ALPINE PLANTS

MAPS OF DISTRIBUTION OF NORWEGIAN VASCULAR PLANTS

EDITED BY

ROLF Y. BERG · KNUT FÆGRI
OLAV GJÆREVOLL

Volume II

TAPIR PUBLISHERS
TRONDHEIM 1990

ALPINE PLANTS

BY

OLAV GJÆREVOLL

DET KONGELIGE NORSKE VIDENSKABERS SELSKAB
THE ROYAL NORWEGIAN SOCIETY OF SCIENCES AND LETTERS

Det Kongelige Norske Videnskabers Selskab

The Royal Norwegian Society of Sciences and Letters

Printed in Norway by TAPIR

This volume is dedicated to the memory of Rolf Nordhagen, late Professor of Botany at the University of Oslo

CONTENTS

GENERAL PART

Introduction

An account of the early history of plans for Maps of the Distribution of Norwegian Plants is given in Vol. 1, Coast Plants (Fægri 1960). Following the publication of that work, a self-appointed committee consisting of Rolf Nordhagen, Johannes Lid, Knut Fægri, Per Størmer and Olav Gjærevoll decided that mapping should continue, priority being given to alpine plants and species belonging to the southern and southeastern floral elements. Responsibility for alpine plants was given to Olav Gjærevoll. It was also decided which species should be mapped.

Criteria used in this selection process were as follows:

1. The age, distribution and immigration history of the Scandinavian alpine flora have been discussed continuously since the 1880's, and a number of species were included with this in mind.

2. The Scandinavian flora is to a great extent isolated. Species occurring in both Scandinavia and the mountains of central Europe have a gap in their distribution of 900 km from Scandinavia to the Carpathians and 1200 km from Scandinavia to the Alps. Many circumpolar species have a gap between northern Scandinavia and northern Russia/Siberia; this varies from one species to another.

3. The northern amphi-Atlantic species which, in Europe, are found in Scandinavia (and in some cases also in adjacent areas) are of special phytogeographical interest. The same applies to the endemic taxa.

4. A considerably larger number of species (taxa) display a disjunct distribution in the Scandinavian mountain range, and attract special interest.

More than one of these criteria will frequently apply. Of the ubiquitous species, which do not comply with the criteria listed here, it was decided to map only a few.

Registration of the alpine plants started in 1961, cand. real. Sigmund Sivertsen being engaged to do the work.

Following the practice used when preparing the volume on coastal plants, he put all data on filing cards. In addition to the material in Norwegian herbaria, he also registered material from Norway kept in herbaria in Copenhagen, Lund, Gothenburg, Stockholm and Uppsala. Corresponding data from herbaria in Finland have been obtained later.

In 1970, Sivertsen was appointed curator of the herbarium of the Museum of Natural History and Archaeology at the University of Trondheim. He nonetheless continued to play a central and coordinating role in the continuing registration. The revision of the material has also largely been carried out by him. A continuous registration process has taken place in that all the other herbaria have sent new data concerning the relevant species to the museum in Trondheim.

As with the coastal plants, publications and field notes have also been extracted; but only some of the data have been used. Because many alpine species belong to critical genera (e.g. *Draba, Papaver*), or are very closely related, to be certain only herbarium material has been taken into account. The registration work revealed a necessity for carrying out comprehensive revision.

Morphologically similar species may easily be confused if field notes alone are used, especially when their ecological conformity is taken into consideration, e.g. *Carex rupestris, C. glacialis* and *C. nardina*. The same applies to tiny species like *Sagina caespitosa* and *S. intermedia*. Furthermore, as the species mapped here are generally collected owing to the great interest attached to them, information in the literature and field notes has proved of limited value for determining their distribution pattern.

For species like *Dryas octopetala*, data from the literature and field notes have been included since there is no risk of confusion. As the map shows, these data have had no real significance for the phytogeographical picture.

The maps have been compiled at the Museum of Natural History and Archaeology in Trondheim under the guidance of the author. A number of persons have participated, mainly advanced students (doctorate students). The completed maps have been sent to the other herbaria for checking.

The maps have been made in a classical manner, as the records are plotted accurately. It should, however, be borne in mind that the size of the standard dot corresponds to ca. 10 km on the ground. The mapping has been very time-consuming, partly because the spelling of many geographical names has varied considerably through time. This applies especially to spelling of Lappish mountain names in North Norway. In many cases, especially when dealing with old records, the labelling was inadequate or even lacking, often given with very local geographical information not found on maps. This has caused a great deal of detective work.

The Norwegian distribution alone generally gives an incomplete and unsatisfactory picture of the species in question in the mountain range. To compensate for this, a small sketch map has been included in the text to show the total distribution in Norway, Sweden and Finland. In addition to the maps in this atlas, these sketch maps are based on Hultén (1971a). Since they also cover some areas east of Finland, including the interesting Rybachiy Peninsula, the term Fennoscandia is used in the captions, even though the whole of Fennoscandia is not included.

Figure 1, showing the alpine region, was drawn by Kari Sivertsen, who also made the small sketch maps.

A single map, that of *Oxyria digyna*, has been produced by computer, based on herbarium data and data obtained from the literature and field notes. The map was processed at the Institute of Geography, University of Trondheim, using a Fortran application program package UNIRAS. The program was written by Axel Baudoin. A laser plotter was used to print the map.

In the years up to 1970, the registration work was financed by the Norwegian Council for Science and the Humanities, which has also given grants for specific purposes subsequently.

Maps showing the distribution of 109 taxa are given in this volume. Several botanists have contributed the text for 27 taxa and their names appear at the end of the relevant descriptions. These contributors are: Simen Bretten (*Draba alpina, D. nivalis, D. oxycarpa,*), Reidar Elven (*Draba cacuminum*), Torstein Engelskjøn (*Draba fladnizensis, D. lactea, Papaver*), Asbjørn Moen (*Carex atrofusca, C. microglochin, Epilobium davuricum, Juncus castaneus, Nigritella nigra, Petasites frigidus*), Leif Ryvarden (*Carex holostoma, Saxifraga paniculata*), Sigmund Sivertsen (*Antennaria, Erigeron politus*), Ola Skifte (*Carex scirpoidea, Scirpus pumilus*), and Bodil Wilmann (*Kobresia simpliciuscula, Pedicularis oederi*).

Olav Gjærevoll has been responsible for the remaining 82 taxa and the introductory chapter, except the section on nunatak theory which has been written by Professor Eilif Dahl.

Richard Binns, M.Sc., has taken care of the language editing.

On the concept of alpine plants

It is scarcely possible to reach an objective definition of the concept of alpine plants, least of all in a country like Norway where alpine conditions are found down to sea level in the northernmost part of the country. In the opinion of Nilsson (1986), alpine plants may be grouped in one of four categories: lowland species that ascend to the alpine belts, northern species that clearly have their main distribution in the mountains, arctic species in the northeast, and alpine plants proper. This is a very flexible definition; in my opinion, too flexible.

Traditionally, the treeless area above the birch forest is looked upon as the alpine belt. This corresponds with the boundary between the northern boreal and alpine regions (the terms northern boreal and subalpine are here taken as synonyms).

The upper birch forest limit cannot, however, be used to define the limit between alpine species and others. A considerable number of species are common to the northern boreal region and the low-alpine belt. Furthermore, many species have a very wide altitudinal amplitude; *Polygonum viviparum* and *Sedum rosea*, among others, grow from sea level to 2280 m.

The coniferous forest limit may be used as an alternative to the birch forest limit. E. Dahl (1958) defined an arctic-alpine species as one having its main distribution above the coniferous timber line or north of the polar coniferous timber line. It is quite obvious that species distribution should be viewed broadly, in the way Dahl did, this being particularly the case with the polar coniferous timber line. The alpine belt and the Arctic region have much in common.

In northwest Europe, with its subalpine birch forest, I believe it is most correct to define an alpine species as one that has its main distribution above the birch forest limit and north of the polar birch forest limit. Ecological conditions differ greatly in the birch forest and alpine belts.

Species like *Polygonum viviparum* and *Sedum rosea* have their main distribution in the alpine belt and are accordingly alpine plants. *Antennaria dioica* ascends to 2000 m, but has its main occurrence in the lowlands, and is therefore not regarded as an alpine plant. *Deschampsia flexuosa* which, sociologically, occurs as a dominating species in low-alpine communities, is first and foremost a species of northern and middle boreal regions and consequently is not regarded as an alpine plant.

Nevertheless, the definition given here does not solve every problem; a number of borderline cases remain. Is *Cystopteris montana* an alpine plant? It occurs abundantly in the low-alpine belt, but also in the northern and even middle boreal regions, probably predominantly in the northern boreal region; it is rare in the Arctic region. It was decided to take it as an alpine plant.

Special difficulties are met with in northeastern parts of Norway where a number of eastern species occur. Some have their main distribution in the Arctic region, some belong to the taiga element. *Polemonium acutiflorum* occurs in the low-alpine belt, but is more common at lower altitudes.

Danielsen (1971) prepared a list of what he regarded as alpine species. It contains species which, in my opinion, should not be defined as alpine plants, but as northern or northeastern boreal species, normally growing in bogs in the taiga region. I cannot agree to species like *Carex lapponica*, *C. tenuiflora* and *Eriophorum russeolum* being included among the alpine plants, and Danielsen himself made reservations.

The total distribution of the species in question should be of greatest importance. If the species is arctic, it may also be regarded as alpine, if it belongs to the taiga flora, it should be omitted.

The list drawn up by Danielsen comprised 230 species. E. Dahl (1958) reached 250 arctic-alpine species using his definition. Nilsson (1986) arrived finally at 270 alpine taxa, but it is not quite clear whether this means species or taxa. The number will vary considerably if, for example, *Papaver radicatum* is listed as a collective species, as Danielsen did, or divided into 10 subspecies in agreement with Nilsson.

It is unavoidable that some degree of judgement has to

12

be exercised. In my opinion, the Scandinavian mountain flora consists of approximately 250 taxa.

This atlas includes some species which fall outside the definition given here. These are species that have their main distribution in the northern boreal region, but display the same phytogeographical problems met with in the alpine flora. Consequently, species like *Nigritella nigra* and *Campanula barbata* are included here. Eastern and northeastern species are not included, even though their distribution in Scandinavia coincides with that of many alpine plants.

Figure 1 shows the areas where most of the ground belongs in the alpine region. It is based on Vegetasjonsregion-kart over Norge 1:1 500 000 (E. Dahl et al. 1986).

Fig. 1 Within the vertically hatched areas most of the land belongs to the alpine region

The distribution of the Fennoscandian alpine plants

The Fennoscandian alpine flora contains several phyto-geographical elements. The northernmost part of Norway approaches the Arctic region and, as might be expected, a number of arctic species occur, most of which are circumpolar in their total distribution. Examples of circumpolar arctic species included in this atlas are *Luzula arctica*, *Draba lactea* and *Ranunculus sulphureus*. Some arctic species display an amphi-Atlantic distribution, e.g. *Sagina caespitosa*, *Braya linearis* and *Pedicularis flammea*. Most of the arctic species are found in the northern part of the Scandinavian mountains, but quite a number also occur in South Norway.

Most of the alpine species of Fennoscandia are arctic-alpine, meaning that they are found in the Arctic region as well as in mountain ranges further south in the northern hemisphere. Species from this element that have been mapped include e.g. *Saxifraga cernua* and *S. hieracifolia*, both of which are circumpolar. Some arctic-alpine species are restricted to Eurasia, e.g. *Oxytropis lapponica* and *Pinguicula alpina*, and some are amphi-Atlantic, e.g. *Lychnis alpina* and *Sedum villosum*. A few species are restricted to the central European and Fennoscandian mountains, e.g. *Chamorchis alpina*, *Nigritella nigra* and *Gentiana purpurea*. Finally, there are some endemic taxa, several of which are dealt with in this atlas, e.g. *Primula scandinavica*, *Poa stricta* and *Papaver laestadianum*.

Leaving aside endemic taxa, isolation is a striking feature of the Fennoscandian alpine flora (Hultén 1955). The nearest localities for the amphi-Atlantic arctic species are found in Svalbard, Greenland or North America. For the European species, there is a gap to the central European mountains. The most remarkable feature is the isolation of the majority of circumpolar species. Continuous distribution from northernmost parts of Norway along the Arctic Ocean coast eastwards to the Urals and Siberia might be expected. The pattern of distribution in fact varies greatly. Many species have a gap between northern Norway and the Urals, others have a smaller gap, e.g. to the eastern part of the Kola Peninsula, Kolguyev, or Novaya Zemlya.

Some species display a gap between South Scandinavia and their nearest occurrences to the northeast. *Pedicularis oederi* has a gap between Åsele Lappmark in Sweden and Kolguyev.

Hultén (1955) discussed the probable causes of the isolation of circumpolar species. He pointed out that many of the species which display a gap are calciphilous. In his detailed survey of the geology of the Kola Peninsula and northern Russia he reached the conclusion that ecological conditions did not offer a reasonable explanation for the distribution gap; a calciphilous species like *Dryas octopetala* has a continuous distribution.

Hultén believed that historical reasons must have been crucial, and proposed two possibilities. 1. The species might have survived the last glaciation in ice-free refugia in Scandinavia and consequently been isolated from the occurrences further east for a very long time. 2. There might have been a gradual immigration from the east as the ice melted, resulting in a continuous distribution. During the postglacial climatic optimum the forests expanded northwards and reached the Arctic shores, with the probable exception of the northernmost parts of the Kanin Peninsula and the Urals. This led to the extinction of the arctic-alpine plants, thus creating the present gap.

There is a weak point in this argument. Why were the plants able to immigrate before the postglacial optimum and not afterwards when the climate turned arctic or subarctic in these areas and the forests disappeared?

The distribution in Fennoscandia

Most Fennoscandian alpine plants are found throughout the mountain range. The distribution pattern varies from species to species. Some are highly ubiquitous, being found from the southernmost mountains to North Cape, the Varanger Peninsula and Grense Jakobselv, and from the westernmost mountains to the easternmost ones, e.g. *Oxyria digyna* and *Athyrium distentifolium*.

Saxifraga cotyledon shows a distinct western tendency, descending to the coast in western as well as northern Norway, and being very rare in the Swedish mountains.

Some other species have a pronounced continental tendency that is most distinct in South Norway, e.g. *Carex rupestris*. This may be partly explained by geological conditions as the western mountains mainly consist of acidic rocks.

A considerable number of species to some extent display a very disjunct distribution. Such species are above all found in two widely separated areas, one in South Norway and the other in northern Fennoscandia. The southern area comprises the northern part of the Jotunheimen, and the

14

mountains of the Dovre, Sunndalen and Trollheimen districts. The northern area extends from the Arctic Circle to western Finnmark (Fig. 2). About 30 taxa are found in both areas, and are termed bicentric (Th. C. E. Fries 1913).

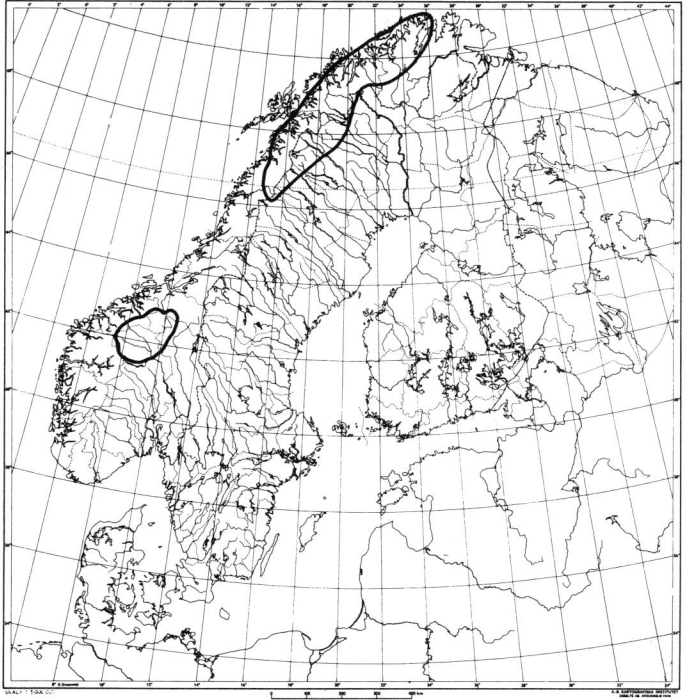

Fig. 2 Map showing the two areas where most of the centric species are found (Gjærevoll 1973)

Great variations are found in the detailed distribution of the bicentric species. *Saxifraga hieracifolia* displays a gap from the Sunndalen mountains to northern Troms, *Campanula uniflora* one from the Trollheimen to Åsele Lappmark. Berg (1963) divided the discontinuous species into 7 groups arranged in a series beginning with those that are highly bicentric, like *Saxifraga hieracifolia*, and ending with nearly ubiquitous ones.

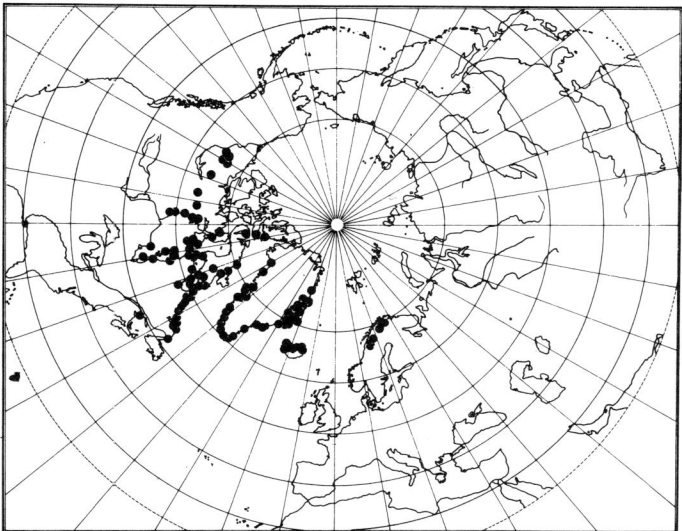

Fig. 3 The distribution of *Pedicularis flammea* (Gjærevoll 1973)

A small number of taxa are restricted to the southern part of the mountain range, e.g. *Artemisia norvegica*. Great variations are met with here, too. *Taraxacum dovrense* occurs from the Jotunheimen to the Trollheimen, *Pedicularis oederi* from the Hardangervidda to Åsele Lappmark. These species are traditionally named southern unicentric (Arwidsson 1928).

About 40 taxa occur exclusively in the northern area (northern unicentric species), but here, too, great variations are found. *Cassiope tetragona* has a continuous distribution from Saltdal to Porsanger. *Arnica alpina* is found from Saltdal to the Varanger Peninsula, *Arenaria humifusa* from Hattfjelldal to Magerøya and *Ranunculus sulphureus* from Lake Torneträsk to the Varanger Peninsula, whereas *Papaver laestadianum* has a very restricted occurrence near Treriksrøysa, the intersection of Norway, Sweden and Finland.

Whatever term is used, centricity or disjunction, many species show considerable congruence in their distributional pattern, others display great deviations. In both cases, centricity offers very interesting and intricate problems. Probably no single explanation is to be found, and attention should be focussed on a number of aspects. Although comprehensive assessment is necessary, each species should also be treated individually, having its total distribution, ecological and sociological behaviour, taxonomical position and dispersal biology examined.

Topographical conditions were brought into the discussion at an early date. Th. C. E. Fries (1913) was the first to draw attention to the bicentric plants, among them many amphi-Atlantic ones. He pointed out that the gap in their distribution contains relatively low mountains compared with those in South Norway and further north in Scandinavia. This must have been important for middle-alpine and high-alpine species.

Many centric species are calciphilous to basiphilous. However, no indication of bicentricity is found in the distribution of *Dryas octopetala* which is usually regarded as the very best calciphilous indicator species. The same applies to *Carex rupestris*. This means that sufficiently calcareous conditions are found in the gap. The low-alpine *Vahlodea atropurpurea*, usually regarded as a calciphobous species (Nordhagen 1943, Gjærevoll 1956), is, on the contrary, slightly bicentric (belonging to the Jævsjø-Frostviken group of Berg (1963)).

The maps show that a number of species display a continental tendency, particularly in South Norway. A. Blytt (1876, 1893) paid great attention to this fact. He emphasised that many alpine species avoid areas influenced by mild, moist oceanic winds, growing preferably on the eastern sides of the mountain range. However, a definite standpoint cannot be taken in this case either. In one of the richest areas in South Norway, the Gjevilvasskammene-Blåhø mountains in the Trollheimen, the climate is extremely oceanic.There is undoubtedly, at least in part, an oceanic climate in the gap, but hardly more oceanic than in the coastal alpine areas further north where a number of centric species are found.

Temperature conditions unquestionably play an important role in the distribution of alpine species. E. Dahl (1951) studied the relationship between the maximum summer temperature and the distribution of alpine plants. It is well known that high temperatures are lethal to some alpine plants. According to his investigations, the absence of alpine species from low altitudes can be explained by their tolerance limit with respect to high temperatures. He concluded that species whose tolerance limit is an average maximum summer temperature of 22°C (or lower) will not be able to grow in the area between the Sylane mountains and southernmost Nordland.

In every part of the mountain range there are great topographical variations. There are numerous ecological niches, such as ravines, cliffs and screes, where climatic conditions may differ considerably from the isotherm in the area around. *Ranunculus glacialis*, which is a pronounced middle-alpine and high-alpine species, may descend to the northern boreal belt in ravines. Many calciphilous species occur at low levels if geological conditions are suitable. A number of alpine plants occur on south-facing scree slopes where the temperature is often much higher than in the surrounding area. The decisive factor seems to be the degree of competition from other plants.

On the whole, it is difficult to point to a single factor. The climatic, edaphic and ecological factors work together. In conclusion, it may be said that the mountains in the gap do not offer the same good conditions for alpine plants as those further south and north.

The climatic fluctuations after the last glaciation must have considerably influenced the distribution of alpine plants. A. Blytt (1876) was of the opinion that the disjunctions known at that time might be explained by climatic fluctuations. In the first period after deglaciation, the alpine plants were distributed throughout the mountain range, but during the postglacial optimum the arctic vegetation was ousted from some areas and its distribution fragmented. Th. C. E. Fries (1913) mentioned that minor disjunctions might have resulted from the influence of the postglacial optimum. Hultén (1955) maintained that the gap might be explained in the same way as referred to above in connection with the isolation of the Scandinavian alpine flora, namely, the forest ousted the alpine plants during the postglacial optimum. In connection with his contributions to the atlas, Moen drew attention to the ecological problems facing alpine plants during the postglacial optimum, writing (A. Moen, written communication 1988): "It is reasonable to postulate that a number of calciphilous alpine species are absent from the gap because they became extinct during the postglacial optimum when all the vegetational regions were displaced upwards by about one region. After the warm period, a number of species (e.g. *Carex atrofusca*) have not been able to re-occupy potential areas because of geographical barriers and poor dispersal ability. Some species may not yet have finished their migration".

After investigating the serpentinicolous plants in central Scandinavia, Rune (1957) concluded that several species are only represented by their serpentinicolous biotypes, indicating that alpine plants, other than those capable of growing on serpentine- rich bedrock, were made extinct during the postglacial optimum. Those invading during that period were unable to conquer areas with that type of bedrock.

It is, nevertheless, not possible to explain the distribution of all centric plants by postulating their continuous distribution prior to the postglacial optimum, with subsequent development of disjunctions resulting from the fluctuations. A slightly bicentric species like *Astragalus frigidus* may occur in the forest region, e.g. on river banks, *Carex bicolor* is generally a species of inundated river banks, and *Luzula parviflora* does very well in northern boreal birch forests. *Rhododendron lapponicum* may grow in pine forests (Nordhagen 1965b) and seems unlikely to have been made extinct from the whole area between the Jotunheimen and Rana. The same applies to *Saxifraga hieracifolia*, which has a gap between the Sunndalen mountains and northern Troms; it may just as well occur in the forest region.

Since a number of bicentric plants are amphi-Atlantic species, historical explanations with a much longer time perspective should be taken into consideration. The large number of northern unicentric, amphi-Atlantic species supports this suggestion.

The question of long-distance dispersal by birds has been brought up in connection with the disjunct distribution. This will be an accidental type of dispersal, and it is difficult to see how accidental dispersal can have created the marked centricity in the distribution of the alpine plants.

History of the Scandinavian alpine flora

by
Eilif Dahl

During the last two million years, Scandinavia has experienced a number of glaciations, periods when glaciers covered most or parts of the country; the last one had its maximum extent 18 000 years ago. No higher plants or animals could survive in areas covered by ice. When the ice melted, plants colonised the new land by migrating from unglaciated areas in Europe south of the glaciers. Since evidence of glaciation is found nearly everywhere in Scandinavia it was concluded that all our alpine and arctic plants were immigrants from the south. This hypothesis, later called the *tabula rasa* **hypothesis**, found support when Nathorst (1892) discovered fossil remains of typical alpine plants in deposits formed during the deglaciation of South Scandinavia.

However, some facts remained which could not easily be reconciled with the *tabula rasa* hypothesis. A. Blytt (1876) pointed out the presence in Norway of a floral element which is not found in the Alps, but has its main distribution in Greenland and North America; Blytt called it an American-Greenland element. If these species survived south of the glaciers in Europe, one would expect them to also have immigrated to the Alps. However, no comparable element is known from the Alps. If they are immigrants, they must have come by long-distance dispersal; Blytt thought transport by icebergs was possible.

Sernander (1896) did not believe in the long-distance dispersal hypothesis, but concluded, with reference to the findings of Blytt, that a "not inconsiderable" number of alpine and arctic species had survived in unglaciated refugia along the Atlantic coast of Norway. This has been called the **nunatak hypothesis**. A corresponding hypothesis pertaining to Greenland had already been proposed by Warming (1888).

Since then, the alternative hypotheses have been discussed by biogeographers and geologists. The development of the discussion was reviewed by E. Dahl (1955, 1987b).

Amphi-Atlantic distribution patterns

According to the *tabula rasa* hypothesis, the flora of Scandinavia is comprised of immigrants from unglaciated areas south or east of the North European ice sheet, perhaps enriched by some arriving by long-distance transport from the west. This seems to apply to the lowland flora that has altitudinal limits below 1000 m in South Norway. Nearly all the species of this flora have a European, Eurasian or circumpolar distribution and are present in the lowlands of Europe or the Alps. This is, however, not the case with the subalpine and alpine elements in Scandinavia that have a lower altitudinal limit in South Norway above 800 m. This flora includes a large number of **amphi-Atlantic** species, i.e. species occurring on both sides of the North Atlantic, but having large gaps in their distribution east of the Ural mountains in Siberia or west of Hudson Bay in North America. In addition, the alpine flora contains a number of northern circumpolar species absent from the Alps. There is a considerable difference in the alpine flora in the Alps and Scandinavia, whereas the alpine floras of Fennoscandia, Scotland, Iceland and Greenland are closely related (E. Dahl 1987b). If this similarity developed after the maximum of the last glaciation, extensive long-distance dispersal must have taken place.

There is also reason to believe that many Scandinavian alpine species were unable to survive south and east of the North European ice sheet during the glacial maxima. Reconstructions of climatic conditions, supported by fossil evidence, suggest that summer temperatures were 6-8°C lower than today and that the climate was dry, with large amounts of loess blowing around (Manabe & Hahn 1977, Frenzl 1987, E. Dahl 1987b). The snow line and the altitudinal vegetation belts were displaced downwards about 1000 m in the Alps. Under such conditions, there were no ecological niches available for more typical alpine species that rarely grow below the timberline in Scandinavia. In addition, the dry climate and the loess left little room for acidophilous oligotrophic species such as *Phyllodoce caerulea*, *Pedicularis lapponica*, *Vahlodea atropurpurea* and *Carex rufina*.

Various distribution patterns may be found within the amphi-Atlantic elements. One has a restricted distribution in Europe, but a wider one in North America. This is the American-Greenland element of Blytt, now called the west arctic element. An example is *Pedicularis flammea* (Fig. 3). A list of species restricted to Fennoscandia (a few are also in Svalbard) is given below:

Antennaria porsildii	*Epilobium lactiflorum*
Arenaria humifusa	*Erigeron humilis*
Braya linearis	*Leucorchis straminea*
Carex arctogena	*Papaver lapponicum*
C. holostoma	*P. dahlianum*
C. macloviana	*Pedicularis flammea*
C. nardina	*Potentilla chamissonis*
C. rufina	*Rhododendron lapponicum*
C. scirpoidea	*Sagina caespitosa*
C. stylosa	
Draba crassifolia	
D. oxycarpa	

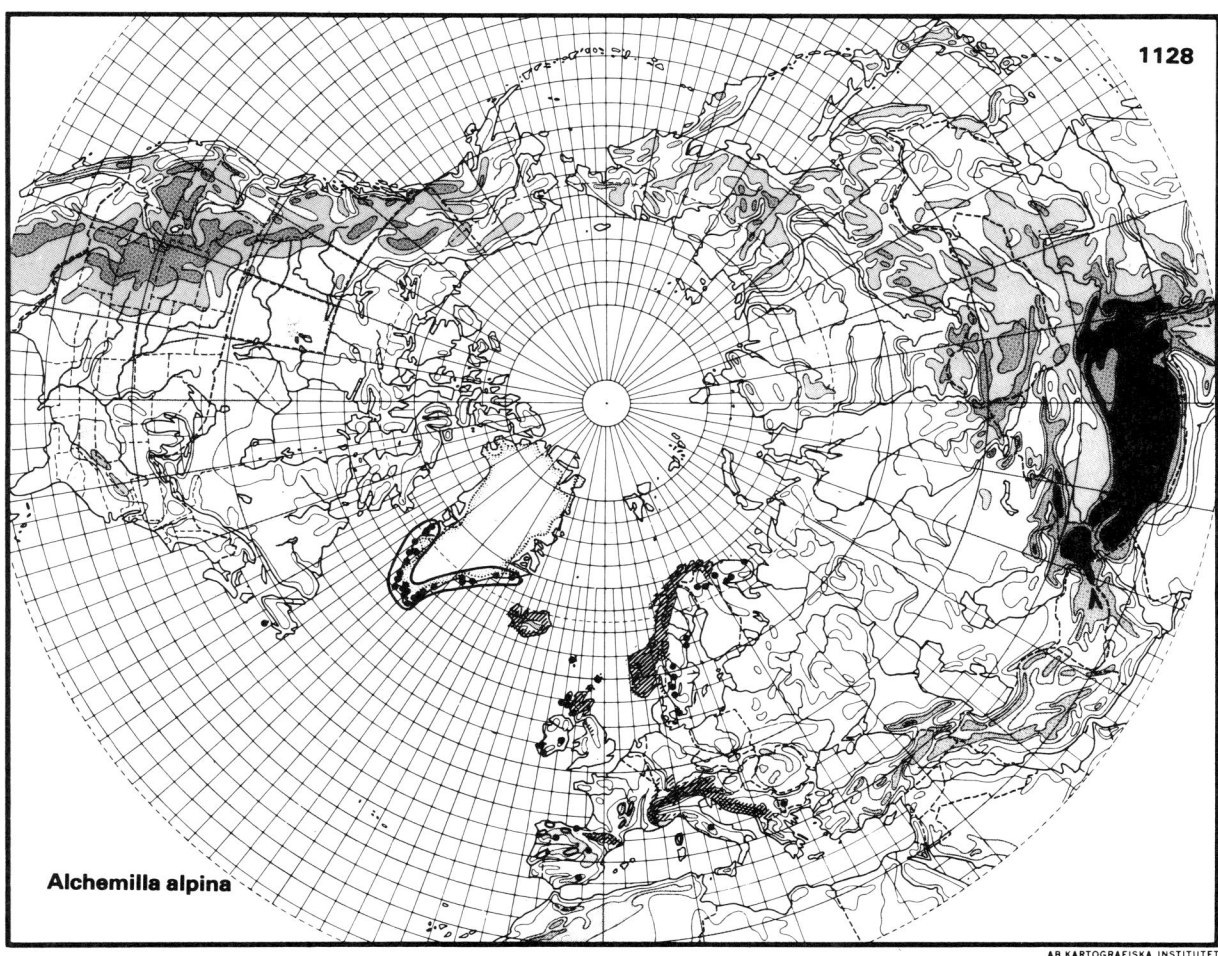

1128

Alchemilla alpina

Fig. 4 The distribution of *Alchemilla alpina* (Hultén and Fries 1986)

Similarly, there is a floral element with a restricted distribution in eastern North America, but a wider one on the European side of the Atlantic. called east arctic in North America. An example is *Alchemilla alpina* (Fig. 4). Then there are species with more equal distribution on both sides of the Atlantic, e.g. *Gentiana nivalis* (Fig. 5). The largest number of amphi-Atlantic taxa on the European side is found in North Fennoscandia, and on the American side the largest number is found in southwest Greenland (E. Dahl 1963b). The characteristics of the amphi-Atlantic element can be summarised thus (E. Dahl 1987b):

1. The amphi-Atlantic element is essentially subalpine – alpine; very few lowland species are amphi-Atlantic.

2. Closely related or identical taxa of polymorphic groups are found on both sides of the North Atlantic and differ from related taxa around the Bering Strait or in the Alps.

3. The amphi-Atlantic taxa are not specially adapted to long-distance dispersal by being equipped with light diaspores for wind-dispersal, having adaptations encouraging animal transport such as edible fruits or fruits with hooks, or being seashore species transportable by ocean currents or limnic ones capable of dispersal by waterfowl. However, such adaptations are prevalent in American taxa that have reached South Greenland and Iceland.

All these factors suggest direct migration over a land connection between North America and northern Europe. Since the amphi-Atlantic element is subalpine – alpine with few temperate species, this suggests a climate on the land bridge that resembled that found in Iceland today. This indicates that the land connection can be dated to late-Tertiary times, since the climate before then was too warm. The close taxonomic relationship between amphi-Atlantic taxa on both sides of the North Atlantic suggest a late date for the connection. Can such a hypothesis be supported by independent evidence?

Important evidence derives from studies of stratification in sediments obtained by drilling holes in the floor of the North Atlantic and North Sea. Sediments of early-Tertiary age are present in such profiles, but those of late-Tertiary age (Miocene and early-Pliocene) are generally missing. Sediments of late- Pliocene and Pleistocene age are present. A likely interpretation of these observations is that in late-Tertiary times a landmass existed west of Scandinavia, probably stretching westwards to Iceland and Greenland. Rokoengen & Rønningsland (1983) depicted

18

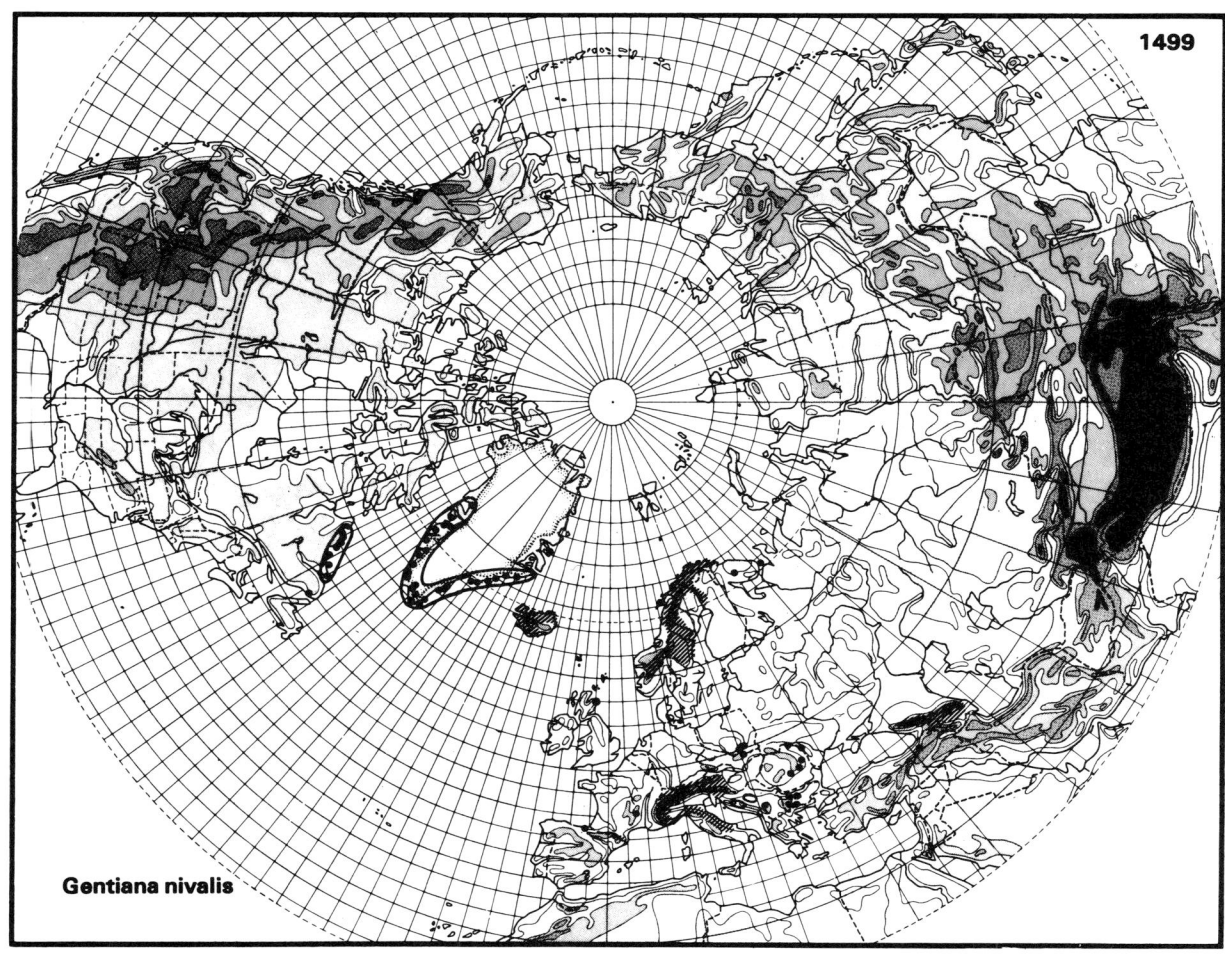

Fig. 5 The distribution of *Gentiana nivalis* (Hultén and Fries 1986)

Centric distributions

a low-lying area of land west of South Norway as late as early-Pleistocene. The floor of the North Sea has subsided since early-Pleistocene, about 600 m in the south and more than 1000 m southwest of Lofoten (Eldholm et al. 1986).

This picture is supported by studies of mollusc faunas in the Polar Basin and North Atlantic made by Strauch (1970, 1983). He found that the faunas in the two basins were separate until the Pleistocene. Elements then began migrating between the basins during the time the Red Crag was deposited in England in the early-Pleistocene about one million years ago. But a barrier to the migration of molluscs must perforce be a bridge for land plants.

However, there is no evidence for a land connection after the maximum of the last glacial event. But the amphi-Atlantic plants had to survive somewhere. If the hypothesis of extensive postglacial long-distance dispersal is rejected, the remaining possibility is refugia along the North Atlantic coast.

As knowledge of the distribution of alpine plants improved, it became clear that some areas were richer in interesting species than others. A large number of amphi-Atlantic, or otherwise disjunct species were found to be concentrated in two areas, one in the south in the Jotunheimen-Dovre mountains, the other in the north in the mountains from the Arctic Circle to northern Troms (Fig. 2). Some occurred in both, but with a gap in between, and were called **bicentric** by Th. C. E. Fries (1913). Others occurred only in one of the areas; they were later called **unicentric** by Arwidsson (1928). The bicentric element has been discussed by Berg (1963). Many amphi-Atlantic species in Fennoscandia are also centric species.

Refugia are areas where plants and animals could survive while they became extinct elsewhere. In or near the refugia one expects to find concentrations of disjunct species. One also expects to find a higher degree of endemism than in other areas, since the plants or animals have a longer history in or near the refugia than elsewhere. Inversely, when a remarkable concentration of disjunct and/or locally endemic species is observed it is tempting to interpret this as indicating the presence of refugia during some past geological period.

Th. C. E. Fries (1913) related the disjunctions in the Scandinavian alpine flora to the nunatak theory. He suggested two areas of refugia in Scandinavia, one along the coast of Møre, another along the coast of North Norway from northern Nordland to western Finnmark. These are areas with an alpine topography with jagged mountains which might have been nunataks above the ice sheet.

However, attempts have been made to explain the centricity in the flora on the basis of environmental conditions. The mountains in the gap between the southern and northern areas are lower than those further north and south, and are open to the influence of westerly winds. A. Blytt (1876) thought that the relevant species were continental in type and were protected against the westerly winds by the high mountains, an explanation supported by Böcher (1951b). E. Dahl (1951) suggested that many alpine species cannot tolerate climates with high summer temperatures and that those which, for this reason, were most demanding were excluded from the gap in Trøndelag and southern Nordland. Gauslaa (1985) pointed out that the palsa mires, typical of climates with cold winters and little snow, also had a bicentric distribution.

On the other hand, the centric species are not necessarily restricted to areas of continental climate. Nordhagen (1963a) was able to find several of the species believed to be continental, far to the west in Møre, e.g. *Euphrasia salisburgensis* near Molde. In the Lofoten and Vesterålen mountains, *Sagina caespitosa*, *Cerastium arcticum* and *Poa arctica* have been found close to the coast.

Even if it were possible to relate the distribution of the centric species to environmental conditions, problems would remain. If their distribution today is limited by environmental conditions, their ecological niche must be narrow and specialised. This is in marked contrast to their behaviour elsewhere. For example, *Carex scirpoidea* has a very limited occurrence in Nordland (Skifte 1988) where it grows on calcareous soils in windswept habitats within a narrow altitudinal zone. In America, it has a wide distribution and descends into the coniferous forest zone, occurring in a wide range of habitats (Böcher 1938). *Rhododendron lapponicum* (Fig. 6) which, in Scandinavia, is a calcicolous plant growing in the subalpine – alpine belts, grows in New York State on the summit of Mount Marcy on granite, along with *Diapensia lapponica* and *Sphagnum* spp. In Wisconsin, in the so-called driftless area, it grows on cliff ledges in forests dominated by temperate deciduous trees such as *Carya spp. Euphrasia salisburgensis*, a rare species not penetrating far down into the forested regions in Scandinavia, grows at 400 m a.s.l. in wine-growing districts near Vienna in Austria.

If a plant population is restricted to a limited refugium for a long time, a loss of alleles must take place. This involves a loss of adaptability to environmental conditions, and the populations become less aggressive. This is a general phenomenon of refugium populations (Braun-Blanquet 1923, Fernald 1925, Hultén 1937). Thus, the narrow ecological niche of many Scandinavian alpine plants suggests that they are relicts which survived at least the last glacial event in refugia in Scandinavia.

Endemism

It was pointed out earlier that the flora of Scandinavia is poor in local endemics compared with areas further south, e.g. the Mediterranean, where almost any large island or high mountain has a number of endemic species often well separated from relatives elsewhere. The lack of endemics in the flora of Scandinavia was, from an early date, taken as support for the *tabula rasa* theory.

This observation must at least be modified following later taxonomic work. Nordhagen (1931a) was able to show that the *Papaver radicatum* complex consisted of a series of different species and subspecies, most of them local endemics. Later taxonomic work has shown that endemic species and/or subspecies are fairly common in the alpine flora. Of the species mapped here, the following are endemic to Fennoscandia:

Antennaria alpina	*Poa stricta*
A. nordhageniana	*P. arctica* subsp.
Papaver laestadianum	*Primula scandinavica*
P. radicatum subsp.	*Taraxacum dovrense*
P. lapponicum subsp.	

In addition, the following alpine – subalpine species and subspecies are endemic.

Arnica angustifolia subsp. *alpina*
Astragalus alpina subsp. *arctica*
Draba cacuminum subsp. *cacuminum*
D. cacuminum subsp. *angusticarpum*
Euphrasia hyperborea
Oxytropis deflexa subsp. *norvegica.*
Pyrola norvegica
Roegneria borealis subsp. *subalpinus*
Silene wahlbergella subsp. *wahlbergella*

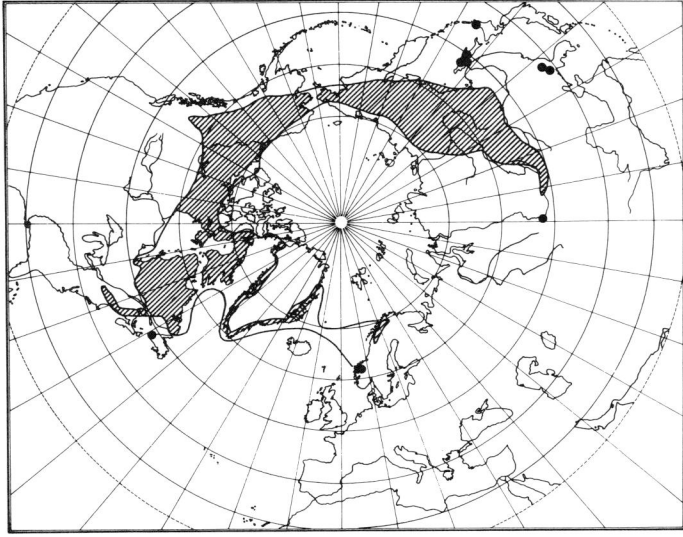

Fig. 6 The distribution of *Rhododendron lapponicum* (Gjærevoll 1973)

In comparison, the lowland flora of Norway is practically devoid of endemics apart from polymorphic apomictic genera like *Hieracium*, *Taraxacum* and *Sorbus*. The exceptions are *Saxifraga osloensis* which is an allopolyploid between *S. adscendens* and *S. tridactylites* (Knaben 1954) and *Arabidopsis suecica* which is an allopolyploid between *Arabidopsis thaliana* and *Cardaminopsis arenosa* (Hylander 1947, Löve 1961). Such allopolyploids can arise as the result of a single crossing. Of the 156 species mapped within the oceanic element by Fægri (1960), only two apomictic *Sorbus* species are endemic, and these are considered to have arisen by hybridisation (Borgen 1987).

Thus, there is a marked difference in the amount of endemism between the alpine – subalpine and the lowland floras in Scandinavia. This indicates that the alpine – subalpine elements have a much older history in Scandinavia than the lowland floras, which is consistent with the nunatak hypothesis.

Geological considerations

The nunatak hypothesis was much discussed after the turn of the century, and was also supported by geologists (e.g. Reusch 1910, Vogt 1913). But later geologists have been reluctant to accept it. Using stratigraphical methods, they were unable to locate areas that had been ice free and have tended to conclude that such refugia did not exist along the Atlantic coast (Mangerud 1973). And, of course, it is circular reasoning from the presence of relict species to conclude that ice-free refugia existed and subsequently explain the distribution patterns by means of refugia.

One difficulty has been the lack of methods for dating sediments. The only method available has been the ^{14}C method, but sediments from the last interglacial are too old to be dated in this way. However, new methods have recently become available.

One such method is amino acid assay. When a mollusc forms its shell, amino acids are included in it. These are initially optically active, but in time they become racemised. This racemisation of certain amino acids is so slow that the degree of racemisation can be used to date shells up to one million years old. The method has been used by Miller (1982) and Forman & Miller (1984) to date shells in beaches along the west coast of Spitsbergen. They found shells up to one million years old. Since the beaches are in situ, they could not have been overrun by an ice sheet after their formation. Hence, ice-free refugia have existed on Vestspitsbergen, probably during the entire Pleistocene. No similar observations are yet available from the coast of Norway.

Thermoluminescence has also been used for dating sediments, and sediments certainly older than 18 000 years and possibly more than 100 000 years old have been dated near Ålesund (Jungner et al. 1989). However, geologists believe that the sites have later been overrun by a glacier (Mangerud et al. 1981, Landvik & Mangerud 1985).

Important results have been obtained by Vorren et al. (1988) from Andøya, North Norway. Here, sediments older than the Weichselian maximum (18 000 years) have been found to contain pollen and other fossils. These indicate that during the period 19 000 – 19 500 years ago the climate corresponded to that of the low to middle arctic zones in Svalbard.

There is, however, other evidence to show that refugia have existed both in South and North Norway, which have been ice free for a very long time, possibly during the entire Pleistocene. A feature often observed on high, flat-topped mountains is block fields. These are large accumulations of blocks weathered from the underlying rock. They have been explained as a result of rapid frost splitting during postglacial times. However, no independent evidence of rapid frost splitting in the high mountains has been presented. It is also difficult to explain how rapid frost splitting could produce large blocks.

Flat-topped mountains capped by block fields are restricted to high altitudes in continental parts of South Norway, above 1700 m in the Jotunheimen mountains. Towards the coast, such features can be found at lower levels, and the altitude of the lowermost block fields corresponds well with what may be expected for the surface of an inland ice sheet. The autochthonous block fields form a readily mappable unit which is shown on Quaternary geological maps.

Towards the Møre coast, and at lower levels, the block fields grade into what British geologists and geographers have termed mountain-top detritus. This is a coarse, but notblocky, sediment which has weathered out in situ. Beneath the mountain-top detritus, and in crevices along cliffs, the underlying parent rocks are chemically weathered, with formation of secondary clay minerals. Vermiculite has been formed by leaching of potassium from biotite; weathering of other minerals has yielded smectite. This weathering is far more advanced than that associated with postglacial soil-profile formation in the lowlands. The mountain-top detritus contains the same minerals as are found in the weathered rocks, and has clearly been derived from the weathered rocks.

Of special interest is the presence of gibbsite in the mountain-top detritus at Stadt and Stemshesten in Møre (E. Dahl 1961 and unpublished data, Roaldset et al. 1982, Longva et al. 1983) and in the Cairngorms in Scotland (Mellor & Wilson 1987). Gibbsite is a laterite mineral which, as far as is known, is only formed under tropical climatic conditions. This indicates that the mountain-top detritus is a remnant of the Tertiary weathering crust that has not been subsequently removed.

The autochthonous block fields and areas of mountain-top detritus cover considerable areas in Norway. Erratics or other conclusive evidence of glaciation have not been observed on such deposits. This indicates that they remained ice free during the last glacial (Nesje et al. 1988), and possibly during the entire Pleistocene (E. Dahl 1987b).

Ecological conditions in the refugia

Even if refugia existed, because of the harsh conditions it is not certain that plants could live there. No vascular plants have been found on nunataks in the Antarctic. These questions must be examined.

The fossils found in sediments dating from the maximum of the last glacial on Andøya suggest a climate comparable to Spitsbergen today. The flora of Svalbard comprises about 140 native species. It is to be expected that refugia further south had better conditions for plant life. The reconstruction of the summer climate 18 000 years ago suggests that the temperature was 6-8°C lower than today along the southern edge of the ice sheet. Plateau-topped mountains in Møre which remained ice free show that the snow line must have been at an altitude of several hundred metres.

An interesting locality is found on Gjevilvasskammene in the Trollheimen at about 1350 m a.s.l. Sørensen (1949) and Grønlie (1953) pointed out the peculiar soil layers found on mountain tops. This is definitely mountain-top detritus resting on chemically-weathered parent rock. Biotite is thoroughly weathered to vermiculite, and smectite is abundant. A number of interesting plants occur in the area.

It is possible to compare the ecological conditions on the Gjevilvasskammen nunatak during the last glacial period with present-day conditions on Jensen's Nunataks in West Greenland. The altitude of the base of Jensen's Nunataks is about 1400 m and the distance to open sea about 75 km. The northernmost birch copses occur a little to the south, so the climatic timberline is approximately at sea level.

In the Trollheimen, the altitude of the timberline is about 1000 m. Assuming, as before, that vegetation zones at the time of the maximum glaciation were 1000 m lower than today, it follows that the timberline was then at 0 m.

The Gjevilvasskammene mountains are situated 80 km from a coastal nunatak at Tustna where there was open sea. Hence, the ecological conditions for plant growth during the coldest period of the last glacial event are thought to be comparable to those on Jensen's Nunataks today. The flora of Jensen's Nunataks has been investigated by Gjærevoll & Ryvarden (1977). They recorded 62 vascular plants and numerous lichens and mosses. Of these, 17 had an amphi-Atlantic distribution and many of the species found on the nunataks have a centric distribution in Scandinavia, e.g. *Papaver* sp., *Campanula uniflora*, *Erigeron humilis*, *Carex scirpoidea* and *C. bicolor*.

There are therefore reasons to believe that the nunataks along the Norwegian coast could have harboured a rich and varied flora of alpine – subalpine vascular plants. Some could also have survived in ice-free areas in mainland Scotland and the Orkneys (Bowen & Sykes 1988) and subsequently, following deglaciation, immigrated across a landmass where the North Sea is now situated.

Some concluding remarks

The nunatak theory has been a storm centre in discussions among biogeographers and geologists for a century. It pertains to fundamental scientific issues and the following points should be emphasised.

Those who favour the *tabula rasa* hypothesis accept the following statements:

1. Evolutionary rates in alpine plant populations are high enough to permit the differentiation of numerous endemic species and subspecies in the course of 18 000 years.

2. Extensive long-distance dispersal of plants lacking pronounced adaptations for such dispersal has taken place since the last glacial.

3. Rapid frost splitting has taken place after the last glacial, leading to the formation of extensive block fields.

Those who favour the nunatak hypothesis accept the following statements:

1. Evolutionary rates in alpine plant populations have been too slow to develop new endemic species or subspecies during the last 18 000 years by purely genetical processes except for alloploidy. Indeed, the separation of Greenland from Europe, which is dated to early-Pleistocene, has not been sufficient to result in any pronounced taxonomic differentiation.

2. Long-distance dispersal has not been an important factor in the formation of floral elements in the North Atlantic region, apart from the immigration of American elements to South Greenland and Iceland.

3. The alpine, autochthonous block fields are remnants of Tertiary weathering and not the result of frost splitting.

The issue therefore has consequences for our understanding of evolutionary processes in alpine plant populations and for our understanding of geomorphological processes. These problems are by no means unique to Scandinavia, but pertain to all the coasts bordering the North Atlantic Ocean.

The ecology of alpine plants

The alpine belt constitutes about half the area of Norway. Figure 1 shows the parts of Norway where most of the area lies above the birch forest. With increasing altitude, both the temperature and the accumulated temperature decrease. Plants with woody stems need a high accumulated temperature. In the lower parts of the alpine belt, the vegetation is characterised by shrubs, dwarf shrubs and ericaceous plants. There are several species of *Salix*, *Betula nana*, *Juniperus communis*, and very frequently *Vaccinium myrtillus*. When the accumulated temperature becomes too low, they no longer grow. Although this varies from species to species, the upper limits of these plants coincide very closely. This common limit is the upper one for the **low-alpine belt**. In central parts of Dovre, it is situated about 1450 m a.s.l. on south-facing slopes and lower on north-facing ones. It drops towards west and north.

Above the low-alpine belt, large areas have a continuous plant cover particularly dominated by grass, sedge and rush, *Festuca ovina*, *Carex bigelowii*, *Juncus trifidus* and *Luzula confusa*. This belt is extensively affected by solifluction and is called the **middle-alpine belt**. Its upper limit is placed where continuous vegetation no longer exists. In the central mountains of South Norway this means 1700-1800 m. Above this belt there is only patchy vegetation. This is the **high-alpine belt**. Solifluction and block fields dominate the landscape, and the vegetation and flora are very similar to those found in snow-beds at lower levels. The number of species may nevertheless be fairly high. In the Jotunheimen, more than 40 species are found above 2000 m.

The Scandinavian mountain range offers very variable geological conditions. Precambrian rocks dominate in the southernmost part. In West Norway, particularly the northwestern part, the mountains mainly consist of granitic or gneissic rocks. From the Hardangervidda in the south to northernmost Finnmark, the range consists of Cambro-Silurian rocks, partly hard schists, partly softer, calcareous rocks which disintegrate easily. Here and there, some small areas of dolomite, peridotite and serpentinite are found, usually only covered by a very thin layer of soil. Due to the special ecological conditions found there, these areas are of considerable phytogeographical interest.

Quaternary superficial deposits are mainly found in the low- alpine belt. Humus production is also chiefly restricted to that belt, and podsol formation rarely takes place above 1200 m in the mountains of central Norway. Since biological activity in the soil declines with increasing altitude, the decomposition rate is low.

Frost heaving and solifluction are important processes in the mountains. In large areas, particularly in the middle-alpine belt, these activities prevent the formation of a stable soil structure.

Because the soil is thin, or even lacking, the vegetation will be characterised by the nature of the bedrock to a greater extent than in most other plant communities. Even in areas where a humus layer is present, most vascular plants will have their roots in the underlying mineral soil which is derived in situ from disintegrating bedrock. In view of the contrast between calcareous and non-calcareous rocks, this leads to pronounced vegetation boundaries, although a gradient is found between the extremes. Mutual competition between species also plays a part in this. Species generally considered calciphilous may grow in exposed localities poor in lime when vegetation is open and scattered.

Snow conditions

When snow falls in a forested area it will stay where it falls. In alpine areas, the wind causes uneven distribution. Snow is blown away from exposed areas and accumulates in depressions and on lee slopes. The amount of snow may vary from year to year, but prevailing winds cause the distribution to remain the same (Nordhagen 1943). The uneven distribution of snow is clearly reflected in the vegetation. In areas that become snow-free early, the vegetation consists of species adapted to low temperatures and strong winds. In winter they are severely exposed to drought because of the wind, and in summer the only supply of water is in the form of rain. The dominating species are therefore xerophilous. They form various chionophobous communities and will have a longer growing season than other plant communities.

Localities where snow accumulates are termed snow-beds. Snow accumulation causes a shorter or longer reduction in the period of growth. Snow-bed plants are therefore adapted to making do with a more or less reduced growing season. They form chionophilous plant communities. They are not exposed to particularly low temperatures as the snow cover offers excellent protection (E. Dahl 1956). They usually have a good supply of moisture.

Due to differences in snow depth there will always be a gradient from the uppermost part of the snow-bed to the foot. Snow-beds that become free of snow early have a lee-side vegetation cover. The soils are usually stable. In those snow-beds that become free of snow late, the soils are often exposed to frost heaving and solifluction.

When the snow has melted, moisture conditions are apt to vary considerably in snow-beds. At the upper margin of a snow drift the soil will dry up rapidly, whereas on the lower side it will be irrigated for a shorter or longer period. Another gradient will therefore be found that varies from dry to seasonally or permanently wet snow-beds.

In effect, there are three main, decisive ecological gradients, the duration of the snow cover, the supply of moisture and the nature of the soil. The snow-cover gradient varies from windswept, almost snow-free areas to extremely late snow-beds, the moisture gradient from very dry to permanently wet, and the soil gradient from acid to calcareous conditions. The constitution of the plant communities is subject to these ecological parameters. Altitude is an additional factor.

This four-dimensional scheme applies above all to the low-alpine belt. A difference is found in the middle-alpine belt as the woody, lee-side species (*Betula nana*, *Salix* ssp. and *Vaccinium myrtillus*) drop out. With increasing altitude, the vegetation gradually assumes a snow-bed character due to reduction in the growing season.

Mires

In the low-alpine belt, there may be extensive flat areas covered by an uneven mosaic of minerogenous and ombrogenous mire elements (minerogenous depressions and ombrogenous hummocks). Most of the mire plant communities in the low-alpine belt are characterised by northern boreal species. Rich fens, however, have a strong element of alpine species. Such fens occur in both the low-alpine and northern boreal belts. Nordhagen (1936a) classified the rich fen communities in the *Caricion bicolori-atrofuscae* alliance, and this classification is followed here (see also E. Dahl 1987a). As mentioned under *Carex bicolor*, it might be sociologically more correct to exclude river-bank communities from this alliance.

Plant communities

In this volume, great emphasis is placed on the sociological position of the species, and several phytosociological terms are used. Species do not occur independently of each other. When a single species is being dealt with it is necessary to study its autecology, but its synecology also needs examining. The distribution of the plant communities must also be taken into consideration.

Where the bedrock and mineral soil consist of granites, sandstones and other rocks which weather into acid soils, the exposed heaths are characterised by xeromorphic species such as *Loiseleuria procumbens*, *Diapensia lapponica*, *Festuca ovina*, *Juncus trifidus*, *Empetrum*, and a number of lichens. The alliance covering these areas is named *Arctostaphylo-Cetrarion nivalis* E. Dahl 1956. It embraces both *Loiseleurieto-Arctostaphylion* Kalliola 1939 and *Juncion trifidi* Kalliola 1939.

The protected lee-side is dominated uppermost by *Betula nana*, further down by *Vaccinium myrtillus*. The various plant communities in this zone are united in the alliance *Phyllodoco-Vaccinion myrtilli* Nordhagen 1936.

With increasing snow depth and later melting, *Vaccinium myrtillus* drops out very abruptly and is replaced by a distinct snow-bed community dominated by grasses and sedges, the alliance *Nardo-Caricion bigelowii* Nordhagen

1936. Prominent species are *Deschampsia flexuosa*, *Anthoxanthum odoratum*, *Nardus stricta* and *Carex bigelowii*. This is the usual community on slopes and in depressions that become snow-free late, but dry up fairly rapidly.

Where the snow melts equally late, but the ground is subject to irrigation for shorter or longer periods, a parallel meadow community is formed that is characterised by hygrophilous species like *Ranunculus acris*, *Rumex acetosa* and *Viola biflora*. This is the alliance *Ranunculo-Anthoxanthion odoratum* Gjærevoll 1956.

When the snow lies so long that the vitality of grasses and sedges is greatly reduced, *Salix herbacea* becomes increasingly dominant. The number of vascular plants decreases substantially, and the soil is often unstable. *Cassiope hypnoides* and *Gnaphalium supinum* show good vitality. At high altitudes, *Ranunculus glacialis* and *Luzula confusa* are additional typical species. The alliance is named *Cassiopo-Salicion herbaceae* Nordhagen 1936.

Areas which become snow-free as late as the last-mentioned alliance and are equally exposed, but irrigated, are characterised by hygrophilous species like *Oxyria digyna*, *Saxifraga stellaris*, *Deschampsia alpina* and *Carex rufina*. They form the alliance *Saxifrago stellaris-Oxyrion digynae* Gjærevoll 1956.

Calciphilous plant communities

In areas where the bedrock consists of carbonate rocks, the most exposed parts are above all characterised by *Dryas octopetala*. This not only dominates exposed ridges, but also the uppermost lee slopes. The *Dryas* communities are very rich in species and display a great variation. Important species are *Kobresia myosuroides*, *Carex rupestris*, *Cassiope tetragona* and *Carex nardina*, the last two being confined to North Norway. The alliance is named *Kobresio-Dryadion* Nordhagen 1936. A large number of species treated in this volume are found in this alliance. With moderate soil humidity and somewhat later snow melting, a low-growing meadow is found with a rich occurrence of, among others, *Silene acaulis*, *Polygonum viviparum* and *Potentilla crantzii*. This is the alliance *Potentilleto-Polygonion vivipari* Nordhagen 1936.

On slopes which dry up quickly after becoming exposed, the *Kobresio-Dryadion* alliance is succeeded by a distinct community dominated by *Salix reticulata*. This is the alliance *Salicion reticulatae-Poion alpinae* Gjærevoll 1956. Parallel to this community, in areas influenced by irrigation or moving soil water, hygrophilous species like *Trollius europaeus*, *Ranunculus acris*, *Petasites frigidus* and *Saussurea alpina* form communities united in the alliance *Ranunculo-Poion alpinae* Gjærevoll 1956.

With very late exposure, *Salix reticulata* drops out and is replaced by *Salix polaris*, which is equivalent to *Salix herbacea* as regards exposure requirement. This alliance is named *Salicion polaris* Gjærevoll 1956.

Where irrigation takes place, a number of hygrophilous species occur, e.g. *Saxifraga oppositifolia*, *S. rivularis*, *Oxyria digyna*, *Ranunculus nivalis*, *R. sulphureus*, *Ce-*

24

rastium arcticum and *Phippsia algida*. This is the alliance *Saxifrago oppositifolio- Oxyrion digynae* Gjærevoll 1956.

In the middle-alpine belt, a special type of snow-bed vegetation occurs in places with late exposure, but not necessarily a thick snow cover. The ground consists of moist, gravelly soil, more or less affected by solifluction, and at least for some time by irrigation. The alliance is named *Luzulion arcticae* Nordhagen 1936, Gjærevoll 1956. Several species belonging to this alliance are dealt with here, including *Luzula arctica*, *Sagina caespitosa*, *Poa stricta* and *Stellaria crassipes*.

A special situation is met with on calcareous scree slopes and some other gravelly localities where the vitality of *Dryas octopetala* is reduced. Species like *Braya linearis*, *Arenaria norvegica*, *Artemisia norvegica* and *Papaver* sp. are characteristic. The communities found in such localities are united in the alliance *Arenarion norvegici* Nordhagen 1936. Some small areas with spring communities are found throughout the Scandinavian mountain range. Communities characterised by eustatic springs on acid soils with water poor in electrolytes are united in the alliance *Mniobryo-Epilobion hornemannii* Nordhagen 1936, and those on calcareous soils with water rich in electrolytes in *Cratoneuro-Saxifragion aizoidis* Nordhagen 1936.

SPECIAL PART

Introduction

This book is organised in the same way as Volume I, Coast Plants (Fægri 1960). It contains maps with explanatory texts. The maps and texts are arranged alphabetically. Generally, the Norwegian distribution alone gives an incomplete and unsatisfactory picture of the distribution of the species in question. To compensate for this a small sketch map is included in the text, showing the total Fennoscandian distribution.

The text has been kept as brief as possible. Compared with Volume I, more attention is paid to the synecology of the species. For alpine plants, sociological behaviour is of special importance when their phytogeographical relationships are being discussed. This is in accordance with Scandinavian traditions.

As in Volume I, the first part of the text is divided into short sections.

1. **Maps**. Previously published maps of the Norwegian or Fennoscandian distribution of the species in question are listed first. Then follows maps covering the total distribution. This information may not be complete, preference usually being given to well-known publications. The "Atlas of North European Vascular Plants" (Hultén & Fries 1986) is so comprehensive that it covers most needs. The subheading, **Local**, often cites maps that are not found in printed publications; many are in reports or theses. Information on these sources is given in the list of references.

2. **First Norwegian record**. With few exceptions, records prior to Flora Norvegica (Gunnerus 1766, 1772) or Flora Danica (Oeder 1761) are not given.

3. **Excluded or doubtful stations**. The revision of the material revealed some erroneous labelling. In several cases, the errors have already been published and corrections were therefore necessary. In other cases, geographical information has been inadequate or indistinct, and misinterpretations may have been published.

4. **Norwegian distribution**. A very brief summary is given here, but some details are usually added when distributions are discontinuous and aberrant.

5. **Altitude limits**. Keen interest has always been taken in the upper limit of alpine species and numerous data exist. However, it is much more difficult to find information about their lower limit. A number of new altitude records are given. In the years 1952-55, Professor Nils Andreas Sørensen made numerous measurements in the central Jotunheimen, but the results were not published. Some have been used in "Fjellflora" (Gjærevoll & Jørgensen, several editions), but are only referred to as "Jotunheimen". When Sørensen passed away in 1987, his detailed survey and diaries were handed over to me. This has enabled accurate geographical data to be given for the new altitude records. In the text, these new observations are followed by NAS. Some new altitude data from Troms have been given by personal communications from Torstein Engelskjøn; these are generally cited as Engelskjøn (1986a). Most of this information is now to be found in the relevant herbaria.

6. **Habitat**. A brief description is given here, more comprehensive information being found in the main text that follows.

The main text that follows gives brief information on the total distribution and somewhat more detailed data on the Fennoscandian occurrence. Both the autecology and synecology of the species are covered, and various views and theories expressed over the years by different authors dealing with immigration and history are referred to.

A number of phytogeographical tables are given in the text. Frequency values and degree of cover are usually given in accordance with the Hult-Sernander scale.

The spelling of Eastern European (mainly Russian) place-names is in accordance with the Times Atlas of the World, 1987 edition.

Herbaria

Standard abbreviations are used to denote herbaria, as follows:

BG	- Bergen	O	- Oslo
C	- Copenhagen	S	- Stockholm
GB	- Gothenburg	TROM	- Tromsø
L	- Lund	TRH	- Trondheim
LE	- Leningrad	UPS	- Uppsala

The species

Antennaria alpina (L.) Gaertn. ♂

Maps: Hultén (1971a, no. 1686a). Local: Benum (1958). Total distribution of the *A. alpina* complex: Hultén & Fries (1986, no. 1721).

Norwegian distribution: Generally in inland mountains. Clearly bicentric, with a somewhat more restricted distribution than the female plants. More frequent in some areas than others; especially in the mountains of Nordland bordering towards Sweden.

Altitude limits: The male plant is known up to about 900 m in North Norway and 1220 m in South Norway. The more common female plants reach much higher, up to 2240 m in the Jotunheimen (Store Memurutind, NAS).

Habitat: In many types of habitat, generally on well-drained soil with only a moderate snow cover, often on screes and in *Dryas* vegetation. Amphicline to calciphilous.

A. alpina is endemic to Scandinavia; the male plant is known only from Norway and Sweden, and is somewhat variable. Selander (1950b) described a new species from Lule Lappmark, *A. lapponica*, based on what he considered to be a possibly sexual species within the *A. alpina* complex. However, since his material appears to have undeveloped pollen, *A. lapponica* is most probably a member of the variable *A. alpina* complex. Male plants are very common in the same area from which Selander described his species. Specimens that may be hybrids also occur, the male gametes probably coming from *A. alpina*. In the type area for *A. lapponica,* some small, very tomentose forms are also found which strongly resemble *A. sornborgeri* of Greenland. Two chromosome numbers are known to occur in the complex in Scandinavia. It is therefore interesting to look more closely at the relationships between different morphological types and their chromosome number, and also at the type of pollen developed in the male plants.

Two types of pollen are found – completely undeveloped pollen (shown as filled circles on the map) and rather well-developed, but somewhat irregular, pollen that may be able to function.

Two main morphotypes are found in the Norwegian material. One is relatively slender and the upper surfaces of its rosette leaves are glabrous or nearly glabrous. The other is somewhat more robust and its leaves are densely hairy on both sides; it has often been called *A. alpina* var. *canescens*. The true *A. canescens* found in Greenland is somewhat similar, but differs among other things in chromosome number. The grey form normally grows highest up in the mountains, the green variety tending to be sub-alpine to low-alpine. The latter appears to be more common in continental areas. No attempt has been made to differentiate the whole material into morphological types,

and the picture is more complicated than is outlined here.

Closer study of the problems found in *A. alpina* should probably start in the area where variation is greatest, i.e. the rich mountains of central Nordland and the adjacent area of Sweden.

Sigmund Sivertsen

Antennaria nordhageniana Rune & Rønning

Maps: Hultén (1971a, no. 1687a), Hultén & Fries (1986, no. 1784), Hämet-Ahti et al. (1984) for Finland.

First Norwegian record: Rune & Rønning (1956) from the Rastigai'si area in Finnmark.

Norwegian distribution: The gai'si area in Finnmark north of the Tana valley, along with an outlier in Navitdalen in Troms. In view of the Finnish find in Enontekis Lappmark the species can also be expected to be discovered in inner parts of Nordreisa and possibly Storfjord in Troms.

Altitude limits: Up to about 800 m in Finnmark (Rune & Rønning 1956), down to about 400 m. In Navitdalen, it has been reported from the forest limit, which generally lies between 500 and 600 m in this area.

Habitat: *A. nordhageniana* chiefly occurs on large solifluction lobes, its long runners making it especially suited for such habitats. At its upper limit, conditions can be judged to be middle-alpine, even though it does not grow above 800 m in the area we are considering. It is a somewhat hygrophilous, amphicline to rather acidophilous species.

In addition to the Norwegian area, *A. nordhageniana* is known to occur in Enontekis Lappmark and Enare Lappmark in Finland. As an endemic species in Fennoscandia, *A. nordhageniana* is of considerable interest. However, it is certainly very close to *A. dioica*, being a normally sexual species like that. There has therefore been some discussion as to whether its taxonomic status would instead be better placed at the infraspecific level. Hultén & Fries (1986) retained it as a subspecies under *A. dioica*.

The somewhat species-poor habitats in which *A. nordhageniana* grows attract few botanists. New finds may provide the necessary information to deal with this interesting taxon and decide whether it should be considered merely as an ecotype of *A. dioica* or as a species in its own right. *A. nordhageniana* occurs outside the main area for centric species and can therefore be suspected of having a somewhat different history.

Sigmund Sivertsen

Antennaria porsildii Elis. Ekman

Maps: Hultén (1971a, no. 1688). Total: Hultén (1958, no. 165), Hultén & Fries (1986, no. 1782). Local: Benum (1958).

First Norwegian record: Nordhagen (1935) mentioned the species from Troms. Neither Th. C. E. Fries (1919) nor Ekman (1927c), when she described *A. porsildii*, mentioned Norwegian material, only material from Sweden. The oldest herbarium collection dates back to 1895 – "Målselv, Alappen og Storfjeldet", A. Landmark (O).

Norwegian distribution: A northern unicentric species growing on inland mountains from Grønfjellet in Rana, just south of the Arctic Circle, to Lohtana in Nordreisa.

Altitude limits: Mostly upper low-alpine to middle-alpine, up to considerable altitudes – Mannfjell in Storfjord 1500 m, and reported by Engelskjøn (1984) up to the same altitude on Gai'bavarri in Bardu. Downwards, it approaches the birch limit in Storfjord (author's observation), but has hardly ever been observed below the forest limit.

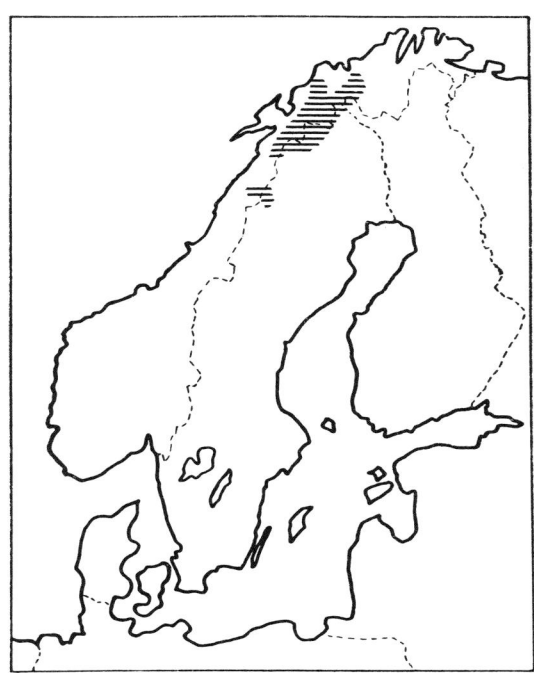

Fig. 7 The Fennoscandian distribution of *Antennaria porsildii*

Habitat: More or less stony, fresh habitats with a good snow cover. Usually ascends higher than *Antennaria alpina* and often prefers localities that tend towards moderate snow-beds. It is rather local and rare, but may be abundant in some rich alpine habitats in Troms. Decidedly calciphilous.

A. porsildii was first described by Ekman (1927c) from Greenland and northern Sweden. It is still an amphi-Atlantic, arctic-alpine species known only from Fennoscandia and Greenland. It is morphologically very similar in its Greenland and Fennoscandian habitats, even though the chromosome numbers are reported to be different. Since it is an apomictic taxon belonging to the critical *Antennaria alpina* complex, its status as an independent species has sometimes been disputed. When Th. C. E. Fries (1919) discussed a find of it from northern Sweden he called it *A. glabrata* (J. Vahl) Porsild. The true *A. glabrata*, however, belongs to a group of species lacking runners. In Sweden, *A. porsildii* is known from Pite Lappmark, and in Finland, from Enontekis Lappmark.

Male plants are very rare. They were previously known only from Torne Lappmark in Sweden, but have recently been reported from central Troms (Engelskjøn 1984). Specimens with pink pappus occur now and then, as in *A. alpina*.

Sigmund Sivertsen

Antennaria villifera Boriss.

Maps: Urbanska-Worytkiewicz (1967), Hultén (1955, 1971a, no. 1685). Total: Hultén (1968), Hultén & Fries (1986, no. 1780), Urbanska (1986). Local: Benum (1958), Ryvarden (1969), Aune (1980).

First Norwegian record: Apparently in 1821 by Zetterstedt (1822) on the mountain Bæskades between Alta and Kautokeino (LD).

Norwegian distribution: Northern unicentric. Inland mountains from Saltdal to Tana and Måsøy in Finnmark.

Altitude limits: Upper low-alpine to middle-alpine. Troms: Målselv, Kirkestinden 1445 m (Engelskjøn 1986a). A find said to be from 1600 m on "Jäggevarre" (Herrmann 1939) should be viewed with some scepticism. When his text is compared with the modern map, the locality cannot be Jäggevarre (Jiek'kevarri) but instead Bredalsfjellet which is some 300 m lower (Jiek'kevarri 1833 m, Bredalsfjellet ca. 1500 m). The species rarely grows below 600-700 m – Karasjok, Giellanjav'rit 430 m, Måsøy, Ryggefjord, probably less than 400 m.

Habitat: Fresh to wet habitats, often in moderate snow-beds. Moderately calciphilous.

A. villifera was first described by Wahlenberg (1826) as *Gnaphalium carpaticum* ß. It was usually called *Antennaria carpatica* until Borissova (1959) coined the name *A. villifera* for North Scandinavian – Euro-Siberian material, retaining the name *A. carpatica* for populations having 2n = 56 chromosomes in more southerly mountains in Europe. (Chrtek & Pouzar (1960) used *A. lanata* (Hooker) Greene for what is being called *A. villifera* here, claiming that Hooker, with direct reference to Wahlenberg's *G. carpaticum* ß, formally based his name on Scandinavian material.)

As it is taken here, *A. villifera* embraces strains with 2n = 28 and 42, in accordance with Urbanska (1986). The species has a divided distribution area. In addition to northern Fennoscandia, it occupies a wide area in Arctic Eurasia

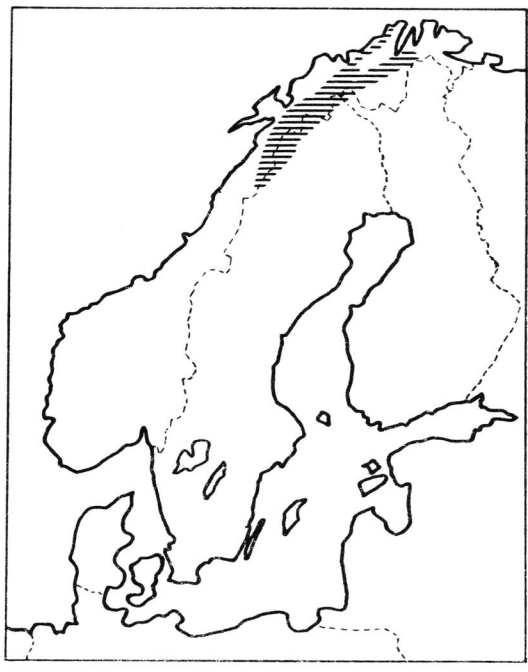

Fig. 8 The Fennoscandian distribution of *Antennaria villifera*

from Kolguyev Island eastwards to the Chukotskiy Peninsula, but does not quite reach the Bering Strait. In Sweden, it is found from Pite Lappmark to Torne Lappmark, but in Finland it only occurs in Enontekis Lappmark.

The two chromosome races in the area we are dealing with do not differ notably in morphological characters. Where male plants are involved, the greater degree of ploidy is usually combined with more irregular pollen. There is nothing to indicate that the cytological races differ in ecological demands.

A. villifera grows in a fairly wide variety of habitats, but always with a good snow cover. The habitats range from *Cassiope hypnoides* vegetation to middle-alpine flushes, fresh vegetation with *Dryas*, and very wet polygon soil (a moss sample from a station of the last-mentioned type included *Blepharostoma trichophyllum*, *Campylium stellatum*, *Cephalozia connivens*, *Drepanocladus intermedius*, *Hylocomium splendens* and *Sphagnum warnstorfii*). The species seems to benefit from solifluction since it is very often found in such habitats, particularly on north-facing slopes. It usually appears to be more or less hygrophilous, but can also be found in well-drained, .well-consolidated grass turf. On the other hand, it does not readily grow on regularly inundated river banks.

A. villifera usually occurs in small, isolated colonies. Its propagation seems to be mostly vegetative, especially in the hexaploid populations. Its competitive power seems very limited.

Sigmund Sivertsen

Arenaria humifusa Wahlenb.

Maps: Nordhagen (1935), Hultén (1971a, no. 718), Rune (1954, 1955), Engelskjøn (1965). Total: Nordhagen (1935, 1954b), Hultén (1937, 1958, no. 169, 1968), Nannfeldt (1947), Raymond (1950), Rune (1955), Hultén & Fries (1986, no. 710). Local: Nordhagen (1935), Aune (1980).

First Norwegian record: Nordkapp: Magerøya 1916, O. Dahl (det. *A. norvegica*) (O).

Norwegian distribution: Very disjunct, from the mountain of Krutvassrøddiken in Hattfjelldal to Magerøya.

Altitude limits: Low-alpine to subalpine. Hamarøy: north of Slædovaggjav'ri 900 m (1050 m on Rakkovarde in Lule Lappmark). Almost down to sea level at Ytre Kåven, Alta.

Habitat: Fairly exposed, moist and open gravelly soil on olivine-rich and calcareous rocks.

In 1807, the Swedish botanist Göran Wahlenberg collected an *Arenaria* on Unna Tuki, a mountain in Lule Lappmark. He described it as *Arenaria humifusa* (Wahlenberg 1812), and denoted it in his Flora Suecica (Wahlenberg 1824). The first Norwegian discovery dates from 1916 when Ove Dahl collected it on the mountain of Duken on Magerøya, but the material was determined by Ostenfeld & Dahl (1917) as *Arenaria norvegica*. In 1934, Nordhagen collected *A. humifusa* on Duken and shortly afterwards revised the determination of Ostenfeld and Dahl (Nordhagen 1935).

The Norwegian distribution is very disjunct. Apart from the localities shown on the map, *A. humifusa* has only been recorded locally in Europe, from western Lule Lappmark where Wahlenberg discovered it and where several localities are now known, the Rybachiy Peninsula, and Svalbard where it is very rare. *A. humifusa* is also known from eastern and western Greenland and Arctic North America westwards to the Yukon. The species therefore belongs to the northern amphi-Atlantic floral element.

The phytogeography of *A. humifusa* was thoroughly investigated by Nordhagen (1935). He thought the species had survived the last Ice Age in Norway and was very old and of North American origin. He pointed out that the localities in Finnmark are concentrated in coastal areas where refugia might have existed, whereas inland occurrences are only found in Nordland. He inferred that the evacuation of alpine species from outer coastal areas in Nordland must have been much more extensive than in Finnmark, due to greater competition.

In 1964, *A. humifusa* was found on Ras'kavarri in Nordfold, only 10 km east of the head of Mørsvikfjorden, one of the innermost arms of Foldafjord (Engelskjøn 1965). The geologist O. T. Grønlie (1927) wrote: "During the last Ice Age the ice that flowed over the Folla district was not thick enough to cover all the land, but did only form a network of glaciers between mountainous parties more or less free of ice". The only locality in Troms, situated close to the innermost part of Lyngenfjord, is interesting in this connection. Nordhagen (1935) was not aware of it.

The ecology of *A. humifusa* was thoroughly dealt with by Rune (1954, 1955). His investigations showed that *A. humifusa* is strictly limited to olivine-rich rocks such as peridotites and olivine gabbro, and to carbonate-rich rocks such as limestone, dolomite and calcareous sandstone. It requires a fair amount of moisture. The typical locality is open, moist, gravelly soil on flat ground. It is often found on the flat centre of small solifluction terraces. *A. norvegica* may also grow in such places, but unlike *A. norvegica*, *A. humifusa* never occurs on dry, warm scree slopes. It does not grow in snow-beds, but is often accompanied by chionophilous species on ground affected by solifluction. The ecology of the species in Greenland and North America seems to agree closely with that described here.

Nordhagen (1954b) published a number of square analyses from patches containing *A. humifusa* on Magerøya and Sørøya, and an extract of these is given below. I and III are Magerøya, II is Sørøya.

Table 1

| Localities: | I | | | | | II | III | |
Scuare no.: (4 m²)	1	2	3	4	5	6	7	8
Arenaria humifusa	+	1	+	+	+	1	2	2
A. norvegica	1	1	1	1	1	.	.	.
Cerastium alpinum	1	+	.	+	1	.	1	.
Pinguicula alpina	1	1	1	1	1	.	1	1
Polygonum viviparum	1	1	1	1	2	+	1	1
Saxifraga oppositifolia	1	1	1	1	2	1	.	-
Silene acaulis	2	2	1	3	2	2	2	3
Thalictrum alpinum	2	2	2	1	1	2	1	1
Tofieldia pusilla	1	1	1	1	1	.	1	1
Carex bigelowii	1	1	1	1	1	1	1	1
C. capillaris	.	.	1	2	1	.	1	2
C. rupestris	2	1	2	2	2	1	.	.
Festuca ovina	1	1	1	2	1	+	1	1
Dryas octopetala	2	1	3	1	2	2	.	.
Empetrum hermaphroditum	1	1	.	1	+	1	1	.
Salix reticulata	2	2	1	1	1	1	.	.
Bare gravel, stones	4	4	4	4	4	4	4	4

Nordhagen characterised *A. humifusa* as a seasonal hygrophilous and sub-neutrophilous species.

Similar analyses were given by Engelskjøn (1965) from Ras'kavarri in Nordfold.

When these sets of analyses are compared with the list of species published by Rune (1954) from localities in Lule Lappmark it is clearly seen that *A. humifusa* can be accompanied by numerous species, although very few reach a higher degree of cover than 1. The most common companions are *Carex bigelowii, C. glacialis, C. rupestris, Cerastium alpinum, Festuca ovina, Juncus trifidus, Polygonum viviparum, Saxifraga oppositifolia, Silene acaulis, Thalictrum alpinum, Tofieldia pusilla* and *Lychnis alpina.*

Arenaria norvegica Gunn.

Maps: Nordhagen (1935), Sørensen (1949), Hultén (1971a, no. 719). Total: Ostenfeld & Dahl (1917), Nordhagen (1935), Hultén (1958, no. 66), Hultén & Fries (1986, no. 711). Local: Nordhagen (1936b), Benum (1958), Lid (1959), Nordsteien (1982), Engelskjøn (1986a).

First Norwegian record: Nordland: Steigen, J. E. Gunnerus 1770 (TRH).

Excluded or doubtful stations: Vardø. Lund (1846a) reported *Arenaria ciliata-norvegica* from Vardø, but no material exists.

Norwegian distribution: From Suldal in Rogaland to Magerøya in Finnmark. Very scattered south of Nordland.

Altitude limits: Mainly low-alpine in South Norway, northern boreal in North Norway. Jomfrunuten, Finse 1450 m. Røragen 720 m. Troms: Målselv, Rostafjell 717 m. Down to sea level in North Norway.

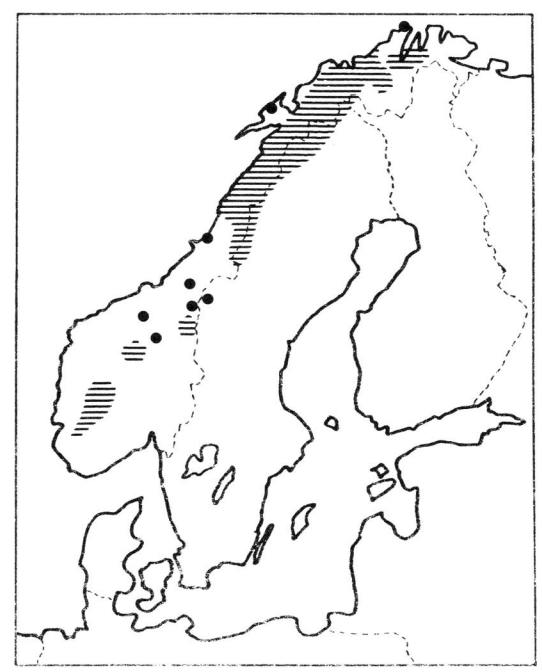

Fig. 9 The Fennoscandian distribution of *Arenaria norvegica*

Habitat: Screes and gravel derived from limestone, dolomite, peridotite, serpentinite and calcareous schists.

A. norvegica was discovered in 1770 in Steigen, Nordland, and described by Gunnerus (1772, no. 1100) — "Habitat in latere alpis stegensis norlandiae". It is closely related to the European species, *A. ciliata*, and has also been taken as a subspecies of that taxon. *A. norvegica* and *A. ciliata* are treated as separate species in Flora Europaea.

Elsewhere in Fennoscandia, *A. norvegica* is known from several localities in Sweden between Jämtland and Torne Lappmark, and from Enontekis Lappmark in Finland. It also occurs in Shetland and mainland Scotland, where it is very rare, and abundantly in Iceland. The species therefore displays an unusual distribution pattern in the North Atlantic region.

As Nordhagen (1935) emphasised, *A. norvegica* is restricted to areas once regarded as having been glaciated during the Pleistocene.

No coastal localities exist in South Norway, and some occurrences are highly continental, e.g. Røragen, east of Røros (Vogt 1944). It is not found in coastal areas south of Leka, but there are numerous coastal localities further north. The present distribution is obviously determined by its special ecological demands. The competitive ability of *A. norvegica* seems weak, as is often the case with specialised species. It may have had better opportunities in early postglacial times when open localities were more readily available. Afterwards, it gradually became forced back to its present localities.

A. norvegica may grow in similar habitats to *A. humifusa*, such as calcareous gravel and material derived from weathering of peridotite and serpentinite and often affected by frost action. Most Norwegian occurrences are found on screes consisting of limestone, dolomite, mica schist and phyllite. These slopes are usually very dry, *A. norvegica* being resistant to drought. It also grows rarely on *Dryas* heaths. Nordhagen (1936a) described an *A. norvegica* community (*Arenarion norvegici* alliance), taking it as an ecological parallel to the *Thlaspeion rotundifolii* alliance of central Europe. Analyses of an *Arenaria norvegica* sociation have been published by Lid (1959). The *Arenarion norvegici* alliance may have a number of occasional species; in North Norway, *Braya linearis* is usually a faithful companion. Species like *Roegneria borealis*, *Oxytropis lapponica*, *Saxifraga aizoides*, *S. oppositifolia*, *Veronica fruticans* and *Poa glauca* are often present. *A. norvegica* is found more or less occasionally on gravel bars both in South and North Norway. In North Norway, it is also found in crevices on sea cliffs.

Arenaria pseudofrigida
(Ostenf. & Dahl) Juz.

Maps: Nordhagen (1935, 1936b), Hultén (1971a, no. 720). Total: Ostenfeld & Dahl (1917), Nordhagen (1935), Hultén (1958, no. 65), Hultén & Fries (1986, no. 711).

First Norwegian record: Syltefjord, Vesterelv. J. M. Norman 1871 (BG, O, TROM).

Excluded or doubtful stations: Saltdal: Junkerdalen 1904, Peters & Pettersson (S). The collection was determined by Nordhagen in 1935, as *A. ciliata* subsp. *pseudofrigida*, certainly by mistake; Nordhagen has never referred to this locality. Herb. UPS contains a collection by N. Lund from Kåfjord in Finnmark, but the name of the locality has been added later. Lund (1846a) mentioned *A. norvegica* from Kåfjord, and there are numerous collections of *A. norvegica* from Sak'kubadni, a mountain in Kåfjord. As was mentioned under *A. norvegica*, Lund referred to

this species from Vardø, but no material exists.

Arenaria was collected at Vardø by M. Vahl in 1787 (Flora Danica 1792, no. 1269), but the specimens are of such poor quality that they cannot be definitely determined (Nordhagen 1935). It seems likely that both Vahl and Lund collected *A. pseudofrigida* in Vardø, but that the locality was destroyed when the town expanded. The collection in herb. UPS may originate from Vardø. Only a short distance separates Vardø and the localities in Persfjord.

Norwegian distribution: Varanger Peninsula from Båtsfjord to Persfjord.

Altitude limits: Båtsfjord: Jonatoppen ca. 300 m. Most localities are between 100 and 250 m.

Habitat: Exposed, weathered carbonate rocks.

J. M. Norman collected this taxon in 1871 in Syltefjord and called it *A. ciliata*. He realised, however, that the flowers were larger, and remarked (Norman 1894) that the plants were exactly like material from Novaya Zemlya. It was described by Ostenfeld & Dahl (1917) as *A. ciliata* subsp. *pseudofrigida*, and they also demonstrated its strange distribution, the taxon being amphi-Atlantic arctic. Its total distribution comprises the Varanger Peninsula, northern Finland (Kuusamo), Arctic Russia, the southwest coast of Novaya Zemlya, Svalbard (Spitsbergen) and East Greenland from Cape Dalton to Peary Land.

In Flora Europaea, *A. pseudofrigida* is included in *A. ciliata* subsp. *ciliata*, which occurs in the central and eastern Alps and the Carpathians. Since *A. pseudofrigida* has a lower chromosome number (2n = 40, subsp. *ciliata* having 80, 120 and 160) and is an amphi-Atlantic arctic plant, I prefer to regard it as a separate species on phytogeographical grounds.

A. pseudofrigida resembles *A. norvegica* ecologically, but its main habitat is exposed dolomite gravel and calcareous shales on gentle slopes. The gravel is often affected by frost. Like *A. norvegica*, though more rarely, it may also occur on loose scree slopes. It is a much more vigorous plant than *A. norvegica*, usually forming large mats. In my experience, it behaves in the same way on Svalbard.

Table 2 gives an extract of 7 analyses (1 m²) from typical localities at Melkar in Båtsfjord made by the author in 1952.

In other places, it was accompanied by, among others, *Chamorchis alpina* and *Carex glacialis*.

This seems to agree well with the lists given by Nordhagen (1954b).

The phytogeographical problems related to *A. humifusa*, *A. norvegica* and *A. pseudofrigida* were dealt with by Nordhagen (1935, 1936b, 1954b). The amphi-Atlantic *A. humifusa* is widely distributed in Arctic North America and is obviously of American Arctic origin. *A. pseudofrigida* is also an amphi-Atlantic, arctic species, but does not reach further west than East Greenland. *A. norvegica* displays a more southerly distribution, deviating from the typical amphi-Atlantic pattern by occurring no further west than Iceland. The last two species are of European origin. It is interesting that they do not meet.

A. Blytt (1893) discussed the probability of Tertiary

Table 2

| Localities: | | | I | | | III | II |
Square no. (1 m²)	1	2	3	4	5	6	7
Antennaria alpina	1	1	1	1	1	.	.
A. dioica	1	1	.	.	.	.	.
Arenaria pseudofrigida	1	1	2	1	2	2	2
Campanula rotundifolia	1	1	1	1	1	1	1
Dryas octopetala	2	2	1	1	2	2	1
Euphrasia frigida	1	1	1	1	1	1	1
Empetrum hermaphroditum	.	1	.	.	1	.	.
Festuca ovina	1	1	1	1	1	1	1
Polygonum viviparum	.	1	.	.	.	1	1
Potentilla nivea	.	.	1	.	1	1	1
Saussurea alpina	1	.	1	.	.	.	.
Saxifraga oppositifolia	1	1	1	1	1	1	.
Silene acaulis	.	.	.	1	.	1	1
Crustaceous lichens	3	3	4	4	3	2	3
Bare gravel	5	5	5	5	5	5	5

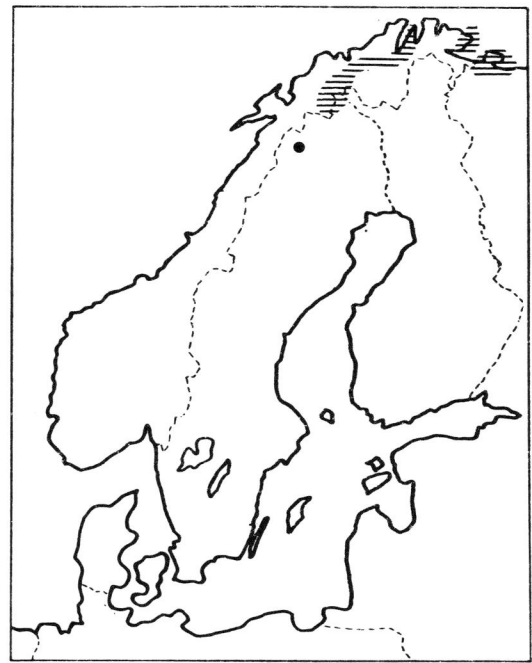

Fig. 10 The Fennoscandian distribution of *Armeria scabra*

land bridges having existed in the northern Atlantic Ocean, a southern one connecting Norway, the Orkneys, Scotland, the Faeroes and Iceland, and a northern one linking northernmost Europe, Svalbard and Greenland. Nordhagen (1935) drew attention to the distribution of *A. norvegica* and *A. pseudofrigida* in relation to the bridge theory. He also pointed out that the entire geographical distribution pattern of *A. pseudofrigida* and *A. norvegica* must have developed during the last interglacial, and that both these and *A. humifusa* must have survived near the areas where they now occur.

Armeria scabra Pall.
(Armeria maritima (Mill.)
Willd. subsp. *sibirica* (Boiss.) Nyman;
A. sibirica Turcz., *A. maritima* (Mill.) Willd.
subsp. *labradorica* (Wall.) Hultén)

Maps: Hultén (1971a, no. 1408). Total: Hultén (1958, no. 88), Meusel et al. (1978), Hultén & Fries (1986, no. 1486). Local: Benum (1958).

First Norwegian record: Troms: Lyngen, Čacca, Læstadius 1830 (UPS). The collection must have been obtained in 1828 or 1829.

Norwegian distribution: Northern unicentric. From Bardu to the Varanger Peninsula and Grense Jakobselv.

Altitude limits: Low-alpine. Målselv, Garanasgai'si 1050 m. Down to sea level at Vardø.

Habitat: Damp heaths, solifluction lobes and moist gravelly soil. Basiphilous.

A. scabra is a member of the very intricate *A. maritima* complex, which has been the subject of numerous taxono-

mical investigations. Hultén (1958) mapped six subspecies which, together, represent a circumpolar distribution. *A. scabra* is apparently also circumpolar.

A. maritima is a seashore plant, but may have some inland occurrences. *A. scabra* displays a different ecological behaviour, growing as a low-alpine species accompanied by a number of centric species. It is fairly common in the mountains of inner Troms, but usually grows dispersed. The main centre is the Storfjord-Kåfjord-Nordreisa district. At Vardø, it occurs at a seashore locality in an eroded dune along with *Festuca rubra* and *Saxifraga oppositifolia* (Elven & Johansen 1983). It has also been reported from dry habitats at sea level in Porsanger, and has been found at sea level at Grense Jakobselv, too (O. Dahl 1934). In Sweden, it is only known from Torne Lappmark. It has not been reported from Finland.

A. scabra occurs in various damp, plant communities, preferably where there are solifluction terraces. It may also grow in shallow alpine fens. Nordhagen (1943) reported it with a relatively high degree of cover on a rich fen dominated by *Scirpus cespitosus* on Sil'bacåk'ka, a mountain in Porsanger comprised of dolomite. He also considered it a characteristic species of his *Caricion atrofuscae-saxatilis* (*Caricion bicolori-atrofuscae*) alliance.

Arnica angustifolia Vahl subsp. *alpina*
(L.) I. K. Ferguson

Maps: Arwidsson (1943), Hultén (1971a, no. 1734), Engelskjøn (1986a). Total: Hultén (1943, 1958, no. 183), Maguire (1943), Hultén & Fries (1986, no. 1850). Local: Benum (1958), Ryvarden (1967), Engelskjøn (1984).

First Norwegian record: The oldest collections in Norwegian herbaria date back to 1841 – Fløyfjellet and Tromsdalstind in Troms, N. Lund (TRH) and M. N. Blytt (O).

Norwegian distribution: From Skaitiaksla in Saltdal to Båtsfjord on the Varanger Peninsula.

Altitude limits: Mainly low-alpine. Målselv, Garanasgai'si 1200 m. 1440 m in Torne Lappmark. Båtsfjord 150 m.

Habitat: Dry exposed heaths on carbonate rocks.

Linné described this plant as *A. montana* var. *alpina*. In 1799, it was given species rank, *A. alpina* (L.) Olin. However, Ferguson (in Heywood 1973) has shown that the name is invalid as it is a later name than *A. alpina* Salisb., which is a *Doronicum* species. It has been com-

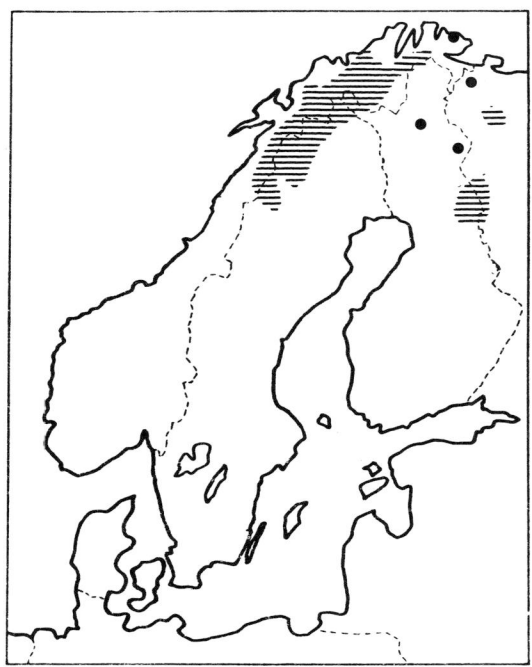

Fig. 11 The Fennoscandian distribution of
Arnica angustifolia subsp. *alpina*

monly accepted that the plant we are concerned with belongs to a variable, circumpolar species. According to Ferguson, the valid name should be *A. angustifolia* Vahl, based on material from Greenland, and the name of our race is therefore subsp. *alpina*. According to Ferguson and Flora Europaea, subsp. *alpina* is known from Sweden (Pite Lappmark, Torne Lappmark), Finland (Enontekis Lappmark, Enare Lappmark, Kuusamo), Russian Karelia, the Kola Peninsula and Svalbard, and is therefore a race that is endemic to northwesternmost Europe. Hultén (1958) regarded the Svalbard plant as subsp. *angustifolia*, a view repeated by Hultén & Fries (1986) and followed here.

The plant occurring from northeastern Arctic Russia to the Bering Strait is taken by Ferguson to be subsp. *iljinii* (*A. iljinii* (Maguire) Iljin). Since subsp. *angustifolia* and

subsp. *alpina* have the same chromosome number, 2n = 76 (2n = 57 is also reported from Greenland), whereas subsp. *iljinii* has 2n = 56, there is obviously a closer phytogeographical connection between subsp. *angustifolia* and subsp. *alpina* than between the latter and subsp. *iljinii*. Subsp. *angustifolia* and subsp. *alpina*, together, form the northern amphi-Atlantic distribution equivalent of a number of other species.

The Norwegian distribution shows a continental tendency. The species prefers exposed heath communities on calcareous soils dominated to differing degrees by *Dryas octopetala*, *Carex nardina*, *Kobresia myosuroides* and *Rhododendron lapponicum*. It may also occur in abundance on open, weathered carbonate rock, and very vigorously on sheltered ledges in ravines. It grows more rarely in meadows on stabilised screes, but normally avoids luxuriant communities. It is favoured by bird excrements, and is often seen growing on the special substrate built up over a long period of time on bird lookout posts. Lunde (1962) gave the results of 112 pH determinations which showed a wide amplitude of 4.6-7.6, with a mean of 5.93.

Artemisia norvegica Fr.

Maps: Elvstrand (1927), Nordhagen (1929, 1933), Ryvarden & Kaland (1968), Toftaker (1969), Hultén (1971a, no. 1725), Vold (1982). Total: Hultén (1954), Gjærevoll (1973), Hultén & Fries (1986, no. 1823). Local: Toftaker (1969), Hagen (1976), Gjærevoll (1980), Nordsteien (1982), Vold (1982).

Excluded or doubtful stations: Møre & Romsdal: Troldkirken (E. Ryan 1901, TRH), Eikesdalen (P. V. Deinboll). According to Lindeberg (1855), P. V. Deinboll had seen *A. norvegica* in Eikesdalen (ca. 1850), but it has never been collected there. Since *A. norvegica* occurs frequently, but as an occasional plant, on gravel bars, the observation may refer to that type of locality (Nordhagen 1931b). Sør-Trøndelag: Røros (a student, B. Skaarden, O).

Norwegian distribution: Dovre and the Sunndalen and Trollheimen mountains; Rogaland: Skardheii in Hjelmeland.

Altitude limits: Mainly low-alpine to middle-alpine. Oppdal: Skirådalskollen (Snøfjellkollen) 1810 m. Rarely below 1000 m, but is an occasional on gravel bars along the River Driva down to almost sea level.

Habitat: Prefers exposed localities, growing on schists which easily disintegrate (mica schist, hornblende schist); often on scree and gravel, too.

A. norvegica was discovered and described in 1780 (Flora Danica) by O. F. Mueller, being called *A. rupestris* ("In rupibus Opdalensibus Norwegiae"). *A. norvegica* is closely related to the very variable amphi-Beringian species, *A. arctica* (Hultén 1954). The correct name, *A. norvegica*, was introduced by E. M. Fries (1817).

The area in which it occurs in central Norway is very concentrated, from Armodshøin in Folldal in the south to Rinnhatten in Rindal in the north, and from Litle Aurhø in Sunndalen in the west to Såtålia in Oppdal in the east. It is a common species in this area, occurring abundantly in many places, often colouring exposed heaths yellow in late summer. This distribution embraces areas with a continental as well as an oceanic climate. In the Sunndalen mountains, *A. norvegica* grows both on mountainsides influenced by an oceanic climate and in valleys occupying precipitation shadows, which are typical topographical features for that district. The Trollheimen also has a highly oceanic climate, whereas Dovre is continental. This concentrated occurrence is strange. The geological conditions should be favourable towards the east in particular, but the species usually has an abrupt limit both there and in the other directions.

In 1966, *A. norvegica* was very surprisingly discovered in Ryfylke (Ryvarden & Kaland 1968).

The total distribution of *A. norvegica* is very disjunct. Apart from the Norwegian areas, it is a very rare plant occurring at a couple of localities in Scotland and a few in the northern Urals. The main distribution area is therefore central Norway. Hultén (1954) drew attention to the differences between the populations in the Urals, Norway and Scotland. Unlike the others, Ural plants (var. *uralensis* Rupr.) have glabrous flowers and leaves. The Scottish plants (var. *scotica* Hult.) differ from Norwegian ones by the shape of their leaves, having fewer capitula and a dwarfed growth.

The total, and also the disjunct, distribution of *A. norvegica* has made this a key species in discussions concerning the age and immigration of the Norwegian alpine flora. Nordhagen (1935) held postglacial migration from the east to be unlikely and presumed that the plant survived in refugia on the Møre coast. E. Dahl (1951) supported this opinion, emphasising that *A. norvegica* could not have immigrated from the Urals after Younger Dryas time through areas where maximum summer temperatures had then become too high for the species. The find in Rogaland is interesting in the light of the theory of survival on an ice-free area of land in the North Sea, since long-distance dispersal from central Norway seems rather improbable.

A. norvegica is generally considered a calciphilous species, and within its restricted area in central Norway it is evidently a characteristic species of the *Kobresio-Dryadion* alliance. It is often found together with *Kobresia myosuroides*, and is therefore an important constituent of the most exposed parts of the alliance. The vegetation is usually very open and includes numerous lichens.

An extract of synedric analyses carried out by Vold (1982) is given below.

I = *Dryas octopetala* heaths, II = *Kobresia myosuroides* heaths, III = heath community characterised by *Artemisia norvegica* and *Campanula uniflora*.

The pH values are generally concentrated in the interval 5.5-6.0. The humus content is very low. The ecological range, however, is very wide. Vold (1982) also showed

that *A. norvegica* frequently occurs on more acid soil, too, for example in *Empetrum hermaphroditum* heaths and *Loiseleuria procumbens-Arctostaphylos alpina* heaths with pH values as low as 4.2. In Scotland, *A. norvegica*

Table 3

	I	II	III
Artemisia norvegica	1111111111111311	11111111:21311	2111222312111121
Empetrum hermaphroditum	114..1.11.1..1.	.1.1.........	..1......2....1
Salix herbacea	.11..1..1111.1.	.1.1.1.:.1.11	..2111.1111111.1
S. polaris	.1....1.......1	......1.11.11	21.1.......1.1.1
S. reticulata	1.....1........	......1......	1....11...1.11..
Vaccinium vitis-idaea	111.11.1..11...	..11.1.:......	..........1.1...
Carex rupestris	.1.11..111.111.	1..11.1:11111	111..12.11111111
Festuca ovina	11...1111111.1.	1111.1....1.1	..1..211121...1.
Kobresia myosuroides	1....1.11......	.342222232324	11...11....111..
Antennaria alpina	...111.1111111.	..111....11..	....1.1..11.1.1.
Astragalus alpinus	1....1.11...1..	.21..1.....1	.......1.....2..
Campanula rotundifolia	1.1111...11..11.	111.1..:...1	....1....1......
C. uniflora	1	.......1.11	11111.111.111111
Cerastium alpinum	......1....11..	..1.....1..1	11.1....1..1.1.
Draba fladnizensis	......1.11.11..	.......1:.....	11....1.1......
Dryas octopetala	2111211..2242111	....3....2...	.2....1...1..1..
Pedicularis oederi	1.1.1.......11	...........1.	1.1.1..........
Polygonum viviparum	11111.1.1..1.1	...1..1...111	.111111.111111.1
Potentilla crantzii	1.............1..	....	.1.......1......
P. nivea	......1....1..	1...1.1......	....
Saussurea alpina	..........1..1	...........11.	1...1...11......
Saxifraga oppositifolia	.1.1..111..1211	1..1..1:11..1	1..........1...1
Sedum rosea	1....1.1........	......1..1...	1....1111...1111
Silene acaulis	.1.11.1..11..12	......1:1111.	1.111121111....1
Thalictrum alpinum	..........1.....1	......1:...1.	...11.1......11.
Polytrichum alpinum	11111111111...1	.1.....1...1..	111.21111....1.
P. piliferum	...1.1.1...11.1.	.121111111...	..112..1.1111.1
Gymnomitrion corallioides	.1.1....1111.1.	...1.11.11...	..12...1..13.22
Alectoria ochroleuca	.11.11111.1.111	.111..11.1.1.	1111.1311111111
Cetraria cucullata	111.11111...11.1	.111.1....1.1	11111111111111
C. nivalis	111.11111.1.113	111211111:11111	2221453132311142
Cladonia coccifera	.11.1111.111111	.111.1121.11.	1.1.....1.1.1111
C. gracilis	..1.1111..1.1.1	.11....1.1...	1.11111111.11...
Corniculata aculeata	1...1....1.11.	.11.111:11111	..1.1.1111..111.
C. divergens	.1...1.11....11	.1.1........11	1111.11.1.112111
Parmelia omphalodes	111.1111.1..11.	...1..1...111	1111.1.1.111111.
Pseudephebe pubescens	111.111.11.1.11	.111.11.11111	1111.1.1...1111.1
Sphaerophorus globosus	111.111111..1.1	.1.1.1..1.111	11111.111111111
Thamnolia vermicularis	11111.11111.1.1	.11.111:.1.11	11112111111111..
Umbilicaria hyberborea	.1...1..11.1.1.	1111.1.:..1.1	1.11.11...1......
U. proboscidea	..1.111...1..11	.111..1..1.1.	111..1.1.11.11.1

grows in *Racomitrium* heaths (Clapham et al. 1962). The plant is, moreover, often found on scree slopes, gravel bars and as a pioneer plant in gravelly places, e.g. roadsides.

In conclusion, it can be said that *A. norvegica* has a wide ecological range, preferring easily disintegrating calcareous schists, but also growing on acid soils when the humus content is low.

Astragalus frigidus (L.) A. Gray

Maps: Hultén (1971a, no. 1139). Total: Hultén (1968, 1971b, no. 61), Meusel et al. (1965), Hultén & Fries (1986, no. 1187). Local: Benum (1958), Toftaker (1969), Holten (1984).

First Norwegian record: Gunnerus (1772, no. 968) – "In alpe dovrina, vulgo Dovrefjeld,..." (TRH).

Norwegian distribution: *A. frigidus* belongs to the bicentric species which have a relatively narrow gap between their two areas. The southern area stretches from Suletind in Vang to Ramfjellet in Meråker, and the northern area from Varnfjellet in Hattfjelldal to Hornvika near North Cape.

Altitude limits: Mainly low-alpine, often also northern boreal. Up to 1400 m in the Dovre area. Alvdal 480 m. Troms: Målselv, Njunis 1080 m. Down to sea level in Finnmark.

Habitat: Low-alpine meadows and low willow thickets, also open birch forests, preferably on calcareous soils.

A. frigidus s. str. is a Eurasian species. A closely related species, *A. americanus* Jones, occurs in North America and was regarded by Hultén (1971b) as a subspecies of *A. frigidus*. Hultén also viewed the Siberian *A. parviflorus* as a subspecies.

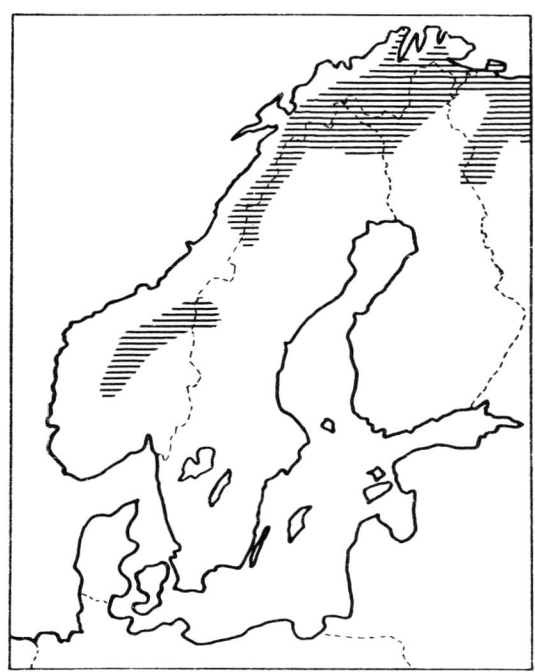

Fig. 12 The Fennoscandian distribution of *Astragalus frigidus*

The Norwegian distribution shows some peculiarities. *A. frigidus* is distinctly continental in both areas. The species is common in the central part of the southern area. It is rare in Nordland, even in the rich parts of Saltdalen. It is again common in inner Troms, as might be expected, and also in inner Finnmark and the area south of Varangerfjord, which contrasts with most bicentric species.

A. frigidus is common in inner parts of Härjedalen and Jämtland, but, in keeping with the situation in Nordland, it is rare in the northern mountains south of Torne Lappmark. The species has a very wide distribution in northernmost Sweden and Finland and on the Kola Peninsula.

The southeasternmost localities are in Kuusamo and Russian Karelia.

In central Europe, *A. frigidus* is widely distributed in the Alps, and is also found in the Carpathians, though rarely. It does not occur in Svalbard. *A. americanus* does not reach either the Canadian Arctic or Greenland. An endemic race, *A. gaspensis* Rousseau, is found on the Gaspé Peninsula, and was regarded by Fernald (1937) as a variety of *A. frigidus*.

A. frigidus grows in a variety of plant communities, preferably in the lowermost part of the low-alpine belt where the vegetation is often an unmappable, intricate mosaic of low willow and *Betula nana* thickets, meadows and heaths. It may occur in abundance in such places. The species requires a good supply of water and a thick snow cover, but needs to be free of snow early.

It is frequently found in the subalpine birch forest and along rivers at lower levels, e.g. on the banks of the River Glåma in Nord-Østerdalen. Occasionally it grows as a pioneer plant on gravelly roadsides. It has been reported growing on shifting sand in North Norway. It is most commonly found in areas with calcareous soils. In Budal, four determinations made by Ouren (1952) showed a mean pH of 5.9. Seven made by the author on Knutshø gave pH = 5.8-6.6.

Astragalus norvegicus Weber

Maps: Hultén (1971a, no. 1141). Total: Meusel et al. (1965), Hultén (1968, 1971b, no. 36), Hultén & Fries (1986, no. 1191). Local: Ouren (1952), Benum (1958), Toftaker (1969), Engelskjøn (1984), Holten (1984).

First Norwegian record: Tydal: Stuedal, M. Wormskjold 1807 (Flora Danica 1810: "Funden ved Stuedalen i Tydalen i Norge af Botanikens utrættelige dyrker M. Wormskiold og derefter af mig paa Dovrefjeld ved Tofte").

Norwegian distribution: From Tyin in the southern part of the Jotunheimen to Polmak in Finnmark.

Altitude limits: Mainly low-alpine, often also northern boreal. Knutshø 1500 m. Alvdal 480 m. On gravel bars at Sel at ca. 250 m and Støren at ca. 70 m. Troms: Bardu, Råkkunbårri 850 m. Down to sea level in Nordland.

Habitat: Meadows, heaths and screes on calcareous soils.

A. norvegicus is a Eurasian species with a very disjunct distribution. In North America, there are two closely related species, *A. eucosmus* Robins. and *A. sealei* Lepage, both regarded by Hultén as subspecies of *A. norvegicus*.

The Scandinavian distribution shows many strange features. In the southern part, the species shows the same continental tendency as *A. frigidus* and may be regarded as being fairly common from the Jotunheimen to Härjedalen and Jämtland. Southern and central parts of Nordland have many localities, some at sea level. Apart from nume-

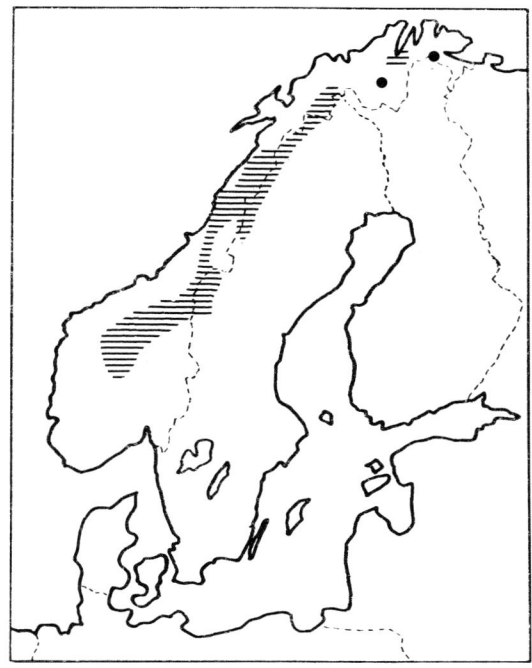

Fig. 13 The Fennoscandian distribution of *Astragalus norvegicus*

Athyrium distentifolium Opiz

Maps: Hultén (1971a, no. 44). Total: Saxer (1955), Hultén (1958, no. 223, 1968), Meusel et al. (1965), Hultén & Fries (1986, no. 50). Local: Toftaker (1969), Kristensen (1981).

Norwegian distribution: Throughout the mountain chain.

Altitude limits: Chiefly low-alpine. Knutsholtind in the Jotunheimen 1870 m. Down to ca. 450 m in southeastern Norway and ca. 250 m in western Norway. Troms: Ibestad 1038 m. Down to sea level in northern Norway.

Habitat: Seasonal-hygrophilous snow-beds, mainly in the low-alpine belt, but also in the middle-alpine and subalpine belts. Also in ravines and along streams. Acidophilous to indifferent.

A. distentifolium was regarded by Hultén (1958) as an amphi-Atlantic species. The species s. lat. has, however, a disjunct circumpolar distribution. It is known from southern Greenland, Iceland, the Faeroes, the British Isles, all

rous localities in the area around Lake Virihaure, the species is relatively rare in the Swedish mountains. It has not been found in the northern part of Nordland, but there are many localities in inner Troms, although it is far less common there than *A. frigidus*. Another gap occurs between Måskogai'si in Storfjord and Masi in Finnmark. It is very rare in Finnmark.

Since *A. norvegicus* is lacking in Finland, the nearest occurrences to Norway are found in the easternmost part of the Kola Peninsula. Hence, whereas the central Scandinavian distribution closely resembles that of *A. frigidus*, their distributions are very different in northern Fennoscandia.

In central Europe, *A. norvegicus* has a similar distribution to that of *A. frigidus*, being found in the Alps and two areas in the Carpathians. No fossil remains of *A. norvegicus* have been recorded between the mountains of central Europe and Scandinavia.

Like *A. frigidus*, *A. norvegicus* is absent from Svalbard. The related plant, *A. eucosmus*, does not reach Greenland.

A. norvegicus has a fairly wide ecological range. It grows in *Dryas octopetala*, *Kobresia myosuroides* and *Rhododendron lapponicum* heaths, and also in low-growing meadows, but avoids communities that are too luxuriant. The plants may become very vigorous in willow thickets and subalpine birch forests. The species is also frequently found on ledges, and unlike *A. frigidus* is a characteristic species on screes. Similarly, it also easily colonises new gravel occurrences, e.g. roadsides. It very commonly occurs on banks of streams and rivers and also gravel bars, descending to low altitudes. It may also grow in basiphilous pine forests where open niches exist. It is found in an *Ulmus glabra* forest at Foldereid in Nord-Trøndelag.

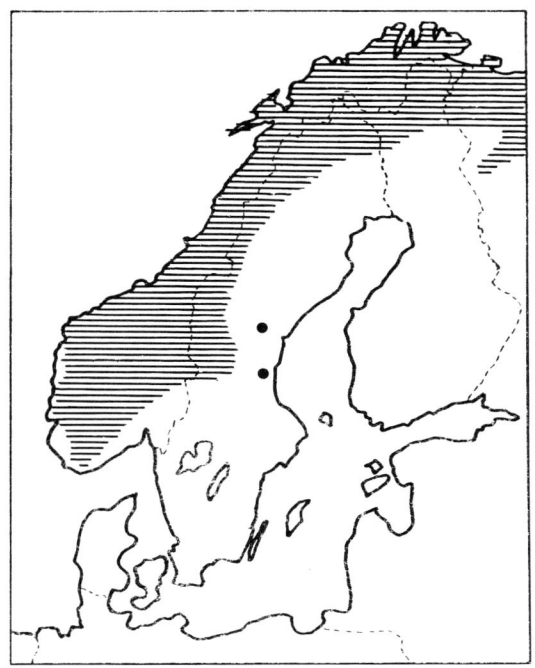

Fig. 14 The Fennoscandian distribution of *Athyrium distentifolium*

the central European mountains and the Caucasus. It has an almost continuous distribution from Scandinavia east to the Urals, and is also known from central Asia, Kamchatka and easternmost Siberia. It is also found in northwesternmost parts of America (var. *americana*) and in easternmost Canada (var. *gaspensis*).

The Norwegian distribution is very wide. In southeast Norway, there are numerous fairly low-altitude localities, e.g. in Nordmarka. It seems to be less common in interior continental valleys (Gudbrandsdalen). It is also rare on the continental plateau of Finnmarksvidda.

In Sweden, *A. distentifolium* is found from Hälsingland to Torne Lappmark, but has relatively few localities away from the mountain chain. In Finland, it is restricted to the Kittilä Lappmark – Enare Lappmark area.

A. distentifolium has its main distribution in the low-alpine belt, but is frequently found considerably lower. In southeastern Norway, it has an optimum between 550 and 1100 m (Odland 1986), but is more alpine in central parts. In western Norway, the main distribution is between 400 and 750 m. In the low-alpine and lower part of the middle-alpine belts, *A. distentifolium* forms a distinct community in the *Cryptogrammo-Athyrion distentifolii* alliance (Nordhagen 1936a, 1943, Gjærevoll 1956), usually dominated by the species itself, which forms dense stands. *A. distentifolium* may be somewhat indifferent as regards soil conditions, but the community occurs on acid soils. The unique humus built up by *A. distentifolium* litter is acidic. Nordhagen (1943) reported 3 determinations showing pH = 4.8-5.1. The author (Gjærevoll 1956) has 30 measurements ranging from pH 3.7 to 5.5.

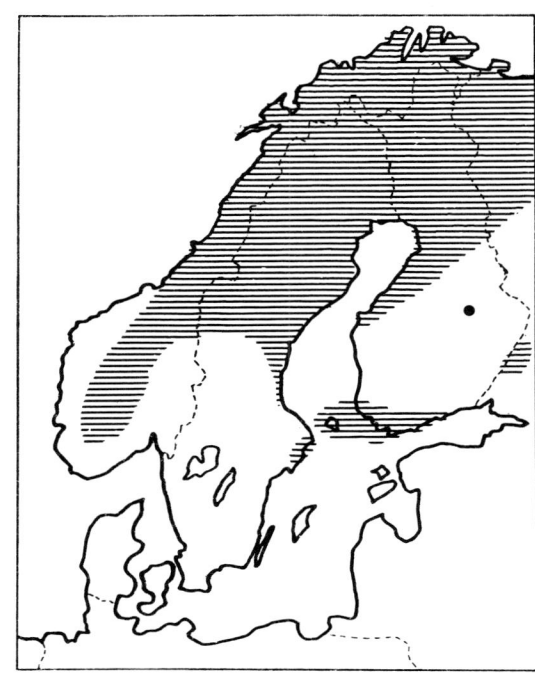

Fig. 15 The Fennoscandian distribution of *Botrychium boreale*

Botrychium boreale Milde

Maps: Hultén (1968, 1971a, no. 22). Total: Hultén & Fries (1986, no. 24). Local: Toftaker (1969), Moen & Selnes (1979), Kristensen (1981), Nordsteien (1982), Holten (1984).

First Norwegian record: Vang, Fillefjell, 1827. S. C. Sommerfelt (O).

Norwegian distribution: From Drakeheii in Suldal to Magerøya and Sør-Varanger.

Altitude limits: Low-alpine to northern boreal. In southern Norway, chiefly between 800 and 1200 m. Trollsteinhøin in the Jotunheimen 1695 m (NAS). Down to 275 m in Orkdal. Frequently down to sea level from Fosen northwards. Ca. 700 m in Troms; according to Benum (1958) it is found at 1300 m on Tromsdalstind (R. E. Fridtz 1902), but since the summit of Tromsdalstind is at 1238 m this height should perhaps be 1300 ft instead of 1300 m.

Habitat: Birch forest, willow thickets and meadows, preferably in fairly open localities with a low-growing grass carpet. Basiphilous.

B. boreale is partly an amphi-Atlantic, partly an amphi-Beringian species. In Europe, it is found from Norway to the Urals and also occurs in northern Iceland and southern Greenland. In Sweden and Finland, there are numerous localities around the Gulf of Bothnia.

Most localities in southern Norway are inland ones; there are no lowland coastal localities. From Fosen northwards, there are many localities on shell sand along the coast.

B. boreale always grows in grassy localities on fairly damp hillsides, in meadows and open birch forest. It prefers habitats with a low-growing grass carpet and avoids luxuriant vegetation. It is very often found on grassy road-

sides and tracks, on previously cultivated land at upland summer farms, and in farmyards. Like other *Botrychium* species it is favoured by grazing cattle which prevent the grass becoming too luxuriant. Places experiencing frequent snow avalanches that prevent willows and birches from maturing are good *B. boreale* habitats.

The amphi-Atlantic distribution presents a special problem as it is rather difficult to regard *B. boreale* as a robust alpine plant. Hultén & Fries (1986) thought it had been overlooked and might prove to be a circumpolar plant, but there are so far no indications to support this assumption; it is extremely rare in Siberia. Hultén (1968) also pointed out that *B. boreale* may be of hybrid origin since it occurs within the distribution areas of *B. lunaria* and *B. lanceolatum*, both as an amphi-Atlantic and amphi-Beringian taxon. It is difficult to accept this as a geographically satisfactory explanation. The present distribution seems to indicate that *B. boreale* is an old species.

Braya linearis Rouy

Maps: Nordhagen (1935), Hultén (1971a, no. 936). Total: Hultén (1958, no. 64), Hultén & Fries (1986, no. 911). Local: Benum (1958), Løkken (1969), Aune & Kjærem (1978).

First Norwegian record: In 1837, J. Ångström collected this species, supposedly for the first time in Scandinavia, in Lule Lappmark and on Båtfjellet in Junkerdalen (UPS). However, the species had probably already been collected by D. C. Solander in Junkerdalen in 1756 (Elisabeth Ekman in Alm 1921). There is a specimen in the Linné herbarium.

Norwegian distribution: Bicentric. In South Norway, known only from Høyrokampen and its vicinity at Lom, and from Jønndalen between Vågå and Dovre. More frequent in North Norway, from Bindal to Magerøya.

Altitude limits: Mostly northern boreal to low-alpine. Høyrokampen 1410 m; down to 900 m in Jønndalen. Troms: Målselv, Gervivarri 1000 m. Down to sea level in North Norway.

Habitat: Screes and gravel derived from carbonate rocks and olivine-rich rocks.

B. linearis belongs to a critical circumpolar group in the *Braya* genus. When E. M. Fries (1839) studied the material collected by Ångström in Lule Lappmark and Junkerdalen he determined it as *B. alpina*, known from the Alps. Rouy (1898) described the Scandinavian plant as a new species, *B. linearis*. When revising a more comprehensive material, Alm (1921) drew the conclusion that the Scandinavian

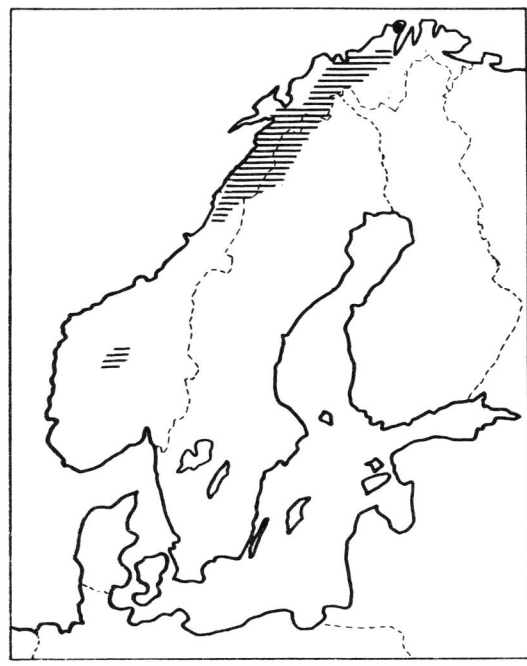

Fig. 16 The Fennoscandian distribution of *Braya linearis*

plant was identical with *B. glabella* from Arctic North America. Later, Schulz (in Engler 1924) concluded that it differed from both *B. alpina* and *B. glabella*, and confirmed Rouy's view. He also stated that *B. linearis* occurs in Greenland.

In addition to Norway, *B. linearis* is found in Scandinavia at a few localities in Lycksele Lappmark, Lule Lappmark and Torne Lappmark. Outside Scandinavia, it is only known from central parts of East and West Greenland. Hence, it shows a similar amphi-Atlantic distribution pattern to that of *Arenaria pseudofrigida*. The ecology of *B. linearis* greatly resembles that of *Arenaria norvegica*, the two often growing together. It is generally found on all kinds of loose, calcareous, south-facing screes, but may also grow on gravel on flatter ground. It also occurs on

gravel bars and as a pioneer plant on roadsides. *B. linearis* grows dispersed in most localities, but is sometimes abundant, e.g. where it occurs as a pioneer plant.

138 determinations (from Troms and Høyrokampen) made by Lunde (1962) showed a pH amplitude of 7.2-8.5. Its competitive ability seems to be weak in Scandinavia, but the Greenland population apparently has a somewhat different ecological behaviour, at least in part. Böcher (1954) described a *Braya linearis* sociation from crevices, where it grows together with *Gentiana detonsa*, *Primula stricta* and *Puccinellia deschampsioides*.

The phytogeographical problems associated with *B. linearis* are apparently like those found in the *Arenaria* species. The very restricted and continental distribution in South Norway implies that this consists of relict occurrences. As Løkken (1969) pointed out, *B. linearis* easily colonises new calcareous mineral soils and might have had favourable conditions in late-glacial and early postglacial times. Since then it has been ousted by other plants except on loose scree slopes and some other places where it has little competition.

Braya purpurascens (R. Br.) Bge.

Maps: Hultén (1971a, no. 937). Total: Fernald (1925), Hultén (1968, 1971b, no. 9), Hultén & Fries (1986, no. 912).

First Norwegian record: Nordkapp: Magerøya, P. V. Deinboll 1822 (O) (*"Draba androsaceae"*).

Norwegian distribution: Known only from Magerøya.

Habitat: Calcareous scree slopes and gravel.

In 1916, Ove Dahl collected *B. purpurascens* on the mountain of Duken believing it to be the first time it had been found in Scandinavia. However, when he was revising material named *Draba androsaceae*, collected by P. V. Deinboll in 1822 in the same area, he discovered that this plant was *B. purpurascens*. The species has subsequently been collected at several places, but within a very restricted area. The southernmost locality is summit 292, west of Skipsfjord, and the northernmost one is Duksfjordfjell, these being less than 10 km apart.

B. purpurascens belongs to a very critical group of *Braya* species. According to Hultén (1971b), three subspecies should be recognised, subsp. *thorild-wulfii* (Ostenf.) Hult. in Arctic Canada and Greenland, subsp. *pilosa* (Hook.) Hult. in Arctic Siberia and western Arctic North America, and the Magerøya plant, subsp. *purpurascens*. The Norwegian plant has a circumpolar, high-arctic distribution which is, however, disjunct. The nearest occurrences are northeastern Arctic Russia, Novaya Zemlya, Svalbard, Iceland and East Greenland. The Norwegian occurrence is therefore very isolated.

The ecology of *B. purpurascens* greatly resembles that of *B. linearis*. The plant occurs fairly frequently on the plateau on Duken and in similar exposed localities. Large areas there can be characterised as wind-eroded *Dryas octopetala* heaths. *B. purpurascens* inhabits gravelly places with an open vegetation and often affected by frost action. It also occurs on loose scree slopes and, like *B. linearis*, as a pioneer plant on roadsides. Such occurrences seem to be decidedly occasional, the species soon being ousted by other plants. *B. purpurascens* obviously has little competitive ability (cf. Høiland 1988).

The bedrock in the area consists in part of limestone and calcareous schists. In Svalbard, *B. purpurascens* is mainly found on calcareous gravel.

Table 4 gives an extract of 6 square analyses made by Nordhagen (1935) on the Duken scree slopes.

Table 4

Square no. (4 m²)	1	2	3	4	5	6
Braya purpurascens	1	1	1	1	1	1
Campanula rotundifolia	1	1	1	2	1	1
Epipactis atropurpurea	1	.	1	.	.	1
Hieracium sp.	1	1	1	1	1	.
Polygonum viviparum	1	1	.	1	1	1
Saxifraga aizoides	1	2	1	3	2	2
S. oppositifolia	2	3	1	1	2	2
Solidago virgaurea	1	.	1	1	.	.
Festuca rubra	1	1	1	2	1	1
Poa alpina	1	1	.	.	1	.
P. glauca	1	1	1	.	1	.
Dryas octopetala	3	1	1	1	2	2
Salix reticulata	1	.	.	.	1	3

The isolated occurrence on Magerøya, combined with the presence there of several amphi-Atlantic species, such as *Braya linearis*, *Arenaria humifusa* and *Sagina caespitosa – Arenaria norvegica* should also be taken into consideration – indicates that *B. purpurascens* survived in the Magerøya area during the last glaciation (Nordhagen 1935). Nordhagen (1963b) drew attention to the occurrence of unsorted, weathered, sharp-edged blocks in roadside exposures on the road to North Cape, taking them to indicate an ice-free refugium.

Campanula barbata L.

Maps: Møller (1966), Hultén (1971a, no. 1656), Nordhagen (1973). Total: Hultén & Fries (1986, no. 1750).

First Norwegian record: Nordre Land: "Synnfjell i Torpa" 1824 – B. M. Keilhau & H. Steffens. Coll. O. Berg (O).

Norwegian distribution: In a very restricted area in Oppland. Numerous localities in Nordre Land and Etnedal, a few in Lillehammer, Gausdal and Nord-Aurdal.

Altitude limits: Northern boreal. Most localities are at 600-800 m a.s.l. Solskiva in Nord-Aurdal ca. 1100 m. Down to 340 m.

Habitat: Open birch forest, grassland, previously cultivated fields, and stabilised scree slopes.

C. barbata is a European, montane species with a strongly disjunct distribution. There is a gap of about 1200 km to the nearest locality in the Sudetes. The main distribution area is the Alps from France to Krain.

C. barbata is a northern boreal species. Its natural habitat seems to be grassland in birch or mixed birch-coniferous forests. According to investigations by Møller (1966), it keeps to birch forests on fairly rich soils, characterised by, among others, *Geranium sylvaticum*, *Ranunculus acris*, *Solidago virgaurea* and a number of grasses. These forests have been the main area for upland, summer farming.

C. barbata seems to be favoured by grazing which keeps the birch forest open. It may invade formerly cultivated land in large numbers, e.g. on upland, summer farms, but also lower-lying farmland. The ecology described here seems to agree well with that found for the species in central Europe. Two recently discovered localities on stabilised scree slopes at ca. 1000 m a.s.l. near Synnfjord (Wesenberg 1988) are very interesting. They are characterised by *Vaccinium myrtillus* and to a certain degree by *Deschampsia flexuosa*, in other words a very common, oligotrophic community. Wesenberg believed these localities to be primary ones and the others secondary.

The distribution is extremely enigmatic. Already A. Blytt (1893) paid great attention to the group of alpine species having one distribution area in the central European mountains and another in Scandinavia, among others, *C. barbata*, *Gentiana purpurea*, *Phyteuma spicatum* and *Nigritella nigra*. These species "kommen bei uns als Relikten vor und sind wahrscheinlich ziemlich früh zu uns gekommen". He, furthermore, supposed that the gap between the central European and Scandinavian occurrences was established during the postglacial climatic optimum. Wille (1905) suggested long-distance dispersal as the most likely explanation.

When dividing the Scandinavian flora into distribution groups, Hultén (1950) placed *C. barbata* in group C, which embraced species that may have survived the last glaciation in Scandinavia.

The history of *C. barbata* has also been dealt with by Nordhagen (1973). He rejected Wille's theory. *C. barbata* is a ballist, unsuited to long-distance dispersal. Taking into consideration both the distribution and the ecology he concluded that "diese Art in Norwegen kein Neueinwanderer ist". Referring to the new localities mentioned above, Wesenberg (1988) supported the views of Blytt and Nordhagen. Glacial survival seems yet very unlikely for a northern boreal species like *C. barbata*.

Campanula uniflora L.

Maps: Arwidsson (1943), Hultén (1971a, no. 1665). Total: Fernald (1926), Hultén (1958, no. 173, 1968), Hultén

& Fries (1986, no. 1759). Local: Benum (1958), Schumacher & Løkken (1981), Engelskjøn (1984, 1986a).

First Norwegian record: Gunnerus (1772, no. 1080) – "In alpibus norvegicis".

Excluded or doubtful stations: Nordkapp: Magerøya, N. Lund; Lund reported *C. uniflora* from here in a list from 1841, but not in a later survey of plants from Finnmark (Lund 1846a). Alta, Norman, Ex. herb. Sommerfelt (O); neither a precise locality nor a year are given, and Norman never referred to this locality.

Norwegian distribution: Bicentric. Southern area from the central Jotunheimen to the southern Trollheimen, northern area from the Junkerdalen mountains to Gjøvarden, a mountain at Skjervøy, Troms.

Altitude limits: Mainly middle-alpine. Kvitingskjølen in the Jotunheimen 1810 m (NAS); down to 1000 m at Rundhø, Tynset. Troms: Målselv, Kirkestinden 1425 m; elsewhere in Troms, down to 400 m.

Habitat: Various heath communities, damp solifluction slopes and terraces, and ledges. Calcareous and base-rich soils.

C. uniflora belongs to the amphi-Atlantic species. Outside Scandinavia, it is also known in Eurasia from the Urals, Vaygach Island and Novaya Zemlya. Then there is a gap of about 1400 km to the mountains east of Lake Baykal. There are only a few scattered localities in central

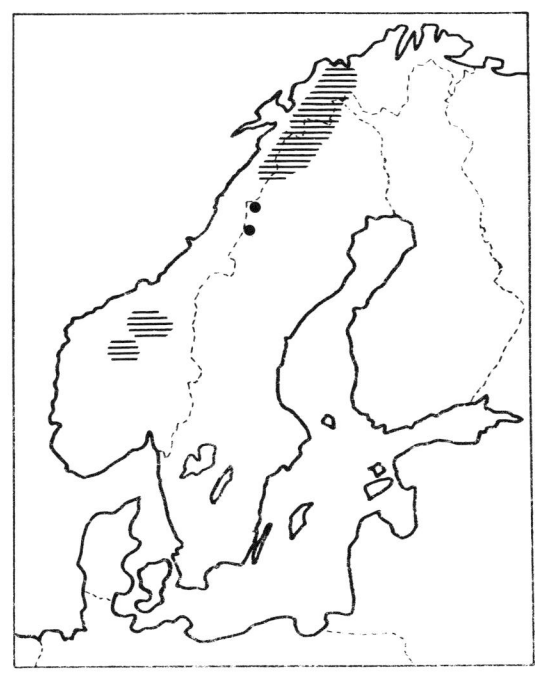

Fig. 17 The Fennoscandian distribution of *Campanula uniflora*

eastern Siberia, but some additional ones occur in the far northeast. The species is widely distributed in Arctic North America and the Rocky Mountains. It also grows in Greenland, Iceland and Svalbard.

In Scandinavia, *C. uniflora* is one of the characteristic

bicentric species. The southern area is divided into two, as there is a distinct gap between the Jotunheimen localities and those at Dovre. *C. uniflora* is very rare in the Trollheimen, only two localities being known (Gjerdhøa and Nyhusfjellet). It is absent from the mountains adjacent to Sunndalen. The westernmost locality in the Dovre area is Drugshø and the easternmost one is Rødalshø in Tynset. *C. uniflora* is rare in Nordland outside the Junkerdalen-Saltdalen area, but fairly common in the mountains of inner Troms. In Sweden, it occurs as far south as Lasterfjället in Åsele Lappmark and is fairly common in Lule Lappmark and Torne Lappmark. It is very rare in Finland and is only known from Enontekis Lappmark. *C. uniflora* is a pronounced continental species in both its southern and northern areas.

Within its restricted area, *C. uniflora* is abundant at many localities. It is very common on some mountains in the Dovre area (e.g. Knutshø), but its flowering varies greatly from year to year. About 80 flowering specimens were counted in a random, 1 m² quadrat on Salefjellet, south of Balvatnet in Nordland.

C. uniflora is found in a variety of plant communities, usually in the *Kobresio-Dryadion* alliance (the *Dryas octopetala*, *Kobresia myosuroides*, *Rhododendron lapponicum* and *Cassiope tetragona* communities). It may also be common on solifluction soil, especially on north-facing slopes where *Dryas octopetala* grows less vigorously. It is very frequent on margins of solifluction lobes. It is a characteristic species of the middle-alpine zone. In central Norway, it is very rare below 1300 m.

Since *C. uniflora* grows in plant communities whose pH values are normally around 7.0, it has always been looked upon as a calciphilous species. 126 determinations made by Lunde (1962) in North Norway showed a mean pH of 5.84, which is somewhat lower than might be expected from the sociological affinity of the species. The amplitude was 4.8-7.5. In exceptional cases, the species grows on acid bedrock where competition is insignificant.

As an amphi-Atlantic plant, *C. uniflora* has been looked upon as a potential survivor of the Ice Age on nunataks. It was found in large quantities on all five of Jensen's Nunataks in Greenland, frequently growing very vigorously, especially on solifluction terraces on amphibolite but also on biotite gneiss (Gjærevoll & Ryvarden 1977).

Cardaminopsis petraea (L.) Hiit.

Maps: Holmboe (1937), Hultén (1971a, no. 920). Total: Meusel et al. (1965), Hultén & Fries (1986, no. 937). Local: Toftaker (1969), Hagen (1976), Nordsteien (1982).

First Norwegian record: Gunnerus (1772, no. 744) referred to Oeder (1768), quoting from Flora Danica VII: "I Leerdalen i Bergens Stift paa Holmene i Elven, paa gruset Grund".

Excluded or doubtful stations: Kristiansund, H. Greve 1865 (BG).

40

Norwegian distribution: From Forsand in Ryfylke to the central Trollheimen.

Altitude limits: Low-alpine to middle-alpine, also northern boreal. Tveråtinden in the Jotunheimen 1730 m (Jørgensen 1933); the text states 1700 m, the list (p. 109) 1730 m. Along rivers and on rocks down to sea level. On the mountains of Gjevilvasskammene and Blåhø the range is 1000-1600 m (Nordsteien 1982).

Habitat: Scree slopes, gravel bars, moraines and other glacial deposits, cliffs and rocks.

C. petraea has a strange, disjunct distribution. In addition to Norway, the species occurs in Ångermanland in Sweden, the Faeroes, Iceland, Shetland and elsewhere in the British Isles, and in central Europe, as well as in a small area beside Lake Onega. From the River Petschora

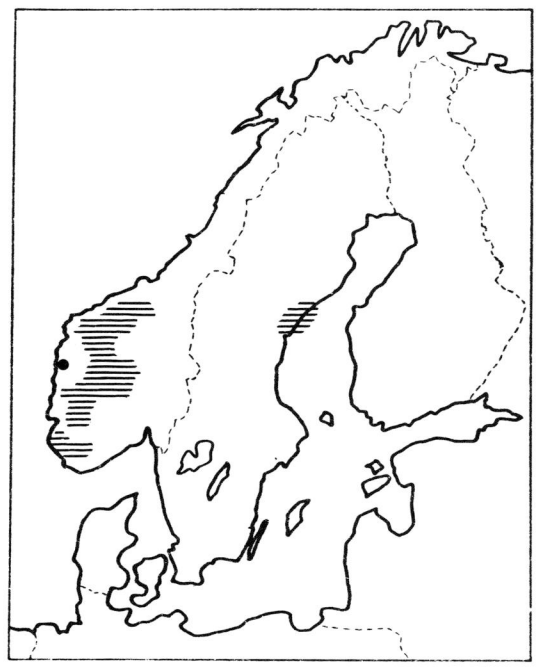

Fig. 18 The Fennoscandian distribution of *Cardaminopsis petraea*

eastwards through Siberia there are two closely related species, *C. septentrionalis* and *C. media*, both mapped as *Arabis petraea* by Hultén & Fries (1986). *C. petraea* s. str. is, however, a European species, and we shall restrict our attention to that.

Its Scandinavian distribution is peculiar. In Norway, it is clearly an alpine species, although there are numerous localities at lower levels. In Sweden, it grows in exposed localities on the skerries in Ångermanland. The distribution of the species reveals a distinct oceanic tendency. There are many localities in coastal mountains, especially in the northwestern part of West Norway. It is also common in the Jotunheimen. In Dovre, it is absent from natural habitats, but occurs as an apophyte at Kongsvoll railway station.

The ecological behaviour of the species is very variable.

It is chiefly found in the low-alpine and middle-alpine belts, on screes and in stony places. It also frequently grows on ledges, in stream and river beds and on moraines. It behaves as a pioneer plant on moraines, roadsides and in gravel pits, and I have also observed it as a pioneer species on the dry floor of a partly drained lake (Sandvatn in Sunndalen).

In central Europe, *C. petraea* is considered a calciphilous species, and it is very closely associated with calcareous schists at alpine localities in Norway. It usually grows dispersed, and I have only observed it growing abundantly at its northernmost locality at Storlifjell in the Trollheimen, which has calcareous schist bedrock. On the other hand, it seems to inhabit localities where it has little competition, such as gravel and bare rock, irrespective of their composition. In Sunnmøre, there are numerous lowland localities, almost without exception on olivine-rich rocks (Bjørlykke 1938). The vegetation on such bedrock is usually very open, consisting of few species.

On screes, *C. petraea* may grow where only a fairly thin snow cover accumulates, but it also frequently occurs on open gravel in snow-beds, accompanied by, among others, *Ranunculus glacialis* and *Oxyria digyna*.

The history of *C. petraea* was discussed by Holmboe (1937) who believed that the species had survived the last glaciation on ice-free refugia in the Møre area. He, furthermore, explained the occurrence in Ångermanland in terms of immigration from Trøndelag: "There will be no difficulty in supposing, that in the course of the Postglacial time it may have been crowded out in districts like Jemtland and the greater part of Ångermanland, while it has been able to persist better in the outer skerries" (Holmboe 1937, p. 24).

Like some other species with a strong affinity to screes and gravel, *C. petraea* may have had favourable conditions for dispersal during late-glacial and early postglacial times.

Carex arctogena H. Sm.

Maps: Smith (1940), Hultén (1971a, no. 325). Total: Hultén (1958, no. 19), Hultén & Fries (1986, no. 430). Local: Bråthen (1973).

First Norwegian record: "Monte Nordcap", P. V. Deinboll 1822 (TRH).

Norwegian distribution: Bicentric. In the southern area, from Vang in Oppland to Grimsdalen. In the northern area, from Gapsfjell in Hattfjelldal to Grense Jakobselv.

Altitude limits: Low-alpine to middle-alpine. The Jotunheimen, Glittertind 1590 m (NAS). Down to about 960 m in Grimsdalen. Troms: Bardu, Rubben 1110 m. Down to sea level in East Finnmark.

Habitat: Wind-eroded heaths, open gravelly soil often affected by solifluction. Indifferent, but tends to be basiphilous.

C. arctogena was described by Smith (1940) from

Moskana, a mountain in Torne Lappmark, close to the Norwegian border. It had previously been included in *C. capitata*. Smith's revision of the Norwegian material

locality. In Arctic North America, *C. arctogena* is found from Labrador westwards to Manitoba. In all, this distribution is like that of a number of amphi-Atlantic species.

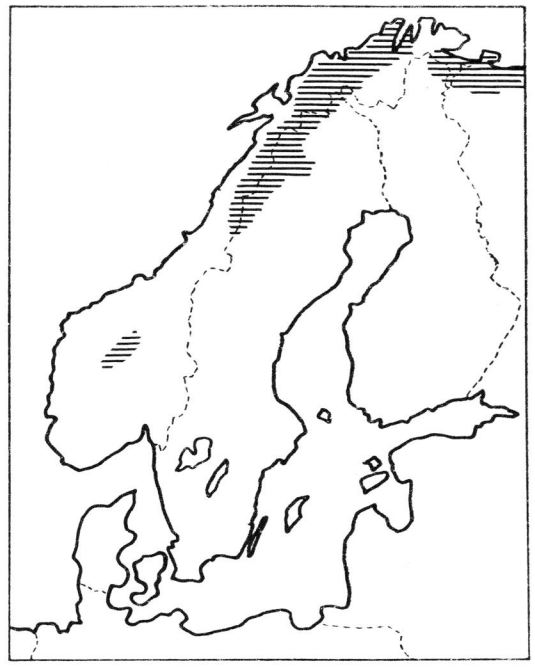

Fig. 19 The Fennoscandian distribution of *Carex arctogena*

Carex atrofusca Schkuhr

Maps: Hultén (1971a, no. 398). Total: Hultén (1962, no. 39, 1968), Hultén & Fries (1986, no. 516). Local: Ouren (1952, 1959, 1961, 1966), Benum (1958), Toftaker (1969), Hagen (1976), Elven (1979), Nordsteien (1982), Holten (1983, 1984), Wilmann (1983), Moe (1985).

First Norwegian record: The oldest herbarium specimens were collected by Kindberg on Dovre in July 1800 (herb. Sommerfelt) (O).

Norwegian distribution: From Hjelmeland in Rogaland to Magerøya in Finnmark.

Altitude limits: Mainly northern boreal, low-alpine and middle-alpine. Bukkeholstindane in the Jotunheimen 1880 m (NAS). Down to 430 m in South Norway (Rindal). Troms: Storfjord, Riep'pigai'si 1040 m. Down to sea level in North Norway.

Habitat: Most common in extremely rich fens and springs; also occurs in damp fen-like heath and grassland, and beside streams and lakes. Only in areas with calcareous soil (pH ca. 6-7).

revealed a large number of localities, mostly in North Norway, but some in the Jotunheimen. The new species, therefore, proved to be bicentric.

In South Norway, the distribution is clearly continental, and the same is on the whole found in North Norway.

Hultén (1958) held *C. arctogena* to be an arctic variety of *C. capitata*, and this view has been followed in Flora Europaea. I do not share this opinion (see also Nilsson 1986). The morphological differences pointed out by Smith are clear. *C. arctogena* and *C. capitata* differ phytogeographically; the latter is ubiquitous in Fennoscandia, whereas *C. arctogena* is bicentric; *C. capitata* is a circumpolar species with numerous localities in central Europe and central Asia, whereas *C. arctogena* is amphi-Atlantic.

The ecological differences between the species are also obvious. *C. capitata* is mainly a subalpine species of shallow rich fens, whereas *C. arctogena* is low-alpine to middle-alpine growing in exposed places, very often in communities on patches of gravel. It prefers calcareous soils, but several reports refer to oligotrophic *Empetrum* heaths, the humus layer being poorly developed in such localities. The species may also grow on peridotite and serpentinite.

In Sweden, *C. arctogena* is known from Lycksele Lappmark to Torne Lappmark, and in Finland it has been found in Enontekis Lappmark and Enare Lappmark. There are also some localities on the Kola Peninsula. *C. arctogena* is widely distributed in the southern half of Greenland, whereas *C. capitata* is known from only a single

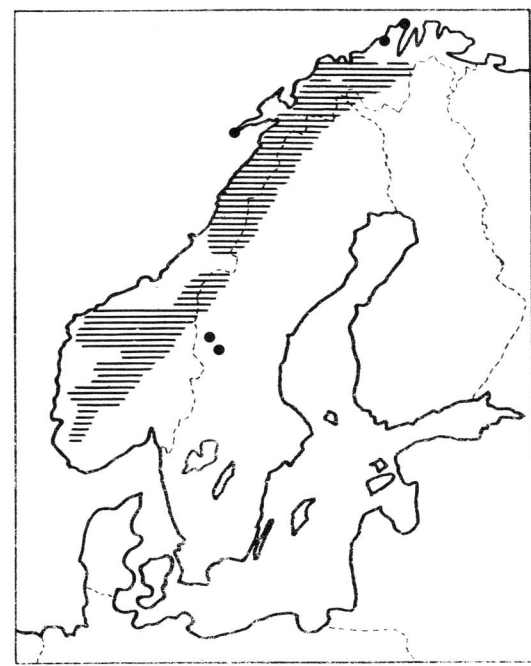

Fig. 20 The Fennoscandian distribution of *Carex atrofusca*

C. atrofusca is a circumpolar, arctic-alpine species, absent from Svalbard and Iceland, but occurring in Scotland, Greenland, North America and Asia. There is a 500 km gap between easternmost Kola and the Urals.

C. atrofusca is rare in the coastal mountains of South

Norway. There are numerous localities on Hardangervidda, but distinctly fewer in the Jotunheimen. The species is very common from Dovre eastwards, thus displaying a continental tendency probably explained by its ecological demands. It is rare in Nord-Trøndelag, but common in Nordland, Troms and West Finnmark. There are many coastal localities in North Norway. It is again rare on Finnmarksvidda and in East Finnmark, being absent from large areas. This is also the case on the Kola Peninsula. In Sweden, *C. atrofusca* occurs from Härjedalen to Torne Lappmark, but in Finland it has only been found in Enontekis Lappmark.

C. atrofusca is first and foremost a species of extremely rich fens, mainly occurring on the margins where the peat layer is thin. It also occurs on the thicker peat of sloping fens in a type of vegetation characterised as mire expanse. It is common in large areas of the Sølendet Fen Nature Reserve, east of Røros (Gaare 1963, Moen 1990). For centuries, this fen was harvested by the local farmers. Management of the nature reserve has included reintroduction of scything, which is clearly beneficial to *C. atrofusca*. In general, the species is favoured in boreal fens by such activities as trampling, grazing and scything which promote the formation of open ground and prevent overgrowth.

C. atrofusca is a characteristic species of the *Caricion bicolori-atrofuscae* alliance (Nordhagen 1936a). In alpine areas, especially oceanic parts, it also grows on rich, moist heathland together with species like *Dryas octopetala*, *Antennaria dioica*, *Campanula rotundifolia* and *Carex rupestris*, e.g. in the Trollheimen (Baadsvik 1974, Nordsteien 1982) and in the Hardangervidda-Voss area (Lid 1959, Moe 1985). *C. atrofusca* also grows in the damp parts of the *Kobresio-Dryadion* alliance.

C. atrofusca is a typical species of the rich spring vegetation of upland areas, mainly growing in springs issuing from mineral soil, and on the margins of springs, accompanied by such species as *Epilobium* spp., *Saxifraga aizoides*, *Juncus castaneus* and *J. triglumis*, and with *Cratoneuron* spp. dominating the bottom layer. It also occurs on the margins of spring-fed streams and on flushed slopes, accompanied by the same species as above (Moe 1985). In Hordaland, steeply sloping flushes are the main habitats for *C. atrofusca*, *C. microglochin*, *C. saxatilis* and *Juncus castaneus*.

Asbjørn Moen

Carex bicolor All.

Maps: Arwidsson (1943), Hultén (1955, 1971a, no. 396). Total: Hultén & Fries (1986, no. 482). Local: Lid (1954), Benum (1958), Moen (1976), Aune & Kjærem (1978).
First Norwegian record: Hedmark: Alvdal, "1/4 mil V-for Sten v. færgestedet i sand". 1854. J. E. Zetterstedt (O).

Norwegian distribution: Bicentric. In South Norway, on the banks of the River Folla and its tributaries, the Grimsa and Einunna. In 1973, numerous localities were discovered by Moen in Innerdalen, Kvikne (Moen 1976), but they were inundated when a large hydro-electricity reservoir was constructed. In 1988, A. Moen (pers. comm. 1988) found several localities in the upper part of the Orkla valley, near Dølvadseter. In North Norway, *C. bicolor* has a dispersed distribution from Rana (Østerdalsisen) to Børselv in Porsanger.

Altitude limits: Dovre: Tverrelvdalen 1050 m. Alvdal 470 m. Nordland: Sørfold, Leirvassfjell 1180 m. Troms: Målselv, Gråhøgda 1050 m. Down to sea level in Porsanger.

Habitat: River banks, shallow eutrophic fens, solifluction soils and snow-beds. Basiphilous.

C. bicolor is a circumpolar species, but has a very disjunct distribution. In Sweden, it is known in Härjedalen and from Pite Lappmark to Torne Lappmark. It is otherwise known in Europe from one locality in the Hebrides, and from the central European mountains and northeastern Russia, as well as Iceland. It occurs from Greenland to Alaska, and has a very dispersed distribution through Siberia. The Scandinavian area is therefore very isolated.

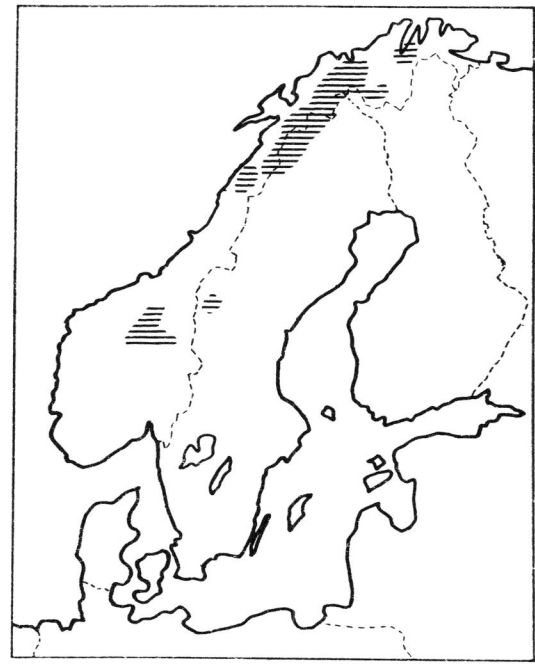

Fig. 21 The Fennoscandian distribution of *Carex bicolor*

The ecological behaviour of *C. bicolor* is somewhat puzzling. In the southern area in Norway, it is only found on river banks, the localities forming a continuous chain along the River Folla and its tributary, the Grimsa. It grows on unstable sandy, clayey and gravelly ground. The localities are normally flooded every year. The vegetation is open and there are numerous accompanying species. Lid (1954) published a series of analyses from the southern area. The most constant and important accompanying species are: *Agrostis mertensii, Carex capillaris, C. microglochin, Festuca ovina, F. rubra, Juncus arcticus, Poa alpina, Astragalus alpinus, Equisetum variegatum, Euphrasia frigida, Oxytropis lapponica, Parnassia palustris, Polygonum viviparum, Primula scandinavica, P. stricta, Saxifraga aizoides, Selaginella selaginoides, Tofieldia pusilla, Salix arbuscula* and *S. hastata*. This list shows that the plant community consists mainly of alpine species. Lid named his community *Caricetum bicoloris*.

Nordhagen (1935) published analyses of a *Carex microglochin-C. bicolor* sociation from Børselv in Porsanger that is very similar to the community described by Lid. The locality is almost at sea level. Nordhagen proposed a *Caricion bicoloris* alliance, but later modified it to *Caricion bicoloris-atrofuscae* (Nordhagen 1936a, 1943). In 1943, he included it in his rich fen alliance, *Caricion atrofuscae-saxatilis*. In my opinion, the river-bank communities should not be included in this rich fen alliance.

In northern Scandinavia, most localities are on alluvial sand and gravel, as in the south. However, the species is also abundant in snow-beds at high altitudes, displaying strong competitive ability and growing in close stands with, among others, *Salix polaris, Carex bigelowii, Saxifraga oppositifolia, S. aizoides* and *Juncus biglumis* in a well-developed carpet of *Campylium stellatum, Drepanocladus uncinatus, D. revolvens, Anthelia juratzkana* and *Nostoc* sp. (Gjærevoll 1950).

Selander (1950b) thought that *C. bicolor* was an alpine species in Lule Lappmark where it grows on moist soil affected by frost and solifluction. The alpine communities mentioned by him do not fit into the *Caricion bicolori-atrofuscae* alliance.

When discussing the presence of *C. bicolor* in Finnmark, Nordhagen (1935) assumed that the species might have survived on a refugium near the mouth of Porsangerfjord, not on a nunatak but on calcareous sandy shores. The inland occurrences in South Norway, being restricted to river banks, cause ecological difficulties if survival on a southern refugium is assumed. It should be taken into consideration that *C. bicolor* also occurs as a pioneer plant on fresh gravel. This is observed near Dalholen in Folldal. Following roadworks, *C. bicolor* rapidly invaded some areas that were laid bare, and formed luxuriant cushions. It might have had favourable dispersal conditions in early postglacial time. In Rana, *C. bicolor* is abundant on the Østerdalsisen moraines (Elven 1978).

In view of the alpine occurrences in North Scandinavia, the southern Scandinavian localities seem likely to be secondary ones. So far, no primary alpine localities have been observed. In 1967, *C. bicolor* was found on the extremely remote Jensen's Nunataks in Greenland (Gjærevoll & Ryvarden 1977), at an altitude of 1350-1400 m. From an ecological point of view, this occurrence is very interesting since it shows that *C. bicolor* may grow as a

nunatak species. When the history of *C. bicolor* is being considered, great emphasis must be placed on the ecological behaviour of the species on Jensen's Nunataks and in the mountains of North Scandinavia.

Carex capitata L.

Maps: Smith (1940), Hultén (1955, 1971a, no. 324). Total: Hultén (1958, no. 20, 1968), Meusel et al. (1965), Hultén & Fries (1986, no. 429). Local: Benum (1958), Hagen (1976), Kristensen (1981).

First Norwegian record: Wahlenberg (1812, no. 419).

Norwegian distribution: From Geitagilet in Vedal, Eidfjord, Hordaland to Brekken (Røros), and from Gapsfjell in Hattfjelldal to Sør-Varanger.

Altitude limits: Mainly northern boreal to low-alpine. Oppdal: Knutshø 1250 m (Lid (1963) reported it up to 1400 m, but this altitude has not been confirmed). Stor-Elvdal 300 m. Troms: Målselv, Tverrelvdal 680 m. Down to sea level in North Norway.

Habitat: Mainly on rich fens and river banks. Basiphilous.

Hultén (1958) thought *C. capitata* was an amphi-Atlantic species, but the map in Hultén & Fries (1986) clearly shows it to be a circumpolar, arctic-montane plant.

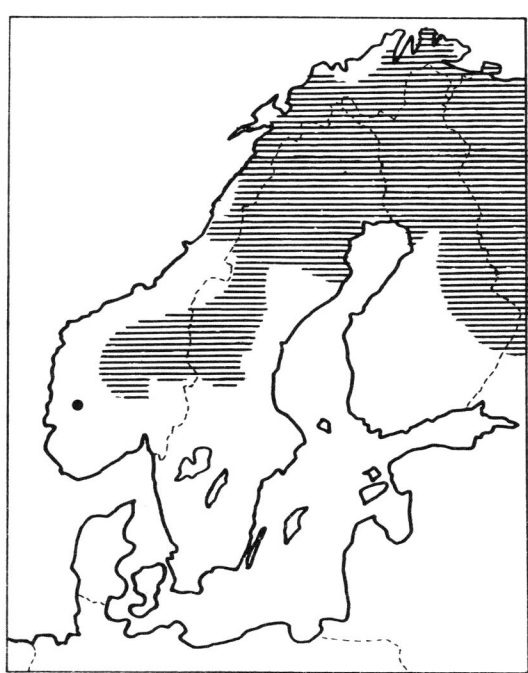

Fig. 22 The Fennoscandian distribution of *Carex capitata*

It is very rare in the southernmost part of the Norwegian mountains. It clearly avoids coastal mountains in South Norway, but there are numerous localities in the continental, eastern mountains. This easterly tendency becomes

44

still more clear when the Fennoscandian distribution is considered. In Sweden, it is found from Dalarna to Torne Lappmark with localities down to the Gulf of Bothnia. If the Norwegian distribution is regarded in isolation it gives the impression of bicentricity, but *C. capitata* has a more continuous distribution through the eastern part of the Scandinavian mountain chain.

In addition to rich fens, *C. capitata* may grow on scree slopes and cliff ledges. There are many coastal localities in North Norway, mainly on shell sand.

It is a characteristic species of the *Caricion bicolori-atrofuscae* alliance (Nordhagen 1936a, 1943), often growing together with *Carex microglochin*, *Kobresia simpliciuscula* and *Scirpus pumilus*. It may be common on river banks, e.g. along the River Folla and its tributaries, where it occurs with *Carex bicolor*, *C. maritima* and *Primula stricta*.

Most localities are in the northern boreal belt, but there are also a number in the middle boreal belt and even lower.

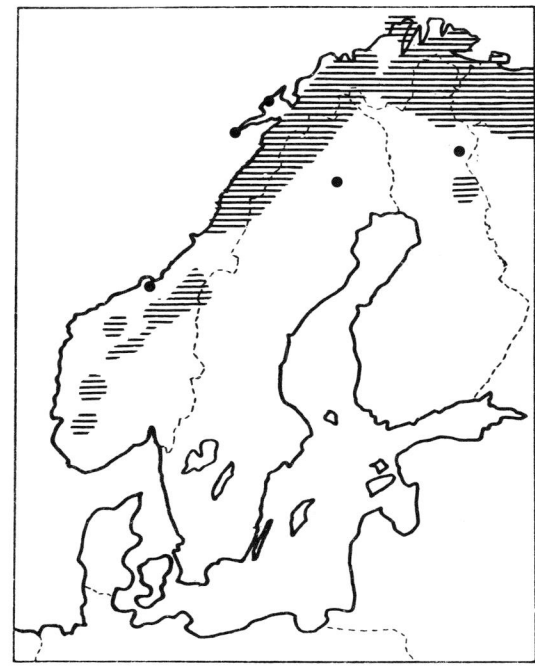

Fig. 23 The Fennoscandian distribution of *Carex glacialis*

Carex glacialis Mack.

Maps: Hultén (1971a, no. 407). Total: Hultén (1962, no. 23, 1968), Hultén & Fries (1986, no. 511). Local: Ouren (1952, 1961, 1966), Benum (1958), Lid (1959), Skogen (1974, 1979), Moen & Kjelvik (1981), Nordstein (1982), Halvorsen & Salvesen (1983), Engelskjøn (1984, 1986a), Moe (1985).

First Norwegian record: Saltdal: Boskarheiene. Sommerfelt 1826 (O).

Norwegian distribution: Slightly bicentric. In the southern area, from Veneheia in Suldal to Hårskallen in Levanger. In the northern area, from Leka to North Cape and Grense Jakobselv.

Altitude limits: Mainly low-alpine. Rusteggji in the Jotunheimen 1500 m (NAS). Troms: Bardu, Råkkunbårri 1180 m. Frequently down to sea level in Finnmark.

Habitat: Windswept ridges and heaths, and open gravel. Basiphilous, growing on calcareous schists, marble, dolomite and peridotite.

C. glacialis is a circumpolar, arctic species. It has a continuous distribution eastwards to Siberia. It is absent from Svalbard, but occurs on Iceland. From Greenland, it has a continuous distribution to the Bering Strait.

The distribution in South Norway is characterised by some rich, but isolated occurrences, e.g. in Suldal, Ullensvang and Tafjord. The species is then rare north to the Trollheimen and the western part of the Dovre area, and also in what is the very centre of the distribution area for bicentric species, the Jotunheimen, Dovre and Sunndalen mountains. The distribution in southern Norway is continental, except for a locality at Lensvik near the mouth of Trondheimsfjord. There are numerous localities on the mountains on Rørosvidda. Most localities in South Norway are between 800 and 1000 m.

In the northern area, there are also some localities on coastal mountains. Localities in Troms range from near sea level to 1180 m. In Finnmark, *C. glacialis* has been reported from a number of localities at 400-500 m, but it also occurs frequently and abundantly in lowland districts, especially on dolomite.

In Sweden, *C. glacialis* is found from Härjedalen to Torne Lappmark, and in Finland it occurs in Kuusamo and from Sompio Lappmark to Enontekis Lappmark.

C. glacialis is one of the most characteristic species of the *Kobresio-Dryadion* alliance, growing especially in the most exposed parts along with *Kobresia myosuroides* and *Carex rupestris*, and, in the northern area, *C. nardina*. *C. glacialis* is definitely a basiphilous species, but may also, like *Dryas octopetala* in Finnmark, grow on weathered gabbro and non-calcareous gravel lacking humus. It is also sporadically found as a pioneer plant on moraines and roadsides.

C. glacialis is one of the dominant species on amphibolite on Jensen's Nunataks in Greenland (Gjærevoll & Ryvarden 1977).

Carex holostoma Drej.

Maps: Nygren (1936), Hultén (1971a, no. 392). Total: Nygren (1936), Hultén (1958, no. 206, 1968), Hultén & Fries (1986, no. 489). Local: Rønning (1956a), Benum (1958), Ryvarden (1969), Bråthen (1973), Engelskjøn (1986a).

First Norwegian record: Finnmark: Alta, Tverrelvdalen, J. M. Norman 1861 (O, BG, TRH).

Norwegian distribution: From the mountains of Aksla and Sildviktind, near Narvik, to Grense Jakobselv.

Altitude limits: Mainly low-alpine. Troms: Storfjord, Paras mountain 1000 m. Finnmark: Sør-Varanger, Gandvik, ca. 50 m.

Habitat: Most common on seasonally to occasionally irrigated gravelly soil derived from various types of bedrock. The hydrological amplitude is rather narrow.

In Norway, *C. holostoma* was first reported by Norman in 1861 from Tverrelvdalen in Alta. Norman described it as a new species, *C. cryptandra*. Later, it proved to be identical with *C. holostoma*, previously described from Greenland.

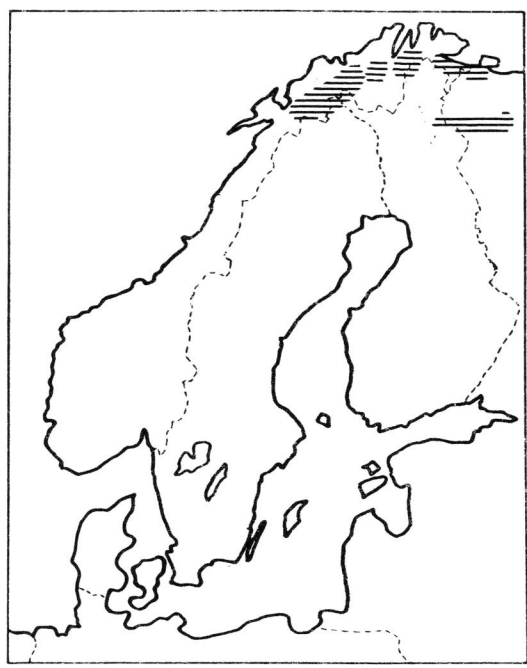

Fig. 24 The Fennoscandian distribution of *Carex holostoma*

C. holostoma belongs to the northern unicentric group of alpine species. Its distribution is somewhat disjunct. Three areas have numerous localities, namely northern Troms, the "gai'sa" mountains in Finnmark, and eastern Finnmark. However, it is an easily overlooked species. It is very rare in Sweden, being known only from Torne Lappmark. In Finland, it has been reported from Sompio Lappmark, Enare Lappmark and Enontekis Lappmark. Some localities on the Kola Peninsula and one in Iceland complete the European distribution.

Eastwards, there is a wide gap to the nearest, very isolated locality near the mouth of the River Jenisej. From east of Lake Baykal, its distribution is more or less continuous eastwards to East Greenland, but altogether the species is rare.

C. holostoma has usually been regarded as an amphi-Atlantic species (Hultén 1958). Immigration from the east after the last glaciation seems unlikely.

The distribution is slightly continental. The localities in eastern Finnmark should also be regarded as continental since precipitation is low in that area. The substrate

varies quite considerably, from gneiss and other acidic rocks to mica schists.

C. holostoma is most frequent in the low-alpine belt. It may descend into the birch forest, and, in such cases, occurs on cliff ledges that are usually irrigated in early summer. It particularly favours wet sites with open gravel and soil affected by solifluction. It may occur as a pioneer plant in gravelly places, e.g. roadsides. It may also grow in fen-like communities on river banks.

The results of analyses carried out on Rastigai'si (Ryvarden 1969), where the species is fairly common, are given in Table 5. It can be seen that the community is open.

Table 5

Square no. (1m²)	1	2	3	4	5	6	7	8	9
Vaccinium uliginosum	.	.	.	.	.	1	.	1	1
Empetrum hermaphroditum	.	.	.	.	1	1	1	2	
Scirpus cespitosus subsp. austriacus	1	1	.	1	1	.	.	.	.
Carex adelostoma	.	.	.	.	.	.	.	1	2
C. holostoma	1	1	1	1	1	3	5	3	4
C. nigra	1	.	.	1	1	.	1	.	1
Luzula spicata	.	1	.	.	.	.	.	.	.
Polygonum viviparum	1	1	1	1	1	.	.	.	.
Lychnis alpina	.	.	.	1	.	.	.	.	.
Thalictrum alpinum	.	.	.	.	.	1	1	2	1
Saussurea alpina	.	.	.	.	.	.	1	.	1
Stones or water	5	5	5	5	5	5	3	5	4

Squares 1-5 are from the southern slope of Jorbakgai'si (ca. 550 m), and squares 6-9 from beside Lævajåkka (ca. 380 m).

Leif Ryvarden

Carex macloviana D'Urv.

Maps: Alm (1944), Hultén (1971a, no. 342). Total: Du Rietz (1940). Arctic area: Hultén (1958, no. 185, 1968), Hultén & Fries (1986, no. 466). Local: Benum (1958).

First Norwegian record: Tromsø: Tromsdalstind 1841, M. N. Blytt (BG, TRH, O, C).

Excluded stations: Glåmos: Stuedalsvollen 1841, J. Ångström (S). As Alm (1944) pointed out, this occurrence must have been occasional.

Norwegian distribution: From Dunderlandsdalen in Nordland to Tromsø and Lyngen.

Altitude limits: Mainly northern boreal to low-alpine. Troms: Storfjord, Paras mountain 1000 m (Engelskjøn 1986a).

Habitat: Meadows, grassy hills, damp heaths, birch forests and willow thickets. Also as an apophyte in areas carrying domesticated reindeer.

46

C. macloviana is an amphi-Atlantic plant and is also bipolar. The species s. lat. has an extremely disjunct distribution. It occurs in Iceland, Greenland and western Canada, and then in Alaska and western America, Kamchatka, Mexico, the Andes, Hawaii and the Falkland Islands. In Sweden, *C. macloviana* is found in the moun-

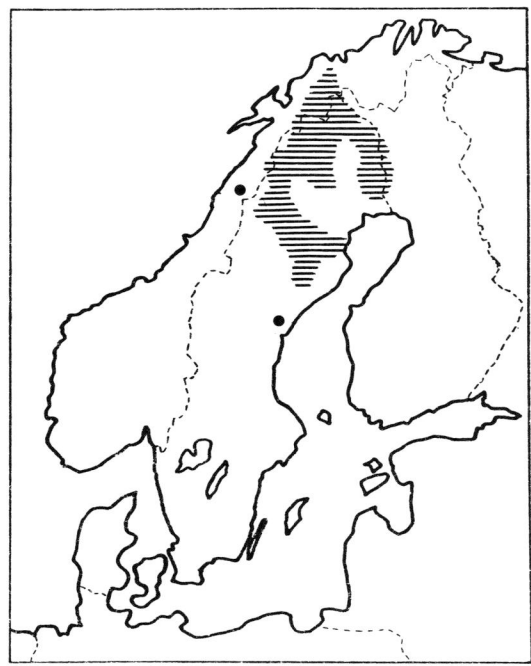

Fig. 25 The Fennoscandian distribution of *Carex macloviana*

tains from Pite Lappmark to Torne Lappmark, and also alongside rivers from Ångermanland to Norrbotten down to the Baltic Sea. In Finland, it occurs in a similar fashion from Enontekis Lappmark to Nordbotten.

The natural habitats of *C. macloviana* are northern boreal and low-alpine meadows and slopes. I have found this to be true in Alaska, too. It may also grow in *Carex bigelowii* communities and in fairly late snow-beds along with *Sibbaldia procumbens* (Gjærevoll 1956). It seems indifferent to soil conditions.

C. macloviana is widely dispersed where domesticated reindeer migrate. As pointed out by Benum (1958), it is frequently found near Lapp summer camps and "near roads and paths in the woodland where Swedish Lapps and reindeer are moving through on their annual passages to and from the summer pastures in Troms fylke". The same was reported by Alm (1944) and Selander (1950a) from Lule Lappmark. Benum also pointed out that the plant "is absolutely unknown in areas of northern Troms and Finnmark where solely Norwegian Lapps have their reindeer pastures".

The isolated occurrence of *C. macloviana* in northern Fennoscandia corresponds well with that of several northern and amphi-Atlantic species. Its history has therefore been considered with this in mind. Alm (1944) was of the opinion that "*C. macloviana* most probably survived the last glaciation in Northern Europe on refuges free from

ice". Selander (1950a) discussed the disjunct distribution of *C. macloviana* and considered that it had "doubtless obtained its distribution, broadly speaking, as early as during the Tertiary, and in Scandinavia it belongs to the glacial survivors that probably lived there through at least two glaciations". These confident statements conflict with the ecological behaviour of the species since it is not reported from any high altitudes and does not appear to be a hardy species. Its disjunct distribution in the northern hemisphere as well as totally, however, indicates that the species must have a long and complicated history.

Carex microglochin Wahlenb.

Maps: Hultén (1971a, no. 322). Total: Hultén (1958, no. 214, 1968), Meusel et al. (1965), Hultén & Fries (1986, no. 433). Local: Ouren (1952, 1959, 1961, 1966), Benum (1958), Lid (1959), Toftaker (1969), Moen (1970), Hagen (1976), Holten (1983), Wilmann (1983), Elven (1984), Moe (1985).

First Norwegian record: Wahlenberg (1812) wrote: "Hab. in uliginosus subalpinis maritimis Norlandiæ rarius copiose ex. gr. in Tromsöa et in Önäset Saltensi frequenter". Specimens were collected at Önäset on 2.7.1807 (UPS). Christen Smith collected *C. microglochin* at Tolga in 1807 (O. Dahl 1895).

Norwegian distribution: From Kortmark on Hardangervidda to Magerøya, and to Skallelv in Vadsø.

Altitude limits: Mainly in the northern boreal and low-alpine belts. Up to 1450 m on Heggjeitlane, Hardangervidda (Lid 1959). Down to 340 m in Nordmøre. Troms: Bardu, Salangsdalen, Veslekletten 870 m. Down to sea level in North Norway.

Habitat: *C. microglochin* grows exclusively on calcareous substrates, extremely rich fens and in types of vegetation that are transitional between fens and heathlands and/or grasslands. It occurs in open vegetation, very often on eroded ground.

C. microglochin was recorded by Hultén (1958) as an amphi-Atlantic species, whereas Hultén & Fries (1986) classified it as a disjunct, circumpolar species. It occurs in Fennoscandia, northern Russia, Scotland (very rare), Iceland, Greenland and North America, and also in the central European mountains, the Caucasus and central Asia. This means that the Fennoscandian area is extremely isolated. *C. microglochin* is bipolar, as it occurs in South America.

The distribution of *C. microglochin* in southern Norway shows a continental tendency. The species is common on the Hardangervidda, but rare or absent in the Valdres-Sogn district. It has a dispersed occurrence in Nord-Trøndelag, but continuous distribution through North Norway, including a number of coastal localities. In Sweden, it is known from Härjedalen to Torne Lappmark, but has a considerable gap in northern Jämtland. The Finnish distri-

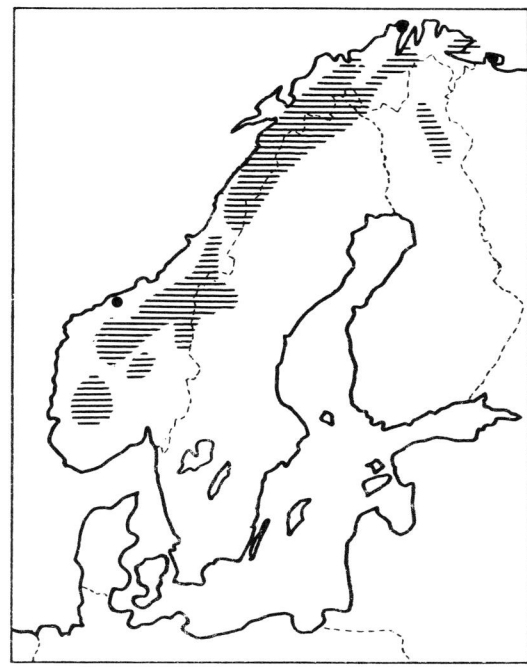

Fig. 26 The Fennoscandian distribution of *Carex microglochin*

bution comprises Sompio Lappmark, Enare Lappmark and Enontekis Lappmark. Its absence or scattered occurrence in many areas can partly be explained by absence or scarcity of calcareous bedrock.

C. microglochin is a weakly competitive species favoured by trampling, grazing, scything and other activities that promote the formation of open areas and prevent over-growth. It occurs in areas with calcareous soil where the water table is at least periodically high. Three main types of habitat can be distinguished.

Moist, sandy or gravelly soil: This habitat is usually found near streams or on seashores, e.g. the *Scirpus pumilus* and *Carex bicolor* communities in Porsanger (Nordhagen 1935) and the *Carex bicolor* communities in Folldal (Lid 1954). According to Nordhagen (1935), it belongs to the alliance then named *Caricion bicoloris*. Nordhagen(1943) later distinguished a separate association, *Caricetum microglochinis*, for communities dominated by *C. microglochin*. Dierssen (1982) also included a number of rather different fen communities, all dominated by *C. microglochin*, in this association.

Fen-like heathlands: This type of vegetation is transitional between rich heathlands and rich fen vegetation in alpine areas, usually with some open ground due to local erosion, for example. Nordhagen (1928) described *C. microglochin* communities of this type from the Sylane mountains that include typical heathland species such as *Carex rupestris*, *Diapensia lapponica*, *Dryas octopetala* and *Festuca vivipara* mixed with *Carex atrofusca*, *Kobresia simpliciuscula*, *Pedicularis oederi* and *Thalictrum alpinum*. Nordhagen (1943) included these communities in his *Caricion atrofuscae-saxatilis* alliance.

Extremely rich fens: *C. microglochin* occurs in the mud bottom, carpet and lawn communities of extremely rich

fens; mainly in mire margin communities, transitional to types described above. These communities are included in the *Caricion bicolori-atrofuscae* alliance.

C. microglochin also grows in fen expanse vegetation on thick peat. It dominates some flarks and flat fen areas in the northern boreal belt and the upper part of the middle boreal belt.

Asbjørn Moen

Carex misandra R. Br.

Maps: Hultén (1955, 1971a, no. 397). Total: Hultén (1962, no. 14, 1968), Meusel et al. (1965), Hultén & Fries (1986, no. 517). Local: Benum (1958), Gjærevoll (1963b, 1980), Toftaker (1969), Hagen (1976), Schumacher & Løkken (1981), Nordsteien (1982), Engelskjøn (1984, 1986a).

First Norwegian record: The oldest collection dates back to the visits of Laestadius to Storfjord in Troms in 1828 and 1829 – "Tjatsa supra Lyngen" (S).

Excluded stations: "Varanger i Østfinmarken" 1822, P. V. Deinboll (BG, O); Rastigaissa 1822, P. V. Deinboll (BG); Saltdal, Solvaagtind, 1.8.1914, D. E. Hylmö (LD); Ringvatsø, -7/1896, ex. herb. O. A. Hoffstad (UPS). The localities in Finnmark (the first two) have been considered very dubious (see O. Dahl 1934); thorough investigations have subsequently been carried out in the Rastigai'si area, but *C. misandra* has not been found. The same applies to Solvågtind in Saltdal, one of the best investigated mountains in the country. Ringvassøy is fairly close to the localities in Tromsø, but no details are known about this supposed locality, and it seems strange that no material is to be found in Norwegian herbaria.

Norwegian distribution: Bicentric. The southern area is very concentrated, from Tron in Alvdal in the east to Store Aurhø in Sunndalen in the west and from Grønhø in Lesja in the south to Mellomfjell in Surnadal in the north. The northern area stretches from Filfjellet in Ballangen to Talvik in Alta and Kågen in Skjervøy.

Altitude limits: Low-alpine to middle-alpine. In the southern area, from ca. 1000 m to 1660 m (Nordre Knutshø). In the northern area, from 200 m (Talvik) to about 1500 m in inner Troms. A sterile specimen has been reported from 1675 m on Kirkestinden in Målselv (Engelskjøn 1986a).

Habitat: Prefers solifluction soil, irrigated or moist cliffs, damp heaths, and meadows which become snow-free late. Calciphilous.

C. misandra is a circumpolar, arctic species. (Flora Europaea 5/1980 classified it as a subspecies of *C. fuliginosa*.) Its occurrence in northern Europe is very disjunct. There is an isolated locality in the central part of the Kola Peninsula, about 500 km east of the Finnmark localities, and then another vast gap eastwards to Novaya Zemlya and the Urals. It is found on Svalbard, but not Iceland. From

48

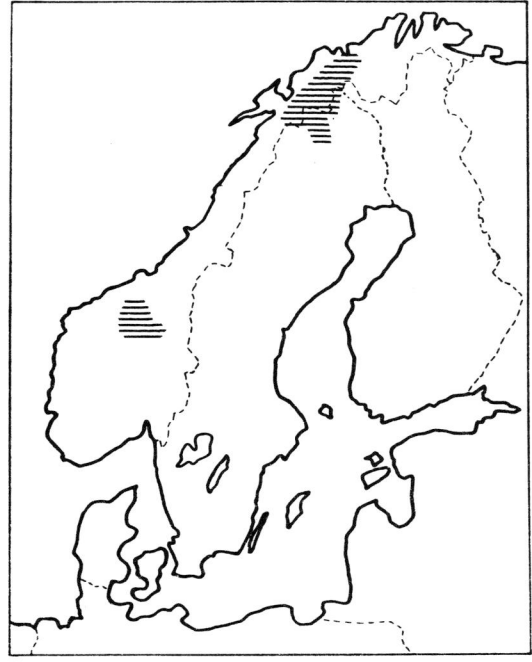

Fig. 27 The Fennoscandian distribution of *Carex misandra*

Greenland there is continuous distribution westwards through Arctic North America and Siberia. The Swedish distribution is restricted to northern Lule Lappmark and Torne Lappmark. In Finland, *C. misandra* is only found in Enontekis Lappmark.

In the southern area, *C. misandra* is very common in the mountains of the Dovre-Sunndalen-Trollheimen area. As ecologically potential habitats exist there, its absence from the Jotunheimen is strange. In North Norway, it is chiefly a continental species, but has some stations in coastal mountains. Here, too, its absence from the mountains of the Junkerdalen area (and southern Lule Lappmark) is strange in view of ecological factors (cf. Selander 1950a).

Its bicentricity and isolated occurrence in Fennoscandia make immigration from the east rather unlikely. If refugia have existed, *C. misandra* may have had possibilities for survival, looked at from an ecological viewpoint.

The ecological range of the species is fairly wide. It may descend into the northern boreal belt on irrigated cliffs and open places. However, it is first and foremost a middle-alpine species found on wet or moist solifluction soil and in various meadow and damp heath communities. Nannfeldt (1940) refers to analyses carried out by Nordhagen from a *C. misandra-Drepanocladus intermedius-Blindia acuta* sociation on Nordre Knutshø. Several species play an important role, e.g. *Deschampsia alpina*, *Juncus biglumis*, *Poa stricta*, *Polygonum viviparum* and *Ranunculus glacialis*. Nordhagen (1954a) placed this community in the *Luzulion arcticae* alliance.

Hatlelid (1980) described two communities from Knutshø, namely a *C. misandra-C. saxatilis*-community and a *C. misandra-Silene acaulis*- community, both from the middle-alpine belt. The former is found in wet, partially irrigated

areas with a long-lasting snow cover. Important companions are *Polygonum viviparum*, *Carex bigelowii*, *Ranunculus glacialis*, and the dominating *Drepanocladus revolvens* in the bottom layer. Hatlelid presumed that this community should be included in the alliance, *Ranunculo-Poion alpinae* Gjærevoll 1956. The latter is found on damp soil with a scanty snow cover. Important constituents are *Salix polaris*, *Polygonum viviparum*, *Pedicularis oederi*, *Saxifraga oppositifolia*, *Juncus biglumis*, and in the bottom layer, *Ditrichum flexicaule* and *Drepanocladus revolvens*. Referring to the above-mentioned analyses carried out by Nordhagen, Hatlelid tentatively included this community in *Luzulion arcticae*.

Carex nardina Fr.

Maps: Hultén (1971a, no. 323), Engelskjøn (1986a). Total: Hultén (1958, no. 168, 1968), Meusel et al. (1965), Hultén & Fries (1986, no. 425). Local: Benum (1958), Aune & Kjærem (1978), Engelskjøn (1984).

First Norwegian record: Saltdal, Junkerdalen, J. Ångström 1837. The species was described by E. M. Fries (1839), based on material collected by J. Ångström in Junkerdalen and Lule Lappmark.

Norwegian distribution: From Čampo, a mountain in Bjellådalen in Rana, to Finskefjell, near Sopnes in Alta.

Altitude limits: Mainly low-alpine. From 235 m on Haukøy, Skjervøy, to 1265 m on Kirkestinden in Målselv. 1440 m in Torne Lappmark.

Habitat: Exposed heaths on calcareous rocks.

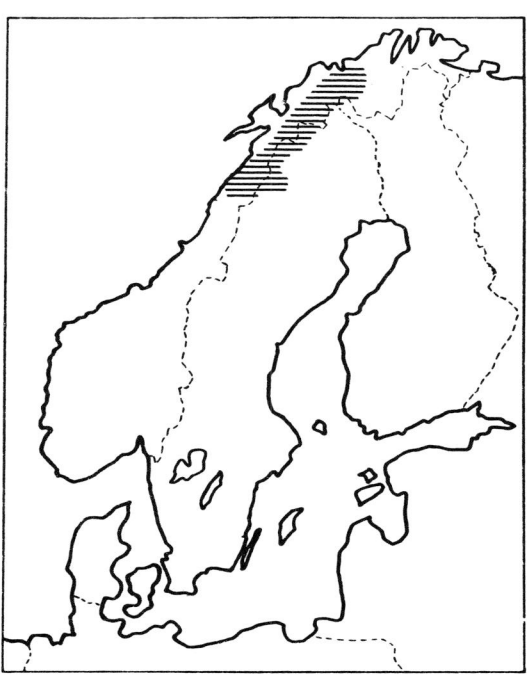

Fig. 28 The Fennoscandian distribution of *Carex nardina*

C. nardina belongs to the northern amphi-Atlantic species. Although there are some localities on coastal mountains (e.g. western Saltfjellet), the distribution is mainly continental. In Sweden, it is known from Pite Lappmark to Lule Lappmark. It also occurs on Svalbard and Iceland. It is widely distributed from East Greenland to the Rocky Mountains and Alaska, and also has a few localities in easternmost Siberia.

The extremely isolated occurrence in northern Scandinavia is one indication of how it may have survived the Ice Age. It was the most common species on Jensen's Nunataks in Greenland (Gjærevoll & Ryvarden 1977), which is interesting since it is otherwise rare in Greenland (Böcher et al. 1966).

The ecological range of *C. nardina* is fairly narrow. It is a highly chionophobous species, inhabitating heaths and ridges on limestone, dolomite and weathered calcareous schists exposed to severe winters with low temperatures and strong deflation. It also grows on scree slopes consisting of calcareous rocks. At its southernmost locality, in Bjellådalen, it grows on serpentinite, the same being found by Selander (1950b) in Lule Lappmark.

It is very abundant at many localities, e.g. on Skjevlfjell in Saltdal where large areas of windswept limestone are dominated by *Dryas octopetala* and *C. nardina*. *C. nardina* is a characteristic species of the *Kobresio-Dryadion* alliance. Hedberg et al. (1952), in their differentiation of this alliance in northernmost Sweden, pointed out an association, *Nardino-Dryadetum*, inhabiting very exposed localities with patches of *C. nardina*, *Dryas octopetala*, *Kobresia myosuroides*, *Saxifraga oppositifolia* and *Silene acaulis*. This is the same community as Nordhagen (1955) described as *Caricetum nardinae*. Table 6 shows an extract of 10 square analyses carried out by the author on two stands on Skjevlfjell (1-10), and 10 analyses by Nordhagen (1955) on Mikalfjell in Kvænangen (11 and 12), Pältsa (13), Skaitiaksla in Saltdal (14-18), Tausa in Saltdal (19) and the Sulitjelma area (20).

Table 6

Square no.	1	2	3	4	5	6	7	8	9	10	11	12	13	14	15	16	17	18	19	20
(1-10: 1m² 11-20: 4m²)																				
Carex nardina	3	3	3	3	3	4	3	4	3	4	4	4	3	3	3	3	2	4	3	3
C. glacialis	1	.	.	.	1	1	1	1	1	1	2	1	.	.	.	.	.	.	.	.
Kobresia myosuroides	1	.	1	.	.	.	1	1	1	.	3	1	1	1	2	1	2	1	1	1
Festuca ovina	2	1	1	1	2	1	2	1	1	2	3	2	.	1	1	1	1	1	.	1
Dryas octopetala	3	2	2	2	3	4	3	3	3	2	3	3	2	2	3	2	3	3	2	2
Salix reticulata	1	.	.	.	.	1	1	.	1	1	.	.	1	1	1	.	1	.	1	1
Antennaria alpina	.	.	.	.	.	1	.	1	1	1	.	.	.	1	.	.	.	.	.	1
Astragalus alpinus	.	.	.	.	.	.	.	.	.	.	.	1	.	1	1	.	1	.	1	1
Campanula rotundifolia	1	1	1	1	.	1	1	1	1	1	1	1	.	.	.	.	1	.	.	1
C. uniflora	.	.	.	.	.	.	.	.	.	.	.	.	.	.	.	1	1	1	.	.
Cerastium alpinum	.	.	.	.	.	.	.	.	.	.	.	.	.	.	.	1	1	1	.	1
Chamorchis alpina	.	.	.	.	.	.	.	1	1	.	.	.	.	.	.	.	.	.	.	.
Euphrasia frigida	1	.	.	.	.	1	.	1	1	.	.	.	.	.	.	.	.	.	.	.
E. lapponica	.	.	.	1	.	1	1	1	1	1	1	1	.	.	.	.	.	.	.	.
Oxytropis lapponica	.	.	.	.	.	.	.	.	.	.	.	.	.	1	1	1	1	1	1	.
Polygonum viviparum	.	.	.	.	.	.	.	.	1	.	.	1	.	.	1	.	1	1	1	.
Saxifraga oppositifolia	2	1	1	1	2	2	2	1	2	1	3	2	.	2	2	2	1	2	2	2
Silene acaulis	1	.	1	2	2	2	.	.	2	1	.	1	.	1	1	1	1	1	1	1
Thalictrum alpinum	.	.	.	.	1	.	.	.	.	.	2	1	.	.	1	.	.	1	1	1
Bare gravel and soil	4	5	5	5	4	3	4	4	4	5	4	4	5	5	4	5	2	4	4	4

In the two stands on Skjevlfjell, the pH was 7.3 and 7.5, respectively. Nordhagen reported values of 7.1 from square 11, 6.7 from square 12 and 8.5 from Skaitiaksla. Lunde (1962) published the results of 125 determinations which gave an amplitude 5.3-8.0 and a mean of 6.92.

Carex parallela (Læst.) Sommerf.

Maps: Samuelsson (1921), Hultén (1971a, no. 317). Total: Samuelsson (1921), Hultén (1958, no. 63), Hultén & Fries (1986, no. 438). Local: Benum (1958), Toftaker (1969), Kristensen (1981), Nordsteien (1982).

First Norwegian record: "Nordlandiæ saltensis copiose"; C. S. Sommerfelt (1826). No herbarium specimens exist.

Norwegian distribution: Bicentric. In the southern area, from the Jotunheimen to Soknedal. In the northern area, from Hattfjelldal to Magerøya.

Altitude limits: Low-alpine to middle-alpine. In the south, Nordre Knutshø 1590 m (Hatlelid 1980), Gynnilfjell in Soknedal 740 m. In Nordland, Skjomen 1213 m. Down to sea level in North Norway.

Habitat: Shallow eutrophic fens, solifluction soil. Basiphilous.

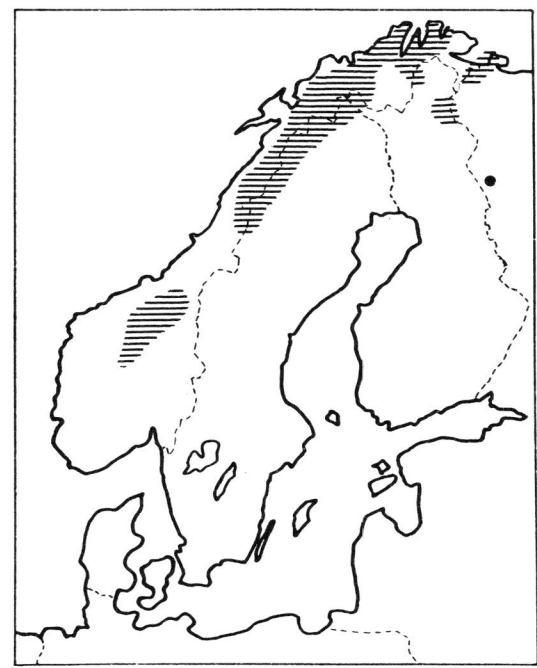

Fig. 29 The Fennoscandian distribution of *Carex parallela*

C. parallela was originally described by Læstadius (1822) as a variety of *C. dioica*. Sommerfelt (1826) gave it species rank. It was later degraded to a subspecies of *C. dioica* (Hartman 1879, Neuman 1901, A. Blytt 1906), but after thorough study by Samuelsson (1921) the rank of species has been maintained.

C. parallela belongs to the amphi-Atlantic species (*C. dioica* is circumpolar). In Sweden, it is known from Åsele Lappmark to Torne Lappmark, and in Finland, it occurs in Sompio Lappmark, Enare Lappmark and Enontekis Lappmark. It is also found eastwards to Novaya Zemlya, and in Svalbard (Spitsbergen) and East Greenland.

C. parallela has a continental distribution in South Norway, and is fairly common in central and eastern parts of Dovre and also in the mountains further northeast. It is very rare in the Trollheimen, the Sunndalen mountains and the Jotunheimen. The continental tendency is also found in North Norway.

Samuelsson (1921) considered immigration from the east unlikely. He also regarded *C. parallela* as belonging to one of the oldest elements of the Scandinavian flora and thought that its present distribution reflects an interglacial pattern. Like *C. dioica*, *C. parallela* may grow in fens, but always shallow eutrophic ones. However, it is most frequently met with on horizontal parts of solifluction lobes in the middle-alpine belt, very often together with, for example, *Carex misandra*. It may also occur on damp *Dryas octopetala* and *Cassiope tetragona* heaths, alluvial soil below snow fields, and irrigated cliffs.

It is definitely a basiphilous species, growing on carbonate rocks and serpentinite.

Carex rufina Drej.

Maps: Arwidsson (1943), Hultén (1971a, no. 363). Total: Hultén (1958, no. 162), Hultén & Fries (1986, no. 475). Local: Gjærevoll & Sørensen (1954), Benum (1958), Lid (1959), Toftaker (1969), Nordsteien (1982), Holten (1984), Engelskjøn (1986a).

First Norwegian record: Oppland: Øystre Slidre, Skjelstøen on Synshorn 1845. N. Moe (O).

Excluded station: Flakstadøy. Lofoten, Ekstrand 18.7.1880 (UPS). According to Engelskjøn (1986a), this record is doubtful and may be due to wrong labelling.

Norwegian distribution: From Bygland (Rustfjell) and Gjesdal to Tønsvikdalen in Tromsø.

Altitude limits: Mainly low-alpine. Ulvik 1600 m, Yset in Kvikne 568 m. In North Norway, from 210 m (Østerdalsisen) to 900 m on Rombaksstøtta, Narvik. Up to 1150 m in Lule Lappmark.

Habitat: Wet snow-beds on sandy soil, mainly acidophilous.

C. rufina is an amphi-Atlantic plant. In addition to Scandinavia, it is known from Iceland, Greenland and a few localities in Keewatin, central Canada, a long way from the Greenland localities. In Sweden, it is found from Härjedalen to Torne Lappmark.

C. rufina has a westerly distribution throughout its range in Scandinavia. The distribution has some strange features. Northwards to the Jotunheimen and southern Sunnmøre, it is known from numerous localities in the

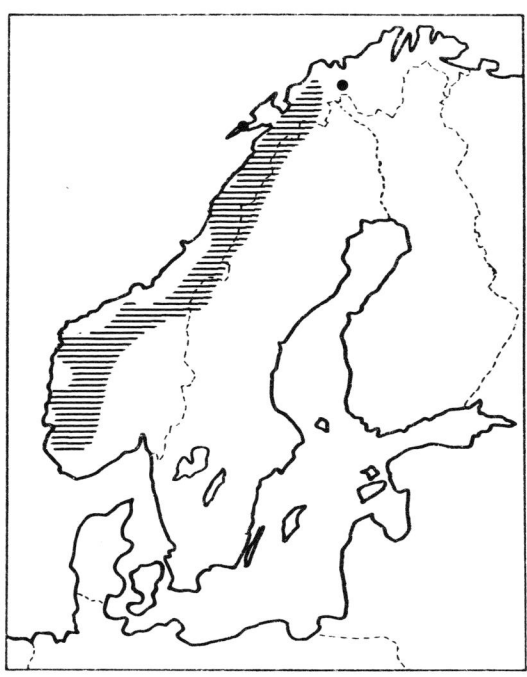

Fig. 30 The Fennoscandian distribution of *Carex rufina*

westernmost mountains as well as in the eastern Jotunheimen. It is then absent from the coastal mountains of northern Sunnmøre, Romsdal and Nordmøre, and is very rare in the inland mountains, too. It appears again in the Trollheimen and east of Dovre. Except for the relatively few inland localities, this distribution pattern corresponds with that of some lowland oceanic species, e.g. *Blechnum spicant* and *Narthecium ossifragum*.

The ecological range of *C. rufina* is very narrow. It is a species of late snow-beds, met with on level ground, on the banks of streams and small lakes, in depressions containing stagnant water, and on alluvial soil; always under wet conditions.

C. rufina forms a distinct community, *Caricetum rufinae* Nordhagen 1943 (Gjærevoll 1956, Lid 1959), in the *Saxifrago stellaris-Oxyrion digynae* alliance, where it is often the only important vascular plant, but always growing in a well-developed mat of various snow-bed mosses.

Table 7 gives an extract from three sociations in *Caricetum rufinae* (Gjærevoll 1956). The differences are first and foremost expressed by the different predominating bryophytes. All the analyses were made in the Trollheimen mountains, and the localities are all low-alpine.

Elven (1978) found *C. rufina* growing as a pioneer plant on the Flatisen and Østerdalsisen moraines in Rana, accompanied by *Pohlia filum*. It may therefore have enjoyed favourable conditions in early postglacial times. The amphi-Atlantic and very restricted distribution indicates that the species is old-established in the areas in which it occurs. Some authors (Selander 1950a, Benum 1958) have regarded *C. rufina* as an amphicline species. In my opinion, it is mainly acidophilous. This at least applies to the *Caricetum rufinae* association. Most of the 44

Table 7

Locality no.	A	B	C
Nos. of squares (1 m²)	10	5	20
Salix herbacea	50-1	60-1	85-3-
Koenigia islandica	50-1	100-1	15-1
Carex lachenalii	60-1	100-1	90-1
C. rufina	100-5	100-5	100-4
Eriophorum scheuchzeri	50-1	80-1+	50-1
Juncus biglumis	60-1	80-1	40-1
Calliergon sarmentosum	100-4	100-2	30-2
Drepanocladus purpurascens	.	100-5-	65-1
Pohlia drummondii	20-1	100-2-	100-1+
Anthelia juratzkana	60-1	80-1	100-5

pH determinations carried out in *Caricetum rufinae* in South Norway lie in the interval 4.5-5.5 (Gjærevoll 1956). In Greenland, the species is characterised as calcifugous by Böcher et al. (1966).

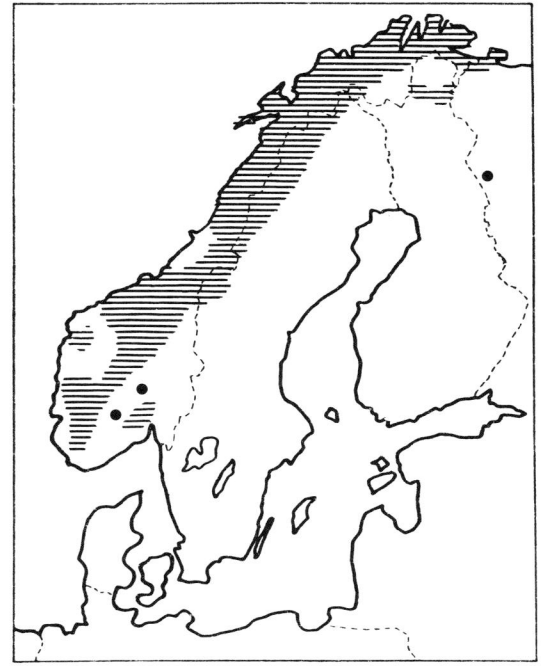

Fig. 31 The Fennoscandian distribution of *Carex rupestris*

Carex rupestris All.

Maps: Hultén (1971a, no. 326). Total: Hultén (1962, no. 20, 1968), Meusel et al. (1965), Hultén & Fries (1986, no. 427). Local: Benum (1958), Ouren (1961, 1966), Toftaker (1969), Moen & Moen (1977), Kristensen (1981), Holten (1983, 1984).

First Norwegian record: Finnmark: Rastigaisa 1802, G. Wahlenberg.

Norwegian distribution: Throughout the mountain range from Gyaoksli in Eigersund to North Cape and Grense Jakobselv.

Altitude limits: Low-alpine to middle-alpine. Kyrkja in the Jotunheimen 2100 m (Jørgensen 1933). Store Memurutind 2100-2200 m (NAS). According to Sørensen, there are numerous localities in the high-alpine belt in the Jotunheimen. Down to 210 m in Lier. Often down to sea level from Fosen northwards. Troms: Målselv, Kirkestinden 1425 m (Engelskjøn 1986a).

Habitat: Dry exposed heaths, cliff ledges. Basiphilous.

C. rupestris is a circumpolar, arctic-montane species. It is widely distributed in arctic and sub-arctic areas. In Europe, it is also known from Scotland and from the Pyrenees to the Carpathians. In Sweden, it is common throughout the mountains from Härjedalen to Torne Lappmark, and in Finland it is found from Sompio Lappmark to Enare Lappmark.

Like *Dryas octopetala*, *C. rupestris* has some interesting localities in South Norway outside the mountain range, at Ytre Sandsvær (Kjørstadelva), Nedre Eiker (Solbergfjellet) and Lier (Glitra). These occurrences are generally regarded as glacial relics from the end of the last Ice

Age (Lid 1958). At the first two localities, *C. rupestris* and *Dryas octopetala* grow together.

The distribution in South Norway is mainly continental and there are few coastal localities in West Norway. There is generally very close conformity between the distributions of *C. rupestris* and *Dryas octopetala*; as a rule they are true companions. The absence from western mountains may have an ecological explanation.

C. rupestris is a characteristic and quantitatively important species of the *Kobresio-Dryadion* alliance (Nordhagen 1928, 1936a, 1943, Malme 1971, Baadsvik 1974, Moe 1985). *C. rupestris* is also abundant and grows commonly on cliff ledges and rock faces where irrigation periodically occurs. The lowland localities referred to above are on steep calcareous bedrock.

Its ecological range is wide, but it is definitely a basiphilous species. From Budal, Ouren (1952) reported pH values of 5.2-7.5, whilst Christophersen (1925) reported 5.5-6.8 from Sylene. On the coastal mountain of Talstadhesten in Møre og Romsdal, Malme (1971) measured 7.9 in a *Ctenidio-Dryadetum* association dominated by *C. rupestris*. Lunde (1962) reported 289 determinations from North Norway with an amplitude of 4.9-7.9 and a mean of 6.47. As might be expected, this is the same as for *Dryas octopetala*.

Carex scirpoidea Michx.

Maps: Hultén (1971a, no. 319), Skifte (1985). Total: Hultén (1958, no. 170, 1968), Gjærevoll (1973), Hultén & Fries (1986, no. 431).

First Norwegian record: Nordland: Saltdal, Solvaagtind, F. Unander and A. Drake 1854 (O).

Norwegian distribution: Only known from Solvågtind and the Frostisen area.

Altitude limits: Low-alpine. On Solvågtind between 550 and 920 m, in the Frostisen area between 830 and 1000 m.

Habitat: Damp meadows, heaths, more rarely on screes. Calciphilous.

C. scirpoidea is an amphi-Atlantic species. The Norwegian localities are the only ones in Europe. It is widely distributed from Greenland throughout Canada and Alaska to easternmost parts of Asia.

Until 1975, Solvågtind was the only locality known in Norway. *C. scirpoidea* occurs fairly extensively on the south slope of that mountain, being abundant in places and in part dominating, particularly at 700-740 m. In 1975, it was found west of Kjelvatnet near Frostisen in Ballangen, and in 1982 two more localities were added east of Kjelvatnet (Skifte 1985, 1988).

The ecological range of *C. scirpoidea* seems to be narrow in Norway, in contrast to the situation in Alaska (Gjærevoll 1963a), for example. On the south slope of Solvågtind, it prefers low-growing meadow vegetation. In the Frostisen area, it occurs in grass and herb meadows, and on ledges. Common companions in both areas are *Selaginella selaginoides, Salix herbacea, S. reticulata, Polygonum viviparum, Silene acaulis, Saxifraga oppositifolia, Cassiope hypnoides, Pinguicula vulgaris, Saussurea alpina, Anthoxanthum odoratum, Carex atrata, C. capillaris, Thalictrum alpinum* and *Tofieldia pusilla*.

Because of its rare and isolated occurrence in northern Europe, *C. scirpoidea* has played an important role in the debate about possible survivors of the last glaciation (Nordhagen 1935, Gjærevoll 1963b). *C. scirpoidea* is obviously of North American origin.

Gjærevoll & Ryvarden (1977) reported *C. scirpoidea* growing in abundance on Jensen's Nunataks in West Greenland, at about 1350-1400 m. The specimens from there are very similar to those found on Solvågtind. Grønlie (1927) believed that nunataks existed in the Saltdal area during the last glaciation. However, investigations of nunataks on the Frostisen glacier have not revealed any occurrences of *C. scirpoidea*.

Analyses of pH from the Frostisen localities gave 5.2-6.5 (19 determinations), and from Solvågtind 5.8-7.0 (6 determinations).

Ola Skifte

Cassiope tetragona (L.) D. Don

Maps: Hultén (1955, 1971a, no. 1372). Total: Hultén (1968, 1971b, no. 17), Hultén & Fries (1986, no. 1448). Local: Benum (1958), Engelskjøn (1984, 1986a).

First Norwegian record: Gunnerus (1772, no. 1077) – "Habitat in norvegia", but no locality was given. Reported by Wahlenberg (1812) from Nordland and Finnmark.

Norwegian distribution: From Solvågtind in Saltdal to Neverfjord in Kvalsund, and inner Porsanger.

Altitude limits: Mainly low-alpine to middle-alpine. Troms: Målselv, Kirkestinden 1520 m. Nissontjärro in Torne Lappmark 1650 m. Down to sea level at some places in northern Troms and Finnmark.

Habitat: Predominates in somewhat sheltered heaths, especially north-facing ones, preferably on calcareous soils.

C. tetragona belongs to the circumpolar, arctic species, but has a disjunct and isolated distribution in northern Europe. Its distribution in Scandinavia is unbroken, but there is then a gap between Porsanger and the mountains in the central part of the Kola Peninsula, and a still larger one from there to the Urals. In Sweden, *C. tetragona* occurs from Pite Lappmark to Torne Lappmark. In Finland, it is confined to Enontekis Lappmark. It is common in Svalbard, but not known from Iceland.

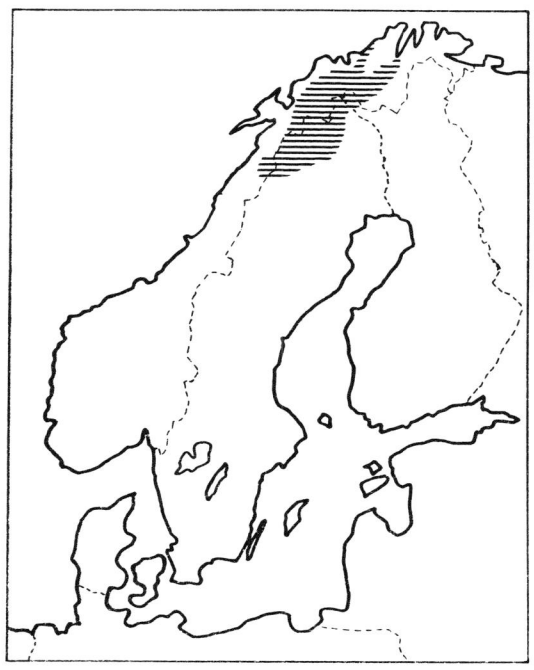

Fig. 32 The Fennoscandian distribution of *Cassiope tetragona*

The isolated occurrence in northern Fennoscandia indicates the same history for *C. tetragona* as for a number of amphi-Atlantic species. Immigration from the east seems unlikely. It is ecologically well adapted to nunatak conditions. In Svalbard, it is abundant on the semi-nunatak, Ossian Sars, near Kongsfjord.

C. tetragona occupies the standard area of a number of bicentric and northern unicentric species. In most of its area it is distinctly continental, but in the northernmost part there are several coastal localities. *C. tetragona* shows great vitality. It is usually very abundant, and its competitive ability is high. In some places there may be some decline towards the margins of its area of distribution, but it is abundant at its southern boundary in the Junkerdalen mountains. The same is reported by Selander (1950a) from its southern boundary in Sweden. According to Engelskjøn (1986a), *C. tetragona* "seems to undergo a concentric areal restriction, caused by a combination of soil chemical, climatical and successional agents". "This process, which may have been acting since early Holocene periods, may be summarized by keywords such as humification, podzolisation and competition by other ericaceous as well as peat-forming bryophyte species".

Within its concentrated area, *C. tetragona* is one of the most conspicuous species and forms a well-defined plant community on gentle slopes with a thin, but stable snow cover. Hedberg et al. (1952) distinguished three associations in the *Dryadion* alliance in North Sweden, one of them being *Tetragono-Dryadetum*. This community has also been described from Svalbard (Hadač 1946, Rønning 1965) where it displays the same ecological behaviour as in Scandinavia, avoiding areas most exposed to wind. In my experience, this is a characteristic circumpolar community occurring mainly on calcareous substrate.

As Th. C. E. Fries (1913) and Nordhagen (1936a) pointed out, *C. tetragona* may also be a dominant species on acid soils and forms a community within the *Arctostaphylo-Cetrarion nivalis* alliance. Engelskjøn (1984) described species-poor *C. tetragona* communities from the middle-alpine region. Many species that are usually characterised as calciphilous are often found growing on more acid substrate where there is little humus production.

C. tetragona may grow in open places in birch forest, and in some places, especially on dolomite, down to sea level. It has also been reported from serpentinite.

According to Lunde (1962), the soil in *C. tetragona* stands is normally alpine humus. The result of 209 determinations was a total amplitude of pH 4.2-7.6, and a mean of 5.76. In Svalbard, Hadač (see Böcher 1954) found an amplitude of 5.5-6.9, whilst Rønning (1965) reported 5.4-7.3.

Cerastium arcticum Lge.

Maps: Hultén (1956, 1971a, no. 688). Total: Hultén (1956, 1958, no. 13), Hultén & Fries (1986, no. 746). Local: Benum (1958), Holten (1984).

Excluded or doubtful stations: According to Lid (1959), the records from Hardangervidda result from misinterpretation of *C. alpinum* specimens. Hultén considered the material to be var. *alpinopilosum*. In my opinion, var. *alpinopilosum* may be a hybrid. The locality has therefore been omitted.

Norwegian distribution: Bicentric. In the southern area, from Gloppen and Jølster to the Trollheimen. In the northern area, from Okstindane to Seiland.

Altitude limits: Mainly middle-alpine, down to the northern boreal belt along streams. Ymisfjell in the Jotunheimen 2245 m (NAS). Troms: Målselv, Njunis 1700 m (Engelskjøn 1986a). On gravel bars down to 400 m in Målselv.

Habitat: Irrigated gravelly snow-beds. Calciphilous.

C. arcticum has been subject to different evaluations by several authors. It was first distinguished in Scandinavia by Lindblom (1837-38), based on material collected near Kongsvoll. Lindblom, however, did not distinguish it from *C. latifolium* L. Based on material from Greenland, Lange (1880) described it as a separate species. The taxonomy of the species was thoroughly dealt with by Hultén (1956) and Böcher (1977).

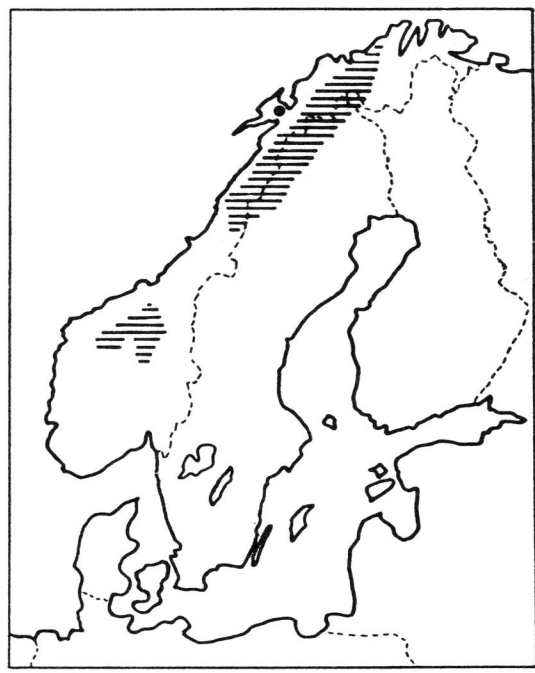

Fig. 33 The Fennoscandian distribution of *Cerastium arcticum*

C. arcticum belongs to the amphi-Atlantic species. In addition to Scandinavia, it is known from the British Isles, Iceland, Svalbard, Greenland and westernmost Arctic Canada. In Sweden, it grows from Åsele Lappmark to Torne Lappmark; in Finland, only in Enontekis Lappmark. *C. arcticum* is a characteristic species of the *Saxifrago oppositifolio-Oxyrion digynae* alliance (Gjærevoll 1956), occurring in wet snow-beds on calcareous soils. It is very often found in snow-beds dominated by *Ranunculus nivalis* and *Phippsia algida*, but may also be the most conspicuous species, especially in gravelly locations where the bottom layer is poorly developed and vascular plants have a scattered growth. This community may cover quite extensive areas. Common companions are *Sagina intermedia*, *Saxifraga cernua*, *S. tenuis*, *S. oppositifolia*, *Silene acaulis* and *Poa alpina vivipara*, and in the bottom layer, *Anthelia juratzkana*. It may also be a pioneer plant on fresh moraines (Elven 1978) and occasionally on river bars.

As it is a very typical amphi-Atlantic species, *C. arc-*

54

ticum may have a history very like that of a number of other species with a similar distribution. Its occurrence in the British Isles corresponds with that of, for example, *Arenaria norvegica*. Taking its ecology into consideration, the species seems well adapted to conditions existing in ice-free refugia.

Chamorchis alpina (L.) Rich.

Maps: Hultén (1971a, no. 529). Total: Hultén & Fries (1986, no. 556). Local: Ouren (1952, 1961, 1966), Benum (1958), Toftaker (1969), Hagen (1976), Nordsteien (1982).

First Norwegian record: Gunnerus (1772, no. 1098) – "Habitat in Hasvig vestfinmarchiæ in solo arenoso inter Melkhouen et Skothouen. Juli 1767".

Norwegian distribution: Slightly bicentric. From the mountains near Gjende to Skjækerfjellene in Snåsa, and from Hattfjelldal to Magerøya and Persfjord.

Altitude limits: Mainly low-alpine. Ryggehø in the Jotunheimen 1570 m (NAS). Down to 750 m in the southern area. Troms: Målselv, Kirkestinden 1150 m (Engelskjøn 1986a). Down to sea level at many places in North Norway.

Habitat: *Dryas octopetala* and *Kobresia myosuroides* heaths, solifluction terraces, cliff ledges and shallow rich fens. Basiphilous.

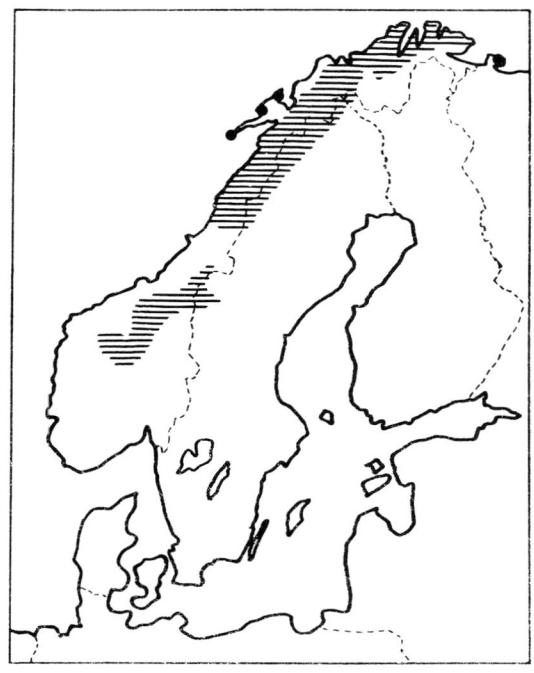

Fig. 34 The Fennoscandian distribution of *Chamorchis alpina*

C. alpina is a European species, restricted to Fennoscandia and the central European mountains. It is rare and very disjunct in central Europe, occurring in the Alps and Carpathians (Visoké Tatry). It has its main distribution in the northern area, by far the majority of localities being found in Norway and Sweden.

In Sweden, it is known from Härjedalen to Torne Lappmark, displaying a gap like that in Norway. In Finland, it is restricted to Enontekis Lappmark. There is also a locality on the Rybachiy Peninsula.

In South Norway, *C. alpina* is mainly continental, the westernmost localities being situated in Norddal, Sunnmøre. In North Norway, there are numerous localities in coastal districts.

C. alpina is mainly a low-alpine plant. In central Norway, most localities are found between 900 and 1100 m (Ouren 1952). In the northern area, it is often found down to sea level on limestone and shell sand.

C. alpina is a representative of the Alpine-North European element. With the proviso that no, or very few, alpine plants have reached Scandinavia from the south in postglacial time Selander (1950a) considered that this element represented glacial survivors. *C. alpina*, like other orchids, has very small seeds that are well adapted for long-range dispersal, so occasional colonisation may have taken place. It is, however, strange that the distribution pattern corresponds with that of species belonging to the centric elements. It is, for example, absent from the Hardangervidda. This weighs against an explanation based on occasional dispersal. *C. alpina* is a characteristic species of the *Kobresio-Dryadion* alliance. I have seen it growing very abundantly in the limestone area west of Saltdalen in several communities belonging to this alliance.

A very typical habitat is the edge of small solifluction lobes. It may also occur in quantities in an entirely different community, on stabilised river banks of rich fen character. Such localities are found in Grimsdalen, Dovre, where *C. alpina* grows together with species like *Carex microglochin*, *C. bicolor* and *C. maritima*, at about 900 m a.s.l. These river banks are inundated during the spring flood and some sedimentation takes place. They are also strongly affected by grazing sheep.

C. alpina is clearly basiphilous, growing on calcareous schists, marble, limestone, dolomite and shell sand. Ouren (1952) reported 5 determinations from Budal with an average pH of 6.3.

Cystopteris montana (Lam.) Desv.

Maps: Hultén (1971a, no. 48). Total: Hultén (1958, no. 226, 1968), Meusel et al. (1965), Hultén & Fries (1986, no. 54). Local: Benum (1958), Ryvarden (1969), Toftaker (1969), Hagen (1976), Moen & Moen (1977).

First Norwegian record: Sørfold in Nordland. Wahlenberg (1812, no. 507).

Norwegian distribution: From Smørslagonuten in Suldal to Hillesøy (northern limit) and Grense Jakobselv.

Altitude limits: Mainly northern boreal. In the Jotunheimen to 1200 m. In southeastern Norway down to 130 m in Krokkleiva; at the foot of Talstadhesten in Fræna at 50-100 m. Inner Troms 900 m. Down to sea level at some places in North Norway.

Habitat: Shady, wet or moist localities, preferably in springs in willow thickets and northern boreal birch forests. Basiphilous.

C. montana is a circumpolar, arctic-montane species with a disjunct distribution. Hultén (1958) classified it as an amphi-Atlantic species. There is a large gap between its distribution areas in western and eastern North America.

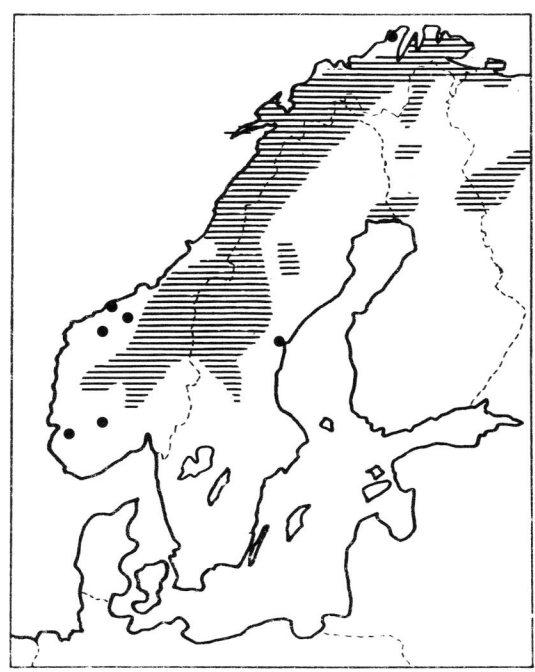

Fig. 35 The Fennoscandian distribution of *Cystopteris montana*

It is very rare in Greenland and unknown in Svalbard and Iceland. In Europe, it is found from northern Spain to the Caucasus and in the British Isles. In Sweden, it is found from Dalarna to Torne Lappmark and there are some coastal localities on the Gulf of Bothnia. Its distribution in Finland is scattered in the northern provinces. It has a continuous distribution from Fennoscandia eastwards to the Urals.

The species has a distinct easterly distribution in South Norway. There are few localities in western Norway, and they are in inner fjord districts except for one at Talstadhesten in Fræna. In southeastern Norway, it is known as far south as Sagelva in Lier, on the western side of Holsfjorden.

C. montana is fairly rare in South Norway apart from some central valleys. In North Norway, it is far more common in Nordland and Troms where there are also numerous coastal localities, but is absent from large areas of Finnmark perhaps partly due to lack of suitable localities (Ryvarden 1969).

C. montana is a characteristic and dominant species of the rich spring communities (*Cratoneuro-Saxifragion aizoidis* Nordhagen 1936a) often found in low-alpine willow thickets and northern boreal birch forests alternating with luxuriant tall herb communities (*Lactucion alpinae*). *C. alpina* may also occur in the latter community, but only in small quantities. *C. montana* may also be frequent in the southern boreal belt where springs are found, and at the base of steep cliffs, especially those facing north, where the soil remains moist. It often grows in shady valleys and gorges where lime-bearing water is found (Berg 1983). Localities at low elevations may be regarded as northern boreal enclaves.

C. montana is a basiphilous species. The pH usually varies between 6 and 7.

Diapensia lapponica L.

Maps: Hultén (1955, 1971a, no. 1387), Tralau (1963). Total: Hultén (1958, no. 204, 1968), Hultén & Fries (1986, no. 1435). Local: Benum (1958), Kristensen (1981), Holten (1984).

First Norwegian record: Oeder (1761) reported it from Fokstua and "Syv Søstre paa Alsten". Gunnerus (1761, no. 118) gave "Habitat in alpe Aalbygfjeldet; 23. Juli 1764". (TRH)

Excluded or doubtful stations: According to A. Blytt (1874), *D. lapponica* grows on Synshorn and Gråhø beside Vinstervatnet, but no specimens from these often visited localities are in any herbaria.

Norwegian distribution: Scattered from Bukollen in Flå to Dovre and Lesja, and then more commonly to Magerøya and Sør-Varanger.

Altitude limits: Mainly low-alpine. Sylane 1600 m. Down to 300 m in Åfjord. Troms: Målselv, Njunis 1480 m. Down to almost sea level at several places in North Norway.

Habitat: Heaths, ridges and plateaus exposed to deflation, gravelly and rocky places. Most common on acid or neutral soil, but also often found on *Dryas* heaths.

D. lapponica s.l. is a circumpolar, arctic species. It comprises two subspecies, the amphi-Beringian subsp. *obovata* and the amphi-Atlantic subsp. *lapponica*. In Eurasia, it is found from Scandinavia to the River Yenisey, and also in northern Scotland and Iceland. It is not known from Svalbard, and generally avoids high-Arctic areas. In Greenland, it has a southerly distribution, and in North America it is chiefly found south of the Canadian Arctic archipelago eastwards to southeast Mackenzie.

In Sweden, it is found from Härjedalen to Torne Lappmark, and in Finland from Enontekis Lappmark to Sompio Lappmark. It is common on the Kola Peninsula, but decreases distinctly further east.

The Norwegian distribution is unique. Apart from some scattered eastern localities, it is not found in the mountain range from the Jotunheimen southwards, or in

56

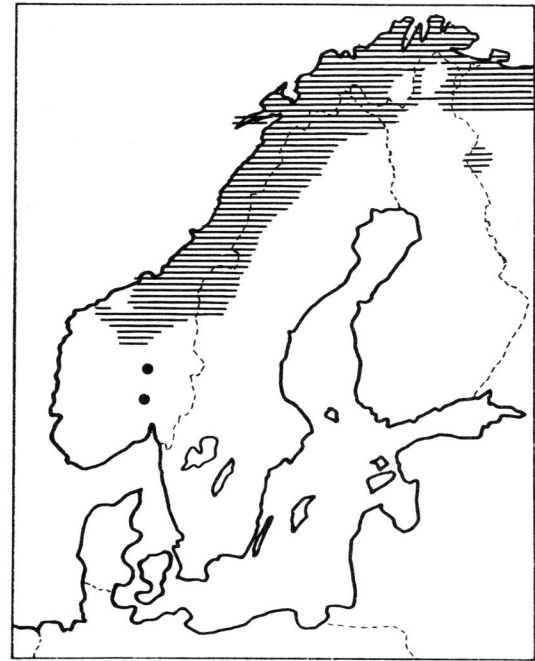

Fig. 36 The Fennoscandian distribution of *Diapensia lapponica*

Fries (1986, no. 948). Local: Benum (1958), Toftaker (1969), Bretten (1973), Hagen (1976), Schumacher & Løkken (1981), Nordsteien (1982).

First Norwegian record: Oeder (1761, no. 56) reported it to be "Ganske sparsommeligen paa de norske Fielde, saasom ved Fogstuen paa Dovrefield". The oldest herbarium specimen dates back to 1828 – Dovre, Boeck (O).

Excluded or doubtful stations: Ekman (1926) reported Ålfjellet, Vang in Valdres, as the southernmost station, but no herbarium specimen is registered.

Norwegian distribution: Slightly bicentric. In South Norway, from Vågå to Meråker. In North Norway, from Hattfjelldal to Alta.

Altitude limits: Mainly low-alpine to middle-alpine. Oppdal, Tythø 1740 m. Troms: Bardu, Kufjell up to 1150 m. Occasionally down in the northern boreal belt in springs, on north-facing cliff ledges and banks of streams.

Habitat: Moderately moist places beside streams and springs, moss-dominated *Dryas* heaths, and snow-beds. Basiphilous.

the coastal mountains, but from Lesja, Stordal and Dovre it occurs continuously to Finnmark. There is no reasonable geological or climatological explanation for this distribution and the plant is far from being an ecological specialist, so its absence from the southern mountains is strange. In South Norway, migration to the continental mountains from north and west seems reasonable. However, it is difficult to look upon the limit in Lesja-Dovre as being related to dispersal, since the seeds are numerous and light. As *D. lapponica* does not occur in the central European mountains, immigration from the south is out of the question. The species must have survived in the North Atlantic area. In this connection, the locality in Scotland is extremely interesting (see *Artemisia norvegica* and *Arenaria norvegica*).

D. lapponica has been recorded from late-glacial deposits at two sites in Scania (Tralau 1963), but the species is not known in glacial deposits in the lowlands of central Europe.

In places where it occurs, *D. lapponica* is a characteristic species of the *Arctostaphylo-Cetrarion nivalis* alliance, a community that is widely distributed throughout the mountain range, on oceanic as well as continental mountains. *D. lapponica* itself grows abundantly on coastal and continental mountains. Although it mainly occurs on soils poor in lime, it may also be an important constituent of *Dryas octopetala* heaths.

Draba alpina L.

Maps: Ekman (1926), Hultén (1971a, no. 891). Total: Marret (1911-24), Hultén (1968, 1971b, no. 53), Hultén &

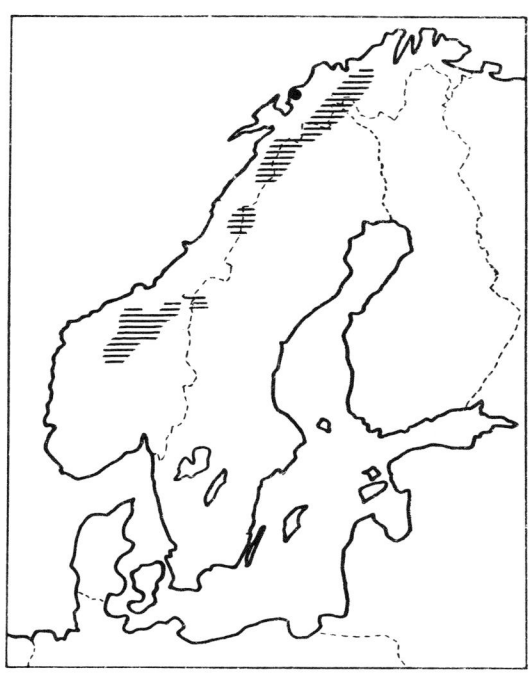

Fig. 37 The Fennoscandian distribution of *Draba alpina*

D. alpina is a circumpolar, arctic, partly arctic-montane, species, widely distributed in the Arctic. In eastern Siberia, there is an isolated population in the Verkhoyanskiy mountains. Closely related taxa have a wide distribution in central Asia, also south of the desert zone. In Sweden, it occurs from Härjedalen to Torne Lappmark, but there is a gap in Jämtland. In Finland, it is only known from Enontekis Lappmark. It also occurs at a few places on the Kola Peninsula.

In South Norway, *D. alpina* is found from Besstrand Rundhø in Vågå to Kjølhaugan in Meråker, with a marked centre on Dovrefjell. The westernmost locality in the

Jotunheimen is Storhø, east of Bøverdalen, and in the Sunndalen mountains it is Storskardhøa. Skrikhø in Rennebu is the northernmost locality in the Trollheimen. East of Dovrefjell, there is a notable locality on Tron, Alvdal.

In North Norway, the distribution extends from Dauingfjell in Hattfjelldal to Did'novarri in Alta.

It is striking that *D. alpina* avoids coastal mountains throughout its range, apart from the isolated station at Senjahesten on the southernmost point of Senja, Troms. The Sunndalen mountains are quite close to the coast, but high mountains result in a rain shadow and a fairly continental climate.

D. alpina is mainly a chionophilous species. It may grow in the protected, moss-rich communities of the *Kobresio-Dryadion* alliance, but is most common in the *Luzulion arcticae*, *Potentillo-Polygonion vivipari* and *Saxifrago oppositifolio-Oxyrion digynae* alliances.

Simen Bretten

Draba cacuminum Elis. Ekman

Maps: Dovre mountains: Gjærevoll & Sørensen (1954), Bretten (1973). Total (Scandinavia): Hultén (1971a, no. 892), Elven & Aarhus (1984).

First Norwegian record: Nordland: Rana, M. N. Blytt 1841 (O).

Excluded stations: (1) Troms: Målselv, Kirkesjordfjell (Notø 1902 (O), Ekman 1917); the material belongs to *D. norvegica* Gunn. (Knaben & Engelskjøn 1967). (2) Finnmark: Alta (Lund 1841 (O)); partly identified as *D. cacuminum* by Ekman, and shown along with a question mark by Hultén (1971a, no. 892); the material belongs to *D. norvegica* (Elven & Aarhus 1984). (3) Oppland: Dovre, Blåhø; referred to by Gjaerevoll & Sørensen (1954), but voucher not found. (4) Sør-Trøndelag: Oppdal, Brattskarven (Nordhagen 1931 (BG)); the material belongs to *D. alpina* (Elven & Aarhus 1984). (5) Sør-Trøndelag: Oppdal, Halsbekkhø (Nordhagen 1946 (O)); incomplete material, may be *D. cacuminum*.

Norwegian distribution: A Scandinavian endemism; bicentric. Finse, the Jotunheimen and Dovrefjell in South Norway (subsp. *cacuminum*). Okstindane area in North Norway (subsp. *angusticarpa*).

Altitude limits: Mainly middle-alpine. Skagsnebb in the Jotunheimen 2000 m (NAS); rarely below 1300 m in South Norway.

Habitat: Exposed ridges and moraines, mostly on calcareous schists. A weak competitor, rarely found in entirely closed vegetation.

D. cacuminum was described by Ekman (1917) from the Dovre and Jotunheimen mountains, but had already been recognised by Kindberg as a new species, *D. ventosa*, a name not published legitimately.

Outside Norway, the species is only known from Sweden, from one mountain in the Okstindane area and two in the Sulitjelma area. The Scandinavian distribution is clearly bicentric and there are apparent disjunctions in both the northern and southern areas. However, the species has tended to be misidentified and the disjunctions may not exist, especially in the north. The species had been repeatedly collected, for example, in the Finse mountains from 1914 and in northern Scandinavia from 1841, but not recognised as such from these areas before 1946 (Selander, northern Scandinavia) and 1980 (Elven et al. Finse). In the Finse mountains, it seems to occur dispersed in most localities, but in large quantities only on recently deglaciated moraines. The distribution seems to correlate well with the presence of large areas of calcareous bedrock in the middle-alpine belt and an abundance of open habitats that are windswept or have recently become deglaciated and have a stable substrate. It seems susceptible to soil movement and cannot exploit open, solifluction habitats or scree.

There is a marked morphological difference between the southern and northern Norwegian populations, recognised at the subspecific level (Elven & Aarhus 1984). There is also some indication that differentiation may be possible between the subareas in South Norway, but the material available is too limited for analysis, especially that from the Jotunheimen. The species is probably an octoploid (2n = 64 (Engelskjøn 1979)). The closest affinities in the northern European area seem to be with the hexaploid, *D. norvegica* Gunn. (common in the Scandinavian mountain range), and the diploid, *D. subcapitata* Simm. (Arctic and Svalbard), but morphological evidence suggests that an alloploid origin from these two species is improbable.

Reidar Elven

Draba crassifolia Grah.

Maps: Arwidsson (1943), Hultén (1971a, no. 894), Rønning (1956b). Total: Hultén (1958, no. 174, 1968), Hultén & Fries (1986, no. 965). Local: Benum (1958).

First Norwegian record: Troms: Fløyfjellet, J. M. Norman 1863 (O, S).

Norwegian distribution: Northern unicentric. From Jørentind in Hattfjelldal to Fløyfjellet in Tromsø and Gætkoai've in Nordreisa.

Altitude limits: Middle-alpine to low-alpine. Troms: Målselv, Kirkestinden 1350 m (Notø 1905). Reported from Bøntuva in Tromsø, the summit of which is 778 m.

Habitat: Snow-beds, gravelly soil and solifluction terraces. Calciphilous.

D. crassifolia is an amphi-Atlantic species, known in Europe only from Norway and Sweden. The distribution in Sweden extends from Pite Lappmark to Torne Lapp-

58

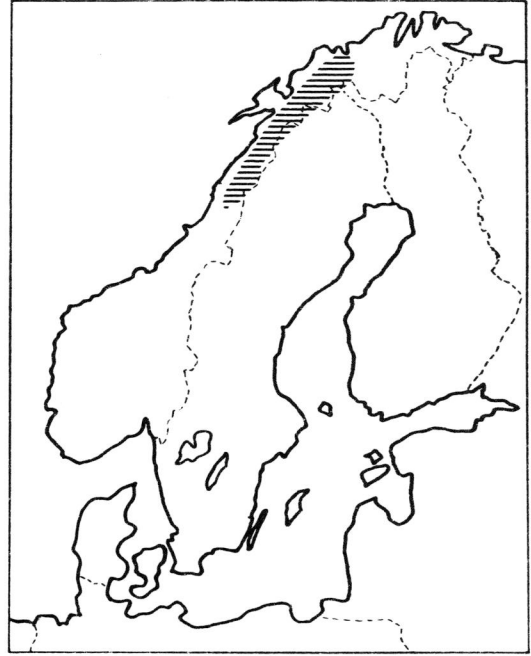

Fig. 38 The Fennoscandian distribution of *Draba crassifolia*

mark. The species is known from two large, but separated areas in East and West Greenland, and also from the eastern Canadian Arctic and from Alaska to the Rocky Mountains.

Compared with the map published by Arwidsson (1943), the recently discovered stations make the distribution more continuous from Hattfjelldal to Nordreisa. More stations will certainly be found, as this tiny species is easily overlooked.

D. crassifolia may grow abundantly in some localities, but its competitive ability appears poor as it is always found in open communities where most of its companions also grow dispersed. It is first and foremost a snow-bed plant, and occurs as an exclusive, but quantitatively insignificant species within the *Saxifrago oppositifolio-Oxyrion digynae* alliance (Gjærevoll 1956).

Its distribution area conforms with that of several other northern unicentric and amphi-Atlantic species. It has therefore been regarded as a potential survivor of the last Ice Age. Ecologically, it seems well fitted for extreme conditions.

Draba fladnizensis Wulf.

Maps: Ekman (1926), Knaben (1966c), Hultén (1971a, no. 895), Bretten (1973). Total: Löve & Löve (1947), Hultén (1958, no. 207), Hultén & Fries (1986, no. 958). Local: Gjærevoll & Sørensen (1954), Benum (1958), Ouren (1966), Bretten (1973), Nordsteien (1982), Engelskjøn (1986a).

Excluded stations: *D. fladnizensis* has often been misidentified, even being confused with *D. crassifolia* (Tromsø, Ramfjord: Durmålsfjell, Benum (TROM), see

also Benum 1958:292 and p. 234). A specimen of genuine *D. fladnizensis* may have been mixed into a *D. norvegica* sheet – Nord-Trøndelag, Lierne, Storfjeld, Wennberg (O). Because of its location in central Norway, this station is considered unlikely and requires confirmation.

Norwegian distribution: Bicentric. In the southern area from Hårteigen on Hardangervidda to Soknedal, in the northern area from Krukki on Saltfjell northeastwards to Gandvikfjellene in Nesseby.

Altitude limits: Mainly low-alpine and middle-alpine, but also reaching high-alpine elevations. In South Norway, Lom: Galdhøpiggen (Keilhau's peak) 2300 m (Jørgensen 1933). In North Norway, Troms: Målselv, Kirkestinden 1320 m (Engelskjøn TROM). Often descends to subalpine shady cliffs or stream banks.

Habitats: On exposed lithosol in gaps between rocks and on ledges, but also enters closed communities belonging to the *Kobresio-Dryadion* alliance. The species is confined to calcareous mica schist, phyllite, dolomite, gabbroid rocks, and schistose mylonite or gneiss.

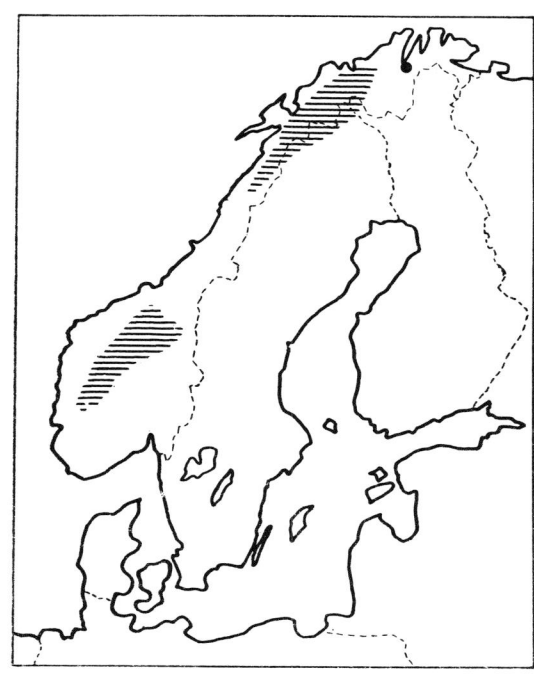

Fig. 39 The Fennoscandian distribution of *Draba fladnizensis*

D. fladnizensis was described from the Austrian Alps in 1778. It is morphologically well defined and uniform in its key characters. It has the diploid chromosome number of 2n = 16 throughout its range, also in Norway (Engelskjøn 1979). Its world distribution is circumpolar, arctic-alpine, and is disrupted but extensive. There is a gap between the central Kola Peninsula and the Urals-Novaya Zemlya. It occurs here and there on Svalbard, and is known from one very restricted area on Iceland.

The conception of Ekman (1932), followed by Hultén (1958) and Benum (1958), that *D. fladnizensis* and *D. lactea* are intergrading species seems untenable. Their distinguishing characters are usually obvious, and their chromosome numbers are 2n = 16 and 2n = 48, respectively.

When the Norwegian material was being revised no interspecific hybrids were seen. An alleged hybrid from North Sweden (Ekman 1932b, Fig. 1) is considered to be a specimen of *D. lactea* with aborted silicles. The same feature, which is not uncommon in *D. lactea*, was observed in a collection from South Norway (Oppdal; Nonshøa, I. Tollan, O), determined as *D. fladnizensis* x *lactea* by S. Arwidsson.

The South Norwegian area of *D. fladnizensis* is centred around the Jotunheimen, Dovre and Trollheimen mountains, with scattered outlying stations. In its northern subarea, the species is only frequent in inner Troms. The marginal stations suggest an areal restriction in postglacial times, due to a secular elimination of suitable edaphic niches.

Torstein Engelskjøn

Draba lactea Adams

Maps: Ekman (1926), Arwidsson (1943), Knaben (1966c), Hultén (1971a, no. 899). Total: Hultén (1958, no. 208), Hultén & Fries (1986, no. 959), Local: Benum (1958), Nordsteien (1982), Engelskjøn (1984, 1986a).

First Norwegian record: Reported by Wahlenberg (1812, no. 317) – "in alpe Lyngentind, 1800" (as "Draba androsacea").

Excluded stations: Some specific delimitation problems are discussed under *D. fladnizensis*. In addition, glabrescent *D. norvegica* has been misidentified as *D. lactea*; hence, some records in Benum (1958) from mountains near Tromsø have not been included on the present distribution map.

Norwegian distribution: The overall pattern is bicentric. In South Norway, there are two stations in the southwest which are marginal and of special chorological interest (Lom: Høyrokampen, Buttler & Gauhl (M & O), and Luster: Myrkrisdalen, A. Skogen (BG)). *D. lactea* is apparently absent from the Dovre mountains, and has a concentration in the Trollheimen and Grøvudalen mountains further northwest. An occurrence at 65° 40' N lat. in Sweden, east of central Norway (Rune 1948, 1950), forms the southern limit of the northern Scandinavian area, the northern limit being in Kvænangen, Troms.

Altitude limits: *D. lactea* is strictly alpine, chiefly middle-alpine. The lowermost stations are usually found on shady, irrigated cliffs close to the forest limit. The Trollheimen, Gjevilvasskammene 1600 m (Nordsteien 1982). Troms: Målselv, Kirkestinden 1565 m (Engelskjøn 1986a).

Habitat: *D. lactea* associates with the *Luzulion arcticae* alliance (Gjærevoll 1956) and the similarly bryophyte-rich association, *Tomenthypno-Dryadetum* (Hedberg et al. 1952), and is thus a moderately chionophilous, eutrophic species. Its preferred substratum is bryophyte turfs covering solifluction soil derived from calcareous mica schists.

In North Norway, it also spreads tơ gabbro, amphibolite and ultrabasic material when irrigation is adequate.

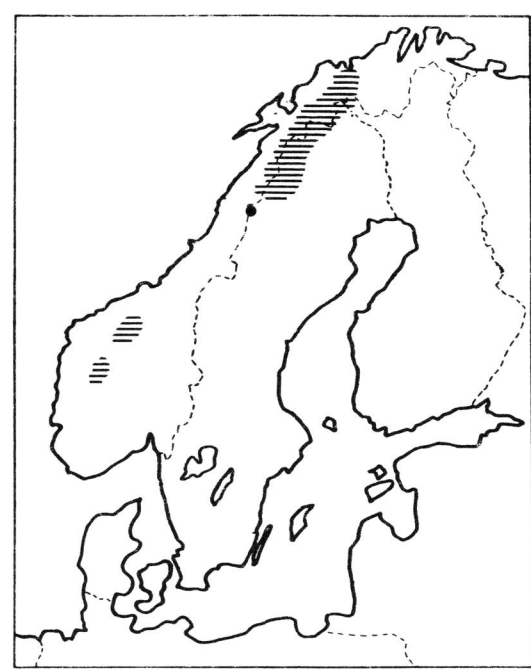

Fig. 40 The Fennoscandian distribution of *Draba lactea*

D. lactea was described from the Lena estuary in eastern Siberia in 1817 and has several synonyms, many of which are ambiguous (Nazarov 1924, Tolmatchev 1975).
It is a circumpolar, arctic species which varies as regards stem pubescence (Knaben 1966c, Fig. 2A); stellate silicles are even seen in high-alpine populations such as Svalbard strains. The species has no outlying, mid-latitude stations south of 61° N lat.

In Sweden, *D. lactea* is known from Lycksele Lappmark to Torne Lappmark, and in Finland it occurs in Enontekis Lappmark. Between Kvænangen and the Urals-Novaya Zemlya only a single locality has been recorded in the central part of the Kola Peninsula. *D. lactea* is common on Jensen's Nunataks in Greenland (Gjærevoll & Ryvarden 1977).

Torstein Engelskjøn

Draba nivalis Liljebl.

Maps: Ekman (1926), Nordhagen (1950-58), Hultén (1971a, no. 902). Total: Fernald (1925), Hultén (1968, 1971b, no. 32), Hultén & Fries (1986, no. 955), Local: Benum (1958), Lid (1959), Knaben (1966c), Bretten (1973), Aune & Kjærem (1978), Schumacher & Løkken (1981).

Excluded or doubtful stations: In the literature, Litlos on the Hardangervidda is repeatedly reported to be the southernmost locality (e.g. Lid 1959), but no herbarium specimen exists.

Norwegian distribution: Bicentric. In South Norway, from Ullensvang to Holtålen. In North Norway, from Hattfjelldal to Berlevåg.

Altitude limits: Mainly low-alpine to middle-alpine. In the Jotunheimen, up to 2100 m on the mountains of Ryggehø, Øystre Bukkeholstindane, Trollsteinhø and Olivinkollen (NAS). In Lom, down to 540 m. Troms: Målselv, Kirkestinden 1530 m (Engelskjøn 1986a).

Habitat: Extremely exposed ridges and peaks, rock fissures, cliff ledges, screes and open gravel. Basiphilous.

D. nivalis is a circumpolar, arctic, partly arctic-montane, species. It has a scattered distribution in northern Asia. In North America, its distribution is continuous throughout the Arctic. It is also known from Greenland, Iceland and

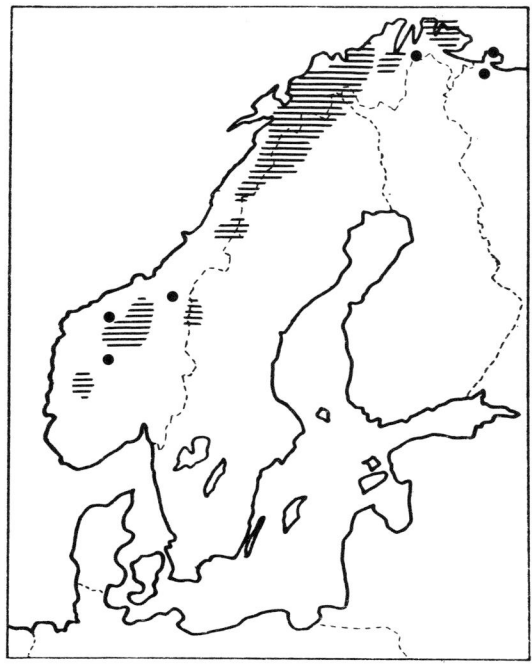

Fig. 41 The Fennoscandian distribution of *Draba nivalis*

Svalbard. In Sweden, a few stations are found in Härjedalen and Jämtland, and there is more continuous distribution from Åsele Lappmark to Torne Lappmark. In Finland, it is restricted to Enontekis Lappmark. One locality on the Rybachiy Peninsula and an inland station just south of there should be added to the Fennoscandian distribution. East of these, a gap of about 650 km ensues to Kolguyev Island, isolating the Fennoscandian distribution.

In South Norway, *D. nivalis* has a rather disjunct distribution from "south of Mt Omnkjelen" in Ullensvang to the mountain Bringen in Holtålen. There is a marked concentration in the northern Jotunheimen, with scattered populations in many directions. To the south, it has a subcentre on northwestern Hardangervidda (Knaben 1966c). West of the Jotunheimen, there are two stations in Tafjord (Norddal). Away from the Jotunheimen, *D. nivalis* declines in the Dovrefjell area where the mountain of Vesle Elgsjøtangen in Oppdal is the northernmost station. Knaben

(1966c) reported the hybrid, *D. curtisiliqua* (= *D. fladnizensis* x *nivalis*), from Sissihø, indicating that both parents may grow there.

Tyrikvamfjell in Oppdal is the only station in the Trollheimen. Bringen, the northernmost locality in South Norway, is rather isolated.

The distribution in North Norway is far more continuous, from Simskarfjell, Hattfjelldal to Kongsfjord, Berlevåg. Northwards to Ofoten, it occurs some distance from the coast. North of there, it also grows on coastal mountains. Contrary to most alpine species, it is rare in lowland areas of Finnmark.

It is striking that in the Jotunheimen-Dovrefjell area, *D. nivalis* grows almost entirely on the east side of the main watershed. Only two localities, Svånådalen and Tverrfjellet in Dovre, are slightly to the west. Ekman (1926) explained the strange distribution by invoking survival through the Ice Age on nunataks in the Jotunheimen-Dovre mountains. Nordhagen (1965b) suggested that the populations in Tafjord and Lesja may indicate a migration route from coastal refugia. Gjærevoll & Sørensen (1954) found it difficult to relate the present distribution to migration from coastal refugia, and believed that *D. nivalis* survived the last glaciation on nunataks. Knaben (1966c) thought it easier to explain the present isolated occurrences of this and other rare mountain species in the area discussed by invoking survival on nunataks rather than coastal refugia alone. It was common on Jensen's Nunataks (Gjærevoll and Ryvarden 1977).

D. nivalis is a chionophobous species sociologically limited to the most windswept parts of the *Kobresio-Dryadion* alliance, together with species like, for example, *Kobresia myosuroides*, *Festuca ovina*, *Carex rupestris*, *C. glacialis*, *C. nardina* (North Norway), *Draba fladnizensis*, *Polytrichum piliferum*, *Gymnomitrion* spp., *Alectoria nigricans* and *A. ochroleuca*. It is often found in patches of open gravel and on screes. It is perhaps most frequently found in fissures in rocks and boulders, and on cliff ledges. An inventory from six fissures at 1300-1320 m on Veslekolla, Dovrefjell, shows exclusively xerophilous and chionophobous species. Species found in two or more fissures are listed here (after Bretten 1973).

Table 8

Carex rupestris			x		x	
Artemisia norvegica		x			x	
Draba fladnizensis	x	x	x	x		
Potentilla nivea			x		x	
Cynodontium sp.		x	x			
Hypnum revolutum	x	x	x			
Alectoria ochroleuca		x			x	x
Caloplaca stillicidiorum		x	x	x		
Caloplaca sp.	x	x				
Cetraria nivalis		x				x
Cornicularia divergens	x				x	x
Ochrolechia sp.		x	x			
Peltigera lepidophora		x	x			

The Norwegian material of *D. nivalis* is fairly homogeneous, but the Tromsø herbarium has several collections from Båtsfjord, Finnmark, and the specimens in these are only sparsely hairy. They mostly have typical stellulate hairs only on the leaf margins. Some even have some cilia and forked hairs on the margins of the fruits (= var. *glabrescens* O. E. Schulz ?).

Simen Bretten

Draba oxycarpa Sommerf.

Maps: Bretten (1973). Local: Bretten (1973), Hagen (1976), Schumacher & Løkken (1981), Nordsteien (1982).
First Norwegian record: Oppdal, Hovdin mountain, 1828, M. N. Blytt (O) (taken as *D. alpina*).
Norwegian distribution: Bicentric. In South Norway, from Vågå to Meråker. In North Norway, very scattered from Fauske to Kåfjord in Lyngen.
Altitude limits: Mainly middle-alpine and upper low-alpine. Oppdal, Namnlauskollen 1840 m. Nordland: Sørfold, Leirvassfjellet 1170 m. Occasionally down to the northern boreal belt on north-facing cliff ledges and in springs.
Habitat: Snow-beds that become exposed moderately late, and moist places beside streams and springs; even on plateaus in the middle-alpine belt.

D. oxycarpa is an amphi-Atlantic, arctic species (*D. gredinii* is being taken as a synonym of *D. oxycarpa*). In addition to Norway, it is known from Svalbard and East Greenland. In Scandinavia, it has so far only been recorded from Norway, but will most likely be found in North Sweden.

D. oxycarpa was described by Sommerfelt (1833) on the basis of specimens from Oppdal and Stans Foreland, Svalbard. Until the 1970's, the taxon was considered synonymous with *D. alpina*.

In South Norway, *D. oxycarpa* has a very concentrated occurrence from Skarshø on the border between Vågå and Dovre in the south to Kjølhaugan in Meråker in the north, and from Storskardhøa in Sunndal in the west to Rødalshø in Tynset in the east. This corresponds highly with the distribution in South Norway of *Carex misandra*. It is remarkable that *D. oxycarpa* has not reached the calcareous, continental parts of the Jotunheimen.

In North Norway, it is known from Austtind in Fauske to Storhaugen in Kåfjord, but future investigations will probably show it to be more frequent.

D. oxycarpa is a calciphilous and somewhat hygrophilous species. Sociologically, it prefers the *Luzulion arcticae* and *Potentillo-Polygonion vivipari* alliances. It is frequently found together with *D. alpina*, but not in extreme snow-beds. In the middle-alpine belt, it may grow in more exposed positions than *D. alpina* and companions are, among others, *Cetraria nivalis*, *Luzula confusa*, *Poa arctica* and *Draba cacuminum*. Beside streams and springs, it

may grow abundantly in moist carpets of *Hylocomium splendens* and *Hypnum lindbergii*, giving the community a bright yellow colour during the flowering period.

Simen Bretten

Dryas octopetala L.

Maps: Hultén (1955, 1971a, no. 1095), Nannfeldt (1958). Total: Hultén (1968, 1971b, no. 45), Tralau (1963), Meusel et al. (1965), Hultén & Fries (1986, no. 1090). Local: Ouren (1952, 1961, 1966), Benum (1958), Lid (1958, 1959), Gjærevoll (1962, 1979, 1980), Toftaker (1969), Bråthen (1973), Hagen (1976), Moen & Selnes (1979), Kristensen (1981), Nordsteien (1982), Holten (1983), Moe (1985).
First Norwegian record: Found at Kongsvoll by Oeder in 1756, drawing in Flora Danica, 1, XXX (Oeder 1761). Finnmark: Måsøy, Hopseidet, Gunnerus 1759, (Gunnerus 1766, no. 106) (TRH).
Norwegian distribution: Throughout the mountain range south to Valle in Aust-Agder; in the lowlands south to Langesund.
Altitude limits: Low-alpine to middle-alpine. Raudhamrane in the Jotunheimen 1830 m (a verbal report of 2275 m in the Jotunheimen is obviously a mistake). Down to sea level at Langesund. Frequently down to sea level from Trøndelag northwards. Troms: Målselv, Kirkestinden 1415 m (Engelskjøn 1986a).
Habitat: Exposed heaths and plateaus, rocks, scree slopes, gravel and shell sand. Basiphilous.

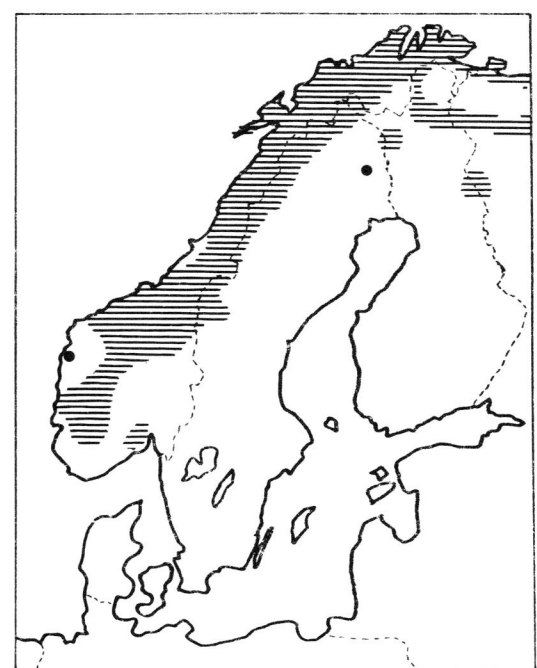

Fig. 42 The Fennoscandian distribution of *Dryas octopetala*

62

D. octopetala is a circumpolar, arctic-montane species. Taken in a wide sense it is polymorphic. There is some variation in the Scandinavian material, but less than, for example, in the Alaskan material (cf. Hultén 1959). During a visit to Trondheim in 1958, A. E. Porsild determined the material from Talstadhesten and its vicinity as being identical with his D. babingtoniana from Ireland, which he took as a separate species.

According to Hultén (1959, 1971b), D. octopetala was already a circumpolar species in preglacial times. During the glaciations, it was apparently isolated in several areas. During the last glaciation, it was limited to the close vicinity of the Scandinavian ice sheet since it has not been found in glacial deposits in the European lowlands. According to Tralau (1963), it was widespread during late-glacial times, especially in the Older Dryas, Allerød and Younger Dryas periods. Abundant fossil finds are known. The postglacial occurrences decrease rapidly north of Scania.

Much attention has been paid to the lowland localities in South Norway. At Langesund, D. octopetala grows abundantly on flat-topped limestone cliffs down to sea level. The present locality was submerged in early postglacial time, suggesting a fairly recent origin (Wille & Holmboe 1903) brought about by long-distance dispersal. Dyring (1911) considered the Langesund habitat to be a secondary glacial relic (see also Fægri 1958). D. octopetala might have grown in the neighbouring area in early postglacial time when the landscape was still very open and steppe-like, and spread later to the present locality. Fægri (1958) also put forward the hypothesis that the plant might have grown on the shore cliffs as soon as they became ice free, gradually invading new areas between the sea and the forest as the land became uplifted, and, thanks to its drought-resistant properties, surviving in this habitat as that gradually changed from a cold steppe climate to a warmer one.

At Solbergfjellet in Nedre Eiker and Kjørstadelva in Ytre Sandsvær, D. octopetala grows on steep limestone slopes where other species have difficulty gaining a foothold. In both places it is accompanied by Carex rupestris, just as on the alpine Dryas heaths (Lid 1958).

At Bergsåsen, Snåsa, Nord-Trøndelag, the steep northwest-facing slope between 50 and 200 m a.s.l. is covered with D. octopetala. This hill was most probably invaded by the plant very early. As the land became uplifted, the plant occupied the steep slope, the rest of the hill being gradually taken over by pine forest. In the present coniferous forest, D. octopetala still occurs on rocks that have a thin soil cover, side by side with, among others, Hepatica nobilis! Carex rupestris is also abundant, in part dominant, on the Dryas slope at Bergsåsen.

In South Norway, D. octopetala is mainly continental. There are, however, some interesting localities in West Norway. The species is able to grow in coastal localities if geological conditions are satisfactory, i.e. if calcareous bedrock is found. The absence of D. octopetala from most of the western Norwegian mountains is apparently due to

lack of geologically suitable habitats (Moe 1985). D. octopetala occupies extensive areas of the marble at Talstadhesten, to such a degree that most other species are prevented from growing (see also Malme 1971).

In Sweden, just as in Norway, D. octopetala is a common species throughout the mountain range from Härjedalen to Torne Lappmark. In Finland, it is found south to Kuusamo.

D. octopetala forms the most conspicuous chionophobous and basiphilous plant community in the Scandinavian mountains, growing on all kinds of calcareous schists, limestone, marble, dolomite and serpentinite (Nordhagen 1928, 1936a, 1943). In the dolomite area at Porsanger, D. octopetala covers large areas down to sea level. Quantitatively, the Dryas associations represent the largest areas in the Kobresio-Dryadion alliance.

Dryas communities have a comparatively wide ecological range as regards snow cover. Hedberg et al. (1952) differentiated three associations of "Dryadion" in Torne Lappmark, Nardino-Dryadetum, Tetragono-Dryadetum and Tomenthypno-Dryadetum. On hills and ridges that have suffered most deflation, D. octopetala is accompanied by, among others, Carex rupestris, C. glacialis, C. nardina (in North Norway) and a number of crustaceous lichens. With decreasing exposure, a number of other species occur, e.g. Rhododendron lapponicum and Cassiope tetragona (in North Norway), and lichens are partly replaced by mosses.

D. octopetala may also occur as a pioneer plant on scree slopes and calcareous gravel. In North Norway, it is also frequently met with on shell sand.

There are numerous measurements of the pH at Dryas localities. The amplitude is fairly wide. It may grow on gravel with a low pH if the humus content is low. Ouren (1952) reported values from Budal of 5.4-7.8. At Vollfjellet in Holtålen, I have measured 8.5. In the Sylane mountains, Christophersen (1925) found an amplitude of 5.9-6.8. Malme (1971) reported values of 6.3-8.1 at Talstadhesten, in his "Ctenidio-Dryadetum" association. From North Norway, Lunde (1962) has 428 tests with an amplitude of 4.5-8.3 and a mean of 6.51.

Epilobium davuricum Horn.

Maps: Hultén (1971a, no. 1279). Total: Hultén (1971b, no. 33), Hultén & Fries (1986, no. 1366). Local: Benum (1958), Lid (1959), Kytövuori (1969), Toftaker (1969), Hagen (1976), Moen & Moen (1977), Aune & Kjærem (1978), Kristensen (1981), Holten (1983), Elven (1984).

First Norwegian record: Reported from Finnmark by Wahlenberg (cf. Hornemann 1821).

Norwegian distribution: From Rauland in Telemark, Hardangervidda and Oslo in the south to North Cape and Sør-Varanger.

Altitude limits: Most frequent in the northern boreal belt, but also occurs in lowland areas as well as the low-alpine belt, ascending to 1370 m north of Peisabotn on the

Hardangervidda (Lid 1959). Troms: Målselv, Njunis 800 m.

Habitat: Springs, flushes, margins of rich fens, and damp grasslands. Calciphilous.

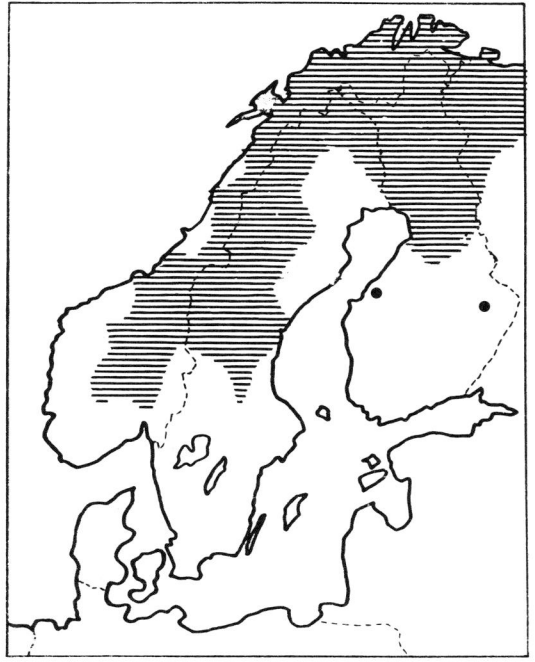

Fig. 43 The Fennoscandian distribution of *Epilobium davuricum*

calcareous soil and peat in damp, open grasslands and fens, very often in eroded localities. The pH of the soil or peat is 6-7.

E. davuricum is a characteristic species of the *Cratoneurion commutati* and *Caricion bicolori-atrofuscae* alliances (Nordhagen 1943).

Asbjørn Moen

Equisetum scirpoides Michx.

Maps: Hultén (1971a, no. 17). Total: Hultén (1962, no. 30, 1968), Hultén & Fries (1986, no. 14). Local: Benum (1958), Toftaker (1969), Moen & Moen (1977).

First Norwegian record: "Norska Nordland wid Qualvik d. 8. juli 1800". G. Wahlenberg (UPS).

Norwegian distribution: From Svarverud, beside Lake Holsfjorden, to North Cape and Sør-Varanger.

Altitude limits: Low-alpine to northern boreal. Grisungknatten in Dovre 1350 m. Down to about 120 m in southeastern Norway. Hustad 20 m. Frequently down to sea level in North Norway. Troms: Målselv, Isdalstind 1102 m. Pite Lappmark 1420 m.

Habitat: Moist sandy soil, gravelly river banks and shallow rich fens. Basiphilous.

E. davuricum s. lat. is a circumpolar species, and can be divided into two subspecies. Subsp. *arcticum* is scattered from Novaya Zemlya through northern Siberia, and also occurs in Arctic North America and Greenland. Subsp. *davuricum* has a more arctic-montane distribution south of where subsp. *arcticum* occurs, but is very rare in Greenland. Neither occur in Svalbard and Iceland, nor in Scotland and central Europe.

In Sweden, *E. davuricum* is found from Dalarna to Torne Lappmark, and in Finland, in the north as far south as northern Karelia. Distribution is continuous eastwards from Scandinavia and the Kola Peninsula.

In South Norway *E. davuricum* is absent from the southern and western parts. From Nordmøre northwards, it also occurs in coastal districts, even in the lowlands. On the island of Frøya, it grows in communities influenced by springs and dominated by *Cratoneuron commutatum* and other alpine species, such as *Equisetum variegatum*, and a number of lowland species, such as *Carex flacca* (Skogen 1970).

E. davuricum is a weakly competitive species, always occurring in open communities with a sparse field layer and, usually, patches of open peat or bare soil. It is somewhat hemerophilous, occurring in ditches and on roadsides, etc. In its lowland localities, it only grows in habitats influenced by springs. In the alpine area, it also grows on

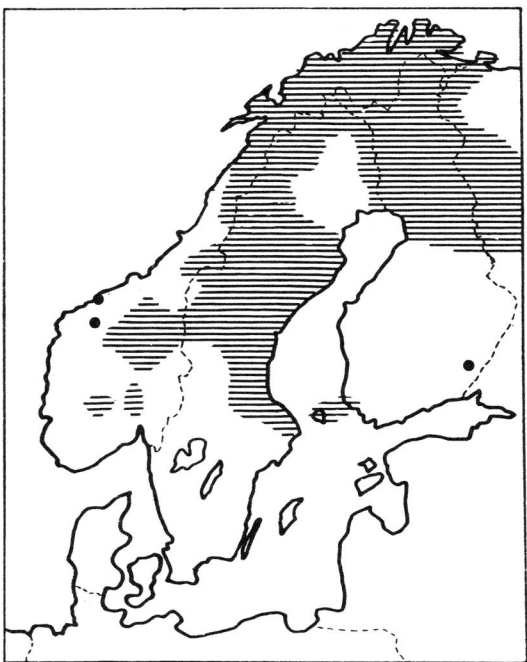

Fig. 44 The Fennoscandian distribution of *Equisetum scirpoides*

E. scirpoides is a circumpolar, arctic to subarctic species, but has a strange gap in the North Atlantic sector, being known from Svalbard but not Iceland or East Greenland. The distribution in South Norway is peculiar in that it is absent from the mountain range south of the northern Jotunheimen, and also from the western mountains (except for one locality at Hustad). In South Norway, it is frequently found at low altitudes, especially on river banks. In North Norway, there are also numerous localities in coastal areas.

The Fennoscandian distribution differs greatly from that of typical alpine plants. In Sweden, it is known from the whole mountain range, but there are also numerous localities along rivers and on the shores of the Gulf of Bothnia. There is continuous distribution from Fennoscandia eastwards.

Phytosociologically, *E. scirpoides* plays a minor role since it rarely occurs in quantity. It may grow in many different habitats, including moist depressions, cliffs, springs, rich fens, scree slopes, pine forests, along streams and rivers, and also as a hemerophile on roadsides. At sea level it keeps to shell sand. It is very common in the dolomite area of Porsanger. Although the habitat range is very wide, it obviously prefers areas with a carbonate substrate.

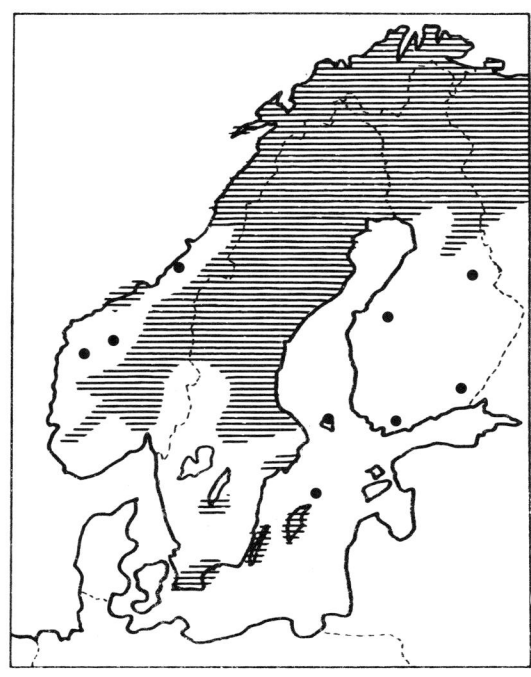

Fig. 45 The Fennoscandian distribution of *Equisetum variegatum*

Equisetum variegatum Weber & Mohr

Maps: Hultén (1971a, no. 20). Total: Hultén (1962, no. 45, 1968), Hultén & Fries (1986, no. 13). Local: Benum (1958), Lid (1959), Toftaker (1969), Moen & Moen (1977), Holten (1983).

First Norwegian record: Reported by Wahlenberg (1812) "in rupibus elatis montium inferalpinum Nordlandiæ passim".

Norwegian distribution: From Sola in Jæren to North Cape and Sør-Varanger.

Altitude limits: Low-alpine to northern boreal. Nordre Knutshø 1530 m (Hatlelid 1980). In southeastern Norway, down to about 60 m; in western and northern Norway, down to sea level. Troms: Målselv, Kirkestinden 1320 m (Engelskjøn 1986a).

Habitat: Rich fens, damp heaths, meadows, and banks of rivers and lakes. Basiphilous.

E. variegatum is a circumpolar, arctic-montane species. It is widely distributed in Europe. Although chiefly an alpine plant in Scandinavia, it also occurs in lowland localities. In Sweden, there are numerous localities along the coast of the Gulf of Bothnia, and the species is fairly common on Öland and Gotland. It is rare in the western mountains of South Norway.

Its ecological range is wide, but it prefers rich fens. Nordhagen (1943) considered it to be a characteristic species of his *Caricion atrofuscae-saxatilis* (*Caricion bicolori-atrofuscae*) alliance. It always plays a quantitatively minor role.

Erigeron humilis Grah.

Maps: Hultén (1971a, no. 1679). Total: Hultén (1958, no. 175, 1968), Hultén & Fries (1986, no. 1773). Local: Benum (1958), Engelskjøn (1967), Aune & Kjærem (1978).

First Norwegian record: The oldest herbarium collection dates back to 1841, Tromsø, Tromsdalstind, M. N. Blytt (O).

Norwegian distribution: Northern unicentric. From Arefjellet in Hattfjelldal to Skarddalen in Alta.

Altitude limits: Low-alpine to middle-alpine. Troms: Målselv, Kirkestinden 1565 m (Engelskjøn 1986a). Down to 540 m in Øverbygd; rarely below the timber line.

Habitat: Prefers snow-beds on north-facing slopes, on solifluction soil. Calciphilous.

E. humilis is an amphi-Atlantic, arctic species. It is known from Svalbard and Iceland, and is also widespread westwards from East Greenland to eastern Siberia. In Sweden, it is known from Pite Lappmark to Torne Lappmark, and in Finland, from Enontekis Lappmark. The distribution in North Norway shows a continental tendency with very few localities on coastal mountains. The main area is inner Troms.

As Engelskjøn (1967) showed, the tetraploid, *E. humilis*, and the diploid, *E. uniflorus*, frequently hybridise, and a number of collections in Scandinavian herbaria that are referred to as *E. humilis* are the triploid hybrid. The hybrid corresponds morphologically to *E. humilis*. Engelskjøn believed that *E. humilis* probably had a Pre-Quaternary origin "evolved in alpine non-Arctic areas, having attained its present arctic distribution in the course of the Quaternary glaciations".

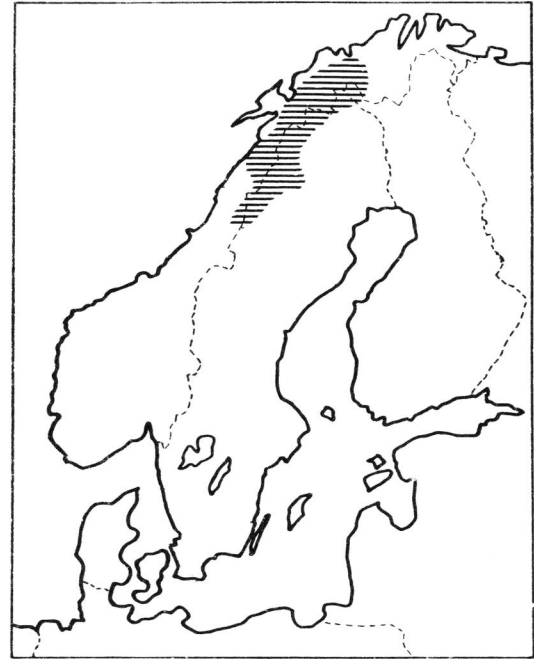

Fig. 46 The Fennoscandian distribution of *Erigeron humilis*

Like a number of other amphi-Atlantic species with an isolated occurrence in Scandinavia, *E. humilis* may have survived the last glaciation in North Norway. Ecologically, it is a hardy species. In Svalbard, it grows abundantly on the semi-nunatak of Ossian Sars in Kongsfjorden. It is the most common *Erigeron* species on Jensen's Nunataks in Greenland (Gjærevoll & Ryvarden 1977).

Its ecological range is fairly wide. It may occasionally grow in gravelly, exposed localities and *Dryas* heaths, but is usually found in damp or moist snow-beds, very often in localities affected by solifluction.

It is a quantitatively unimportant, but fairly frequent constituent of snow-bed communities characterised by, among others, *Salix polaris*, *Saxifraga oppositifolia* and *Cerastium arcticum*, i.e. calciphilous communities. It prefers north-facing slopes with a well-developed moss mat, occurring in company with, among others, *Luzula arctica* and *Antennaria villifera*. It also frequently grows on moist ledges.

Erigeron politus Fr.

Maps: Hultén (1971a, no. 1676). Total: Hultén (1968, 1971b, no. 280), Hultén & Fries (1986, no. 1771). Local: Benum (1958).

First Norwegian record: Botanists prior to Ledebour and Fries usually failed to differentiate the present taxon from *E. acris*, or other glabrescent races such as *E. droebachiensis*, making it difficult to judge which observation might be the first from Norway. For example, "E. acre variet. glabratum" reported from Alta by Zetterstedt (1822) is

probably the present species. However, no substantiating collection has been seen. A report in Flora Danica is likewise somewhat dubious.

Norwegian distribution: Generally found in inland districts from Ringerike and inner Sogn to Finnmark; a few coastal stations, mainly in the northernmost part of the area. There is a limited disjunction in Norway in the Trondheimsfjord area, but distribution is continuous in Sweden.

Altitude limits: From the lowlands to the lower part of the low-alpine belt; down to sea level in western and northern Norway. The upper limit just exceeds 1100 m in southern Norway (in the Jotunheimen and Dovre mountains); 760 m in central Troms.

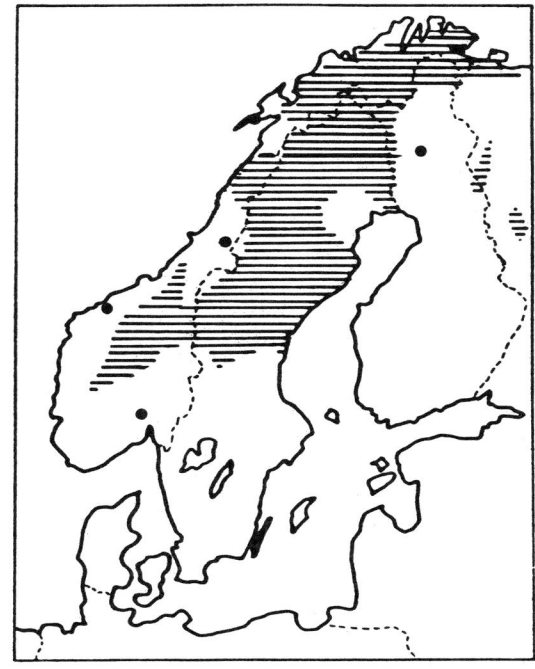

Fig. 47 The Fennoscandian distribution of *Erigeron politus*

Habitat: Generally open. Scree slopes, often with *Fragaria*, rock ledges, stream and river banks; frequently south-facing. Often follows rapidly flowing rivers down to sea level. Most abundant in subalpine areas, but extends in smaller numbers up into the low-alpine and down into the coniferous forest belts, in the latter case usually in canyons containing streams, and similar precipitous habitats. It often invades open, semi-natural habitats around farms and upland summer farms, where it may meet culturally-dispersed *E. acris*. Grows in dry to fresh mineral soil. Moderately basiphilous.

E. politus was described by E. M. Fries (1843). Before then, it was sometimes known as *E. elongatus* Ledeb. (1829, non Moench 1794). *E. politus* is circumpolar, but absent from Iceland and Greenland. In spite of this, it has often been taken as a subspecies of the *E. acris* complex.

E. angulosus Gaudin is very close, and if it is eventually found to be synonymous it will take priority over *E. politus.*

Where *E. politus* grows together with *E. acris*, transitional specimens sometimes occur (the chromosome number is the same). Such dubious specimens have been seen from southern Telemark (not shown on the map). An outlying station at Molde ("Molde, -/7 1907. A. T. Hvass. herb. S") is shown, but requires confirmation.

Sigmund Sivertsen

Euphrasia salisburgensis Funck
(*E. lapponica* Th. C. E. Fr.)

Maps: Hultén (1971a, no. 1576). Total: Meusel et al. (1978), Hultén & Fries (1986, no. 1683). Local: Nordhagen (1952), Benum (1958).

First Norwegian record: Troms: Målselvdal 1841, N. Lund (O). In South Norway, the first record was made in 1930 (Lom, Bøvertunvatnet, F. Jebe).

Norwegian distribution: Bicentric. In South Norway, from the Varahaugane mountains on Hardangervidda to the Trollheimen. In North Norway, from Bindal to Magerøya.

Altitude limits: Mainly low-alpine. Oppland: Lom, Høyrokampen 1430 m (Løkken 1969). Troms: Målselv, Rostafjell 1000 m. Down to almost sea level in Porsanger.

Habitat: Dry exposed heaths and ridges on calcareous soils, calcareous screes.

In the oldest reports of this species from Norway there was uncertainty about its taxonomical position. A. Blytt (1874) regarded the Norwegian material as being different

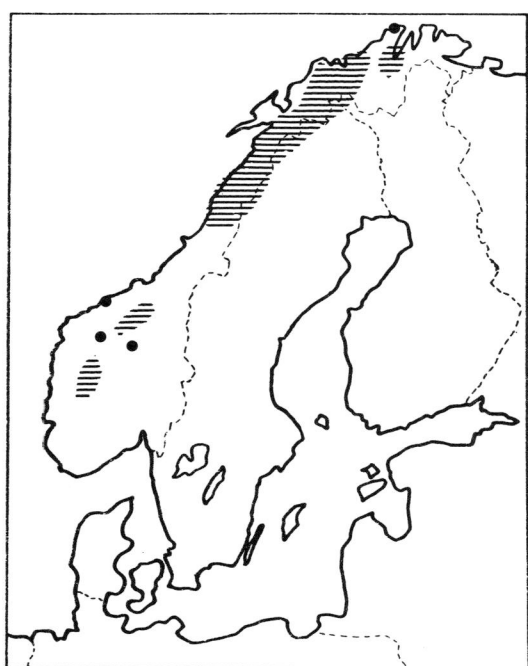

Fig. 48 The Fennoscandian distribution of *Euphrasia salisburgensis*

from *E. salisburgensis*, and named it *E. officinalis* δ *subulata.* However, Hartman (1879) reported it as *E. salisburgensis.* When more material became available from both Norway and Sweden, Th. C. E. Fries (1921) separated the northern Scandinavian populations from *E. salisburgensis* and named the resulting species, *E. lapponica*, endemic to northern Scandinavia. His view was supported by Holmboe (1934) and Nordhagen (1952).

There is considerable variation within the Scandinavian material. Var. *purpureocoerulea* was described by Nordhagen (1952) from some localities in South Norway, and Rune & Rønning (1954) described var. *pallida* from Magerøya. In the Alps, too, a similar situation seems to occur, and Flora Europaea has included *E. lapponica* in *E. salisburgensis*, a view which I accept with some hesitation.

E. salisburgensis is a European species widely distributed in the central European mountains and also occurring in Ireland and on Gotland. Its ecological behaviour seems identical in the Scandinavian mountains and the Alps. However, on Gotland it is completely different, as it occurs on rich *Schoenus* fens. This plant was regarded by Yeo (1978) as var. *schoenicola.*

The distribution in South Norway is very disjunct. Great attention has been paid to the isolated occurrence at Trollkyrkja-Sleppskaret in Romsdal (Nordhagen 1952). He believed this occurrence supported the theory of ice-free refugia in that part of the country. In North Norway, the distribution is more continuous. It is remarkable that the species has not yet been found east of Porsanger.

In the Swedish mountains, *E. salisburgensis* is known from Åsele Lappmark to Torne Lappmark. In Finland, it has only been found in Enontekis Lappmark.

As already emphasised, *E. salisburgensis* has a fairly narrow ecological range, growing in exposed heaths and screes on calcareous soil. Nordhagen (1936a) took it as a characteristic species of his *Kobresieto-Dryadion* alliance.

Gentiana purpurea L.

Maps: Hultén (1971a, no. 1425), Møller (1966), Nordhagen (1973). Total: Meusel et al. (1978), Hultén & Fries (1986, no. 1495). Local: Ouren (1950, 1979), Lid (1959), Halvorsen & Salvesen (1983).

First Norwegian record: As *G. purpurea* has long been a well-known medicinal plant there are numerous old reports of this species in Norway, such as, Absalon Pedersen Beyer (ca. 1570), Jens Nilsson (1595), Gartner (1694), Jonas Ramus (1715) (see Holmboe 1905). It is reported in Flora Danica 1 (Oeder 1761) and Flora Norvegica (Gunnerus 1766, no. 97) – "Søt-fieldet" in Budal.

Norwegian distribution: From Hekkfjell in Åseral to Vang in Valdres, Ringebu and Midtre Gauldal. The map also shows some reliable former localities, where *G. purpurea* is now extinct. The roots were collected in large quantities to be used as a drug and were in part exported to Sweden, perhaps causing this local extinction (Holmboe 1905).

Altitude limits: Northern boreal to low-alpine. Lifjell, Telemark and Storenuten, Odda, 1360 m. On the Hardangervidda, frequently to 1100-1250 m. Down to 325 m at Etne. Mainly between 800 and 900 m in Midtre Gauldal.

Habitat: Birch forests, willow thickets and meadows. Acidophilous to neutral.

G. purpurea is a European, arctic-montane species restricted to two widely separated areas, one in Scandinavia and the other in the Alps and northern Apennines. In Scandinavia, in addition to those shown on the map, there is a single locality in Härjedalen.

G. purpurea is widely distributed from Åseral to Valdres, occurring in large quantities in many places. It was collected by J. B. Barth in ca. 1890 at Ringebu (Bubekkdalen). In herb. TRH, there is a collection by P. Green made in 1947 and labelled Gudbrandsdalen. According to verbal information from the collector, this is most likely Ringebufjellet.

Already in 1694, Gartner reported that *G. purpurea* grew, among elsewhere, in Gauldalen. In 1769, Gunnerus (see O. Dahl 1888-1911) received *G. purpurea*, collected in Budal, from a Støren clergyman, Ole Lie. However, more information on the plant in this area was not obtained until 1922 when it was discovered at Singsås. It is now known from numerous localities in Midtre Gauldal (Budal and Singsås (Ouren 1979).

G. purpurea grows in the upper part of the subalpine birch forest, in willow thickets and low-alpine meadows, always in places well-protected by snow. It seems to avoid birch forests dominated by *Vaccinium myrtillus*, preferring areas where *Deschampsia flexuosa* and *Nardus stricta* are the most conspicuous plants. Such birch forests have been heavily grazed, and sometimes also harvested for hay.

The species seems to prefer acid soils, but avoids the most acid ones. Ouren (1950) published the results of 14 measurements in Budal with a mean pH of 4.6 (range 4.1-5.2), and 3 from Geilo with pH = 4.7, 5.0 and 5.1).

As the distance between the northernmost locality in central Europe and the southernmost one in Norway is about 1300 km, some attention has been paid to the immigration of *G. purpurea* (Nordhagen 1973). Even though it is an old medicinal plant, its present distribution and the history of its immigration can hardly be explained in terms of human import. The plant is very clearly older in Norway than its use as a drug. So far, no fossil evidence has been obtained.

Gentianella tenella (Rottb.) Börner

Maps: Hultén (1971a, no. 1426). Total: Hultén (1958, no. 224), Meusel et al. (1978), Hultén & Fries (1986, no. 1500). Local: Benum (1958), Toftaker (1969).

First Norwegian record: Gunnerus (1772, no. 1077) – "Habitat in gulbrandsdalia" (named *Swertia rotata*) (TRH).

Excluded or doubtful stations: Nordkapp: Magerøya, Kjelvik, P. V. Deinboll, undated (BG). O. Dahl (1934) gave Bossekop as the northern limit for this species and did not mention Magerøya. As there has been a great deal of doubt about several reports from Deinboll, and as there are no other records of *G. tenella* from this very well investigated area, the locality is regarded as doubtful.

Norwegian distribution: Slightly bicentric. In the southern area, there is an isolated occurrence at Finse and then it is found from Kalvåhøgda, north of Bygdin, to the Trollheimen and Røros. In the northern area, from Kappfjellet in Hattfjelldal to Bossekop in Alta.

Altitude limits: Mainly low-alpine. Skarvhø in the Jotunheimen ca. 1600 m (S. Løkken, O). Nordre Knutshø 1555 m. Down to at least 480 m (Alvdal). Troms: Målselv, Maddagai'si 1380 m (Engelskjøn 1986a). Down to sea level in Alta.

Habitat: The ecological range is wide. It may occur on rocks and in exposed heaths as well as grassy meadows, shallow fens and river flats. Basiphilous.

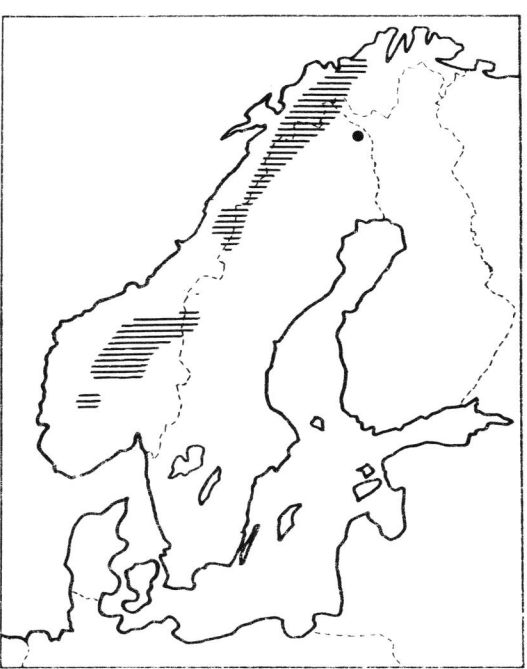

Fig. 49 The Fennoscandian distribution of *Gentianella tenella*

G. tenella coll. is a circumpolar, arctic-montane species with large disjunctions. Several races have been described as separate species. It is common in Iceland, but very rare in Svalbard. The Greenland distribution is divided into two parts, one area in East Greenland and another in West Greenland. It has a wide, but scattered, distribution from Spain to the Carpathians.

In Sweden, it is found from Härjedalen to Torne Lappmark, with the same gap as in Norway. In Finland, it grows only in Enontekis Lappmark. From Alta there is a gap of about 700 km to easternmost Kola.

G. tenella displays a distinctly continental distribution in South Norway, and a similar tendency is found in North Norway.

Sociologically, G. tenella has no pronounced preferences. It usually grows dispersed and in a variety of plant communities. It is most often found in protected heaths and grassy slopes, where it seems favoured by grazing. It is often a hemerophilous plant in farmyards, often together with Gentiana nivalis, Gentianella amarella and G. campestris. It may also occur as a pioneer plant on roadsides. It quite often grows on grassy river banks, tolerating sheep grazing. G. tenella is a basiphilous species growing on calcareous schists, limestone, dolomite and serpentinite gravel.

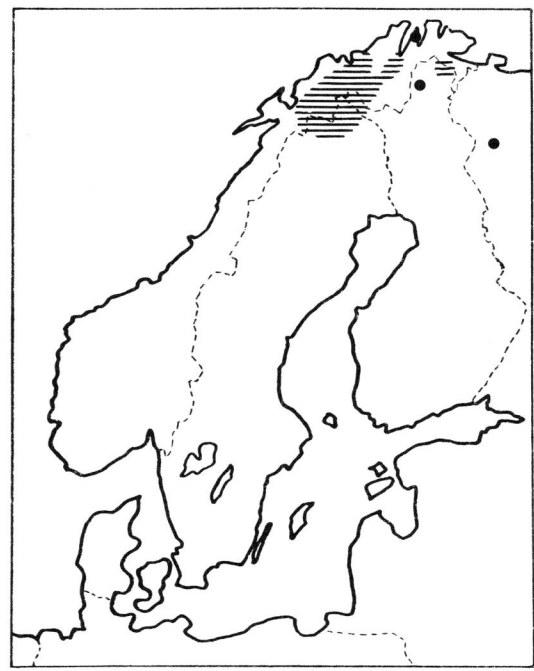

Fig. 50 The Fennoscandian distribution of *Hierochloë alpina*

Hierochloë alpina
(Willd.) Roemer & Schultes

Maps: Th. C. E. Fries (1913), Hultén (1971a, no. 138), Engelskjøn (1986a). Total: Hultén (1962, no. 13, 1968), Hultén & Fries (1986, no. 346). Local: Benum (1958), Engelskjøn (1984).

First Norwegian record: Troms: Storfjord, Čacca ("Tjatsa supra Lyngen Norlandiae 1838"). L. L. Læstadius (GB, O, C, LD). It is mentioned by Wahlenberg (1812, no. 31) from his journey in 1800 "ad Tsaitsekaise".

Norwegian distribution: Northern unicentric. From Frostisen to Stjernøy and Porsanger.

Altitude limits: Low-alpine to middle-alpine. Frostisen 1355 m. Troms: Målselv, Maddagai'si 1340 m. Down to 150 m in Reisadalen.

Habitat: Exposed localities, with a wide range from oligotrophic gravel to fairly eutrophic schists.

H. alpina is a circumpolar, arctic species. In Sweden, it is known from Lule Lappmark and Torne Lappmark. In Finland, from Enontekis Lappmark and Enare Lappmark. There is a gap between the localities in Enare Lappmark and those in central and outer coastal parts of the Kola Peninsula, but it is of minor significance.

H. alpina is mainly an inland species. In inner Troms, it is an important and partly dominating grass, forming a characteristic community in exposed heaths, a community in the *Arctostaphylo-Cetrarion nivalis·* alliance. From northern Sweden, Th. C. E. Fries (1913) described a "Flechtenreiche *Hierochloë alpina*-Assoziation" (cf. also Holmboe 1912). In my experience, this community is circumpolar. Common companions are *Calamagrostis lapponica*, *Carex bigelowii*, *Festuca ovina*, *Empetrum hermaphroditum* and *Betula nana*.

H. alpina inhabits gravel ridges (moraines and eskers) and exposed plateaus. It mainly grows in oligotrophic

habitats, but is also often seen on windswept ridges of schist, sometimes growing together with *Dryas octopetala*. H. alpina has often been regarded as an acidophilous species, but is more correctly viewed as an indifferent one.

Possible immigration from the northeast was discussed by Th. C. E. Fries (1913) and later by Elfstrand (1927) and Weimarck (1976, 1981). As H. alpina grows well on morainic material it might have had favourable dispersal conditions in early postglacial time. The occurrences on Stjernøy (550-650 m) and at Øksfjord (450 m), and especially the recently discovered localities on the Frostisen nunataks at an altitude of 1320-1355 m, show that H. alpina may have been able to survive on ice-free refugia in North Norway.

Juncus arcticus Willd.

Maps: Hultén (1971a, no. 437). Total: Hultén (1962, no. 18), Hultén & Fries (1986, no. 198), Local: Benum (1958), Lid (1959), Kristensen (1981), Nordsteien (1982).

First Norwegian record: Flora Danica (1792) "In Finmarkia orientali ad Kiberg".

Norwegian distribution: Slightly bicentric. From Stavand, Valle in Aust-Agder to the Sylane mountains, and from Hattfjelldal to Magerøya and the Varanger Peninsula. Reported from Hjelmeland (Ryvarden & Kaland 1968), but no voucher specimen exists.

Altitude limits: Mainly low-alpine to northern boreal. Hordaland: Finse, Jomfrunuten 1400 m. On river banks down to 400 m (Jordet in Trysil). Almost down to sea level on the River Orkla at Rømmesmoen (O. Dahl 1920 (O)), but this locality was occasional. Troms: Målselv, Jerdni (a mountain in Dividalen) 890 m. Down to sea level in northernmost Norway.

Habitat: Shallow rich fens, wet gravel, and banks of rivers and lakes. Calciphilous.

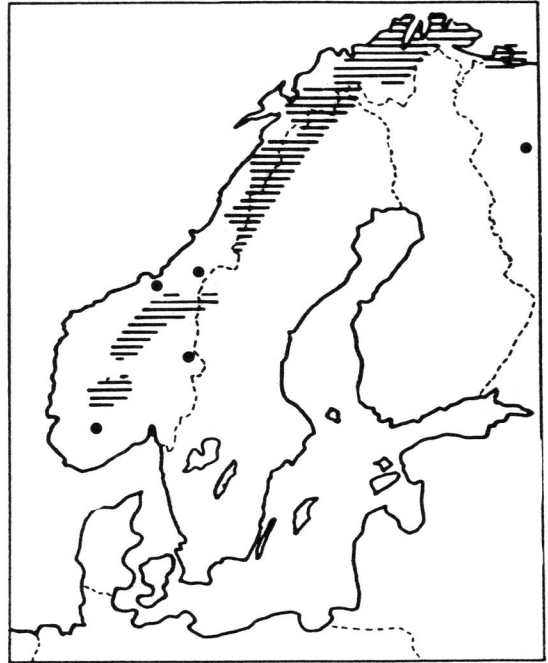

Fig. 51 The Fennoscandian distribution of *Juncus arcticus*

J. arcticus belongs to a critical group. Hylander (1953) thought it impossible to draw a clear distinction between *J. arcticus* and *J. balticus*, the situation in Iceland being particularly obscure, and he took *J. balticus* to be a subspecies of *J. arcticus*. His view was followed by Hultén (1962), who regarded *J. arcticus* as a circumpolar species divisible into a number of races. Flora Europaea, Hämet-Ahti et al. (1984) and Nilsson (1986) retained *J. balticus* as a separate species. As Hylander emphasised, it is usually easy to distinguish between the two taxa in Scandinavia.

In Sweden, *J. arcticus* is known from Härjedalen to Torne Lappmark. It is very rare in Finland, and has only been reported from Enontekis Lappmark and Kuusamo.

J. arcticus is generally found on shallow, rich fens. Nordhagen (1943) regarded it as a somewhat unreliable indicator of rich fens. In my opinion, it is always linked with *Carex atrofusca* fens and fen-like communities on river banks being associated with, among others, *Carex bicolor* and *C. microglochin*, on seasonally inundated sandy and gravelly soil. It is also frequently favoured by human activities. It has invaded the medieval King's Road over Hjerkinnhø in many places where the gravelly substrate offers favourable conditions.

Juncus castaneus Sm.

Maps: Hultén (1971a, no. 445). Total: Hultén (1962, no. 22, 1968), Hultén & Fries (1986, no. 200). Local: Benum (1958), Lid (1959), Toftaker (1969), Moen (1970), Skogen (1974, 1976), Hagen (1976), Moen & Moen (1977), Nordsteien (1982), Holten (1983, 1984), Elven (1984), Moe (1985).

First Norwegian record: Telemark: Rauland, Bossbøen. W. A. Uldahl. Uldahl travelled to Telemark in 1802 together with Martin Vahl. The species was drawn in Flora Danica 23, no. 1332, reference being made to the first record by Uldahl.

Excluded or doubtful stations: A. Blytt (1876) reported localities on river banks in Orkdal (Gjølme) and Trondheim. These apparently occasional stations are excluded.

Norwegian distribution: From Suldal and Hjartdal/Seljord to Korgen in Nordland, and further north from Nordreisa to Vardø.

Altitude limits: Mainly in the northern boreal and low-alpine belts. Between Roksdalskammen and Svorundfjell in the Trollheimen 1500 m. In central Norway down to 230 m; in secondary localities on river banks to 60 m (Ouren 1964). Mainly at low altitudes in Troms and Finnmark. Alta: Haldde 600 m.

Habitat: Extremely rich fens, calcareous springs and flushes.

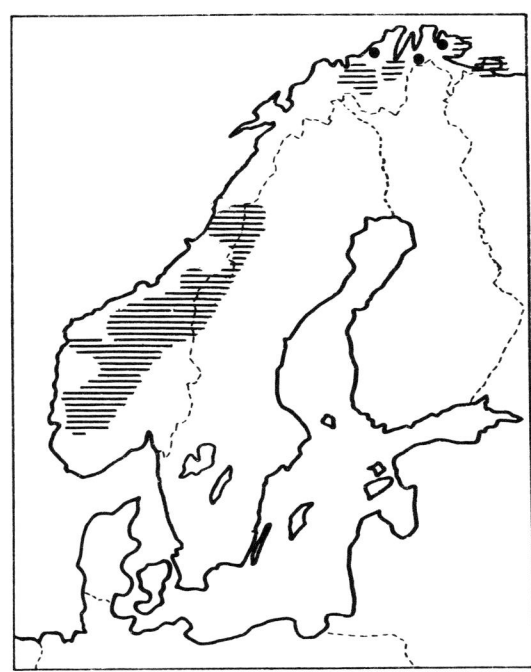

Fig. 52 The Fennoscandian distribution of *Juncus castaneus*

J. castaneus is a circumpolar, arctic-montane species. In central Europe, it occurs disjunctly from the Alps to the Carpathians. In Sweden, it is found in the mountains close to the Norwegian border from Dalarna to east of Rana, but is absent from northernmost Sweden and from Finland. It has a scattered but continuous distribution from Finnmark along the coast of the Kola Peninsula and further eastwards.

Most localities in South Norway are found inland. Moe (1985) classified it as one of the species also occurring in coastal areas, but there are relatively few coastal localities. Being a basiphilous species, unfavourable geological conditions seem to be its limiting factor, not climate. *J. castaneus* occurs in areas where the lawns of extremely rich fens have a thin peat cover, most frequently where there is seepage of groundwater, local erosion, or the fen surface has been disturbed, for example, by hay-making in former times. It also occurs in secondary localities on unstable margins of streams and rivers, and at the edge of ditches, roadsides and tracks.

J. castaneus is a characteristic species of the *Caricion bicolori-atrofuscae* alliance (Nordhagen 1943) and the upland variety of the *Cratoneurion commutati* alliance.

The total absence of *J. castaneus* between Rana and Nordreisa means that the species is disjunct with an "aberrant disjunction" according to Berg (1963), since the gap deviates from that which is normal for bicentric species. In this area, apparently suitable habitats for *J. castaneus* are very common, and its absence cannot be explained by present-day ecological conditions, nor be ascribed to the postglacial warmth optimum (e.g. in the Saltdalen mountains). It seems reasonable to suppose that *J. castaneus* has never grown in this area since the last glaciation. Hence, the species must have either survived the last glaciation in at least two separate areas, or subsequently migrated into Norway from two different directions or areas.

Asbjørn Moen

Kobresia myosuroides (Vill.) Fiori

Maps: Hultén (1955, 1971a, no. 314). Total: Hultén (1962, no. 40, 1968), Meusel et al. (1965), Hultén & Fries (1986, no. 423). Local: Ouren (1952), Benum (1958), Lid (1959), Toftaker (1969), Hagen (1976), Nordsteien (1982).

First Norwegian record: Oppland: Lom, Blakar 1813. C. Smith (cf. O. Dahl 1895).

Norwegian distribution: Slightly bicentric. In the southern area, from Trossovnuten in Røldal to Skjækerfjella in Snåsa. Then from the border between Nord-Trøndelag and Nordland to Magerøya.

Altitude limits: Low-alpine to middle-alpine. Ryggehø in the Jotunheimen, above 1900 m (NAS). Down to 675 m in Drivdalen. Troms: Målselv, Kirkestinden 1100 m (Engelskjøn 1986a). Down to sea level at several places in North Norway.

Habitat: Exposed heaths. Basiphilous.

K. myosuroides is a circumpolar, arctic-montane species. Its distribution in central Europe is disjunct. In Sweden, it

is found from Härjedalen to Torne Lappmark, but has a gap corresponding to that in Norway. In Finland, it is only known from Enontekis Lappmark. A single locality has been recorded on the Rybachiy Peninsula. Between there and the nearest one in the Urals (where it is very rare) is another gap of about 1200 km. East of the Urals there is a ca. 1000 km gap to the nearest northern Siberian occurrences. The gap between the Rybachiy Peninsula and the Urals is a feature found in several other species. Hultén & Fries (1986) recorded *K. myosuroides* in Svalbard, but as no evidence for this is known it seems to be wrong.

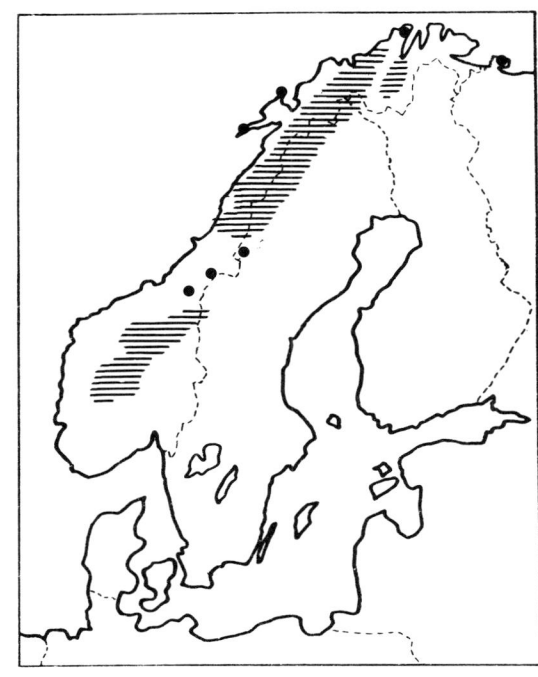

Fig. 53 The Fennoscandian distribution of *Kobresia myosuroides*

In South Norway, *K. myosuroides* is only known in inland areas, and most localities are in the Jotunheimen-Trollheimen region. In North Norway, there are several localities in coastal districts.

K. myosuroides is generally met with in the low-alpine belt. It may also be abundant at lower levels where open ground has been created by man, leading to increased deflation, especially in the vicinity of upland, summer farms. Many valleys have a "low-alpine" segment below a distinct lower birch forest limit, most likely due partly to human activities and grazing, partly to the flow of cold air that takes place in valley bottoms.

K. myosuroides grows on calcareous schists, limestone and dolomite. It is a characteristic species of the *Kobresio-Dryadion* alliance, inhabiting the most windswept ridges which have little snow cover in winter. The tufts may form a continuous mat, but there is more often a mosaic of *Kobresia* tufts and wind-eroded gravel. The vitality of

Dryas octopetala is reduced in such places.

Nordhagen (1955) published a comprehensive survey of *Kobresietum myosuroidis* data derived from various parts of Norway. An extract of this is given in Table 9. Only the most characteristic and dominating species are listed. The community is very rich in species.

Table 9

Locality no.	Sk.	Ha.	Å.	Vå.	K.
Nos. of squares (4 m²)	6	4	10	10	30
Salix reticulata	100-1	100-2	30-1	30-1	20-1
Dryas octopetala	100-4	.	100-1	.	rare
Astragalus alpinus	100-1	100-2	100-1	40-1	30-1
Artemisia norvegica	.	.	100-2	.	30-1
Campanula rotundifolia	33-1	50-1	100-1	100-1	100-1
C. uniflora	100-1	.	100-1	.	.
Oxytropis lapponica	100-1	.	.	.	70-1
Polygonum viviparum	100-1	50-1	100-1	100-1	100-1
Saxifraga oppositifolia	100-1	75-1	100-1	.	30-1
Saussurea alpina	.	50-1	50-1	100-1	50-1
Silene acaulis	100-1	.	100-1	70-1	.
Thalictrum alpinum	50-1	75-1	30-1	100-2	.
Carex rupestris	100-1	.	100-1	100-2	80-1
Festuca ovina	100-2	100-3	100-2	100-1	100-3
Juncus trifidus	.	25-1	100-1	100-1	50-1
Kobresia myosuroides	100-4	100-3	100-5	100-5	100-5
Rhytidium rugosum	.	100-4	.	.	100-5
Gymnomitrion corallioides	100-3	.	100-2	.	.
Alectoria ochroleuca	83-1	.	100-3	100-4	30-2
Cetraria cucullata	100-1	100-1	100-5	100-5	100-1
C. nivalis	50-1	100-1	100-1	100-2	100-1
Ochrolechia frigida	100-3	100-2	100-2	.	80-1
Sphaerophorus globosus	66-1	.	100-1	.	40-1
Thamnolia vermicularis	83-1	.	.	100-1	90-1

Explanation of abbreviations:
Sk = Skaitiaksla, Saltdal (1250 m); Ha = Habavuoppe-bak'te, Kautokeino (400 m); Å = Store Åmotshytten, Oppdal (1483 m); Vå = Raudnebba, Vågå (ca. 1350 m); K = 10 squares from Kongsvoll (ca. 900 m), Svahø in the Trollheimen (ca. 1300 m) and Nonshø in Sunndalen (ca. 900 m).

The table shows that the greatest variations are in the bottom layer. Where domesticated or wild reindeer graze, the vitality of fruticose lichens is greatly reduced, whereas crustaceous cryptograms like *Ochrolechia frigida* and *Gymnomitrium corallioides* increase.

Although *K. myosuroides* is a basiphilous species, the pH amplitude is wide. Nordhagen (1955) reported pH = 5.7-6.5 at Skaitiaksla in Nordland. Lunde (1962) has 130 measurements from North Norway with an amplitude of 4.9-7.9 and a mean of 6.52.

Kobresia simpliciuscula
(Wahlenb.) Mack.

Maps: Tengwall (1913), Hultén (1971a, no. 315). Total: Hultén (1958, no. 213, 1968), Meusel et al. (1965), Hultén & Fries (1986, no. 424). Local: Ouren (1952, 1961, 1966), Toftaker (1969), Hagen (1976), Nordsteien (1982), Halvorsen & Salvesen (1983), Wilmann (1983).

First Norwegian record: Herb. O has two collections leg. C. Smith: 1. "In alp. Hardangeriae et Quiknae" (1814 mis.). 2. "Imellem Tolgen og Opdal" (undated). As Smith visited Kvikne on his journey in 1807, the first collection was obviously made there.

Excluded or doubtful stations: Telemark: Veslefjeld in Drangedal. This locality is marked on the map published by Tengwall (1913), but there is no known collection or source (cf. Holmboe 1932). Finmarkia orientalis, P. V. Deinboll (O). Much doubt has been expressed about the reliability of Deinboll's Finnmark material (see O. Dahl 1934).

Norwegian distribution: From Bjørnaskallen on the Hardangervidda to Rana (between Bøvervatnet and Steintjønna).

Altitude limits: Mainly low-alpine, but also in the northern boreal and the lowest part of the middle-alpine belts. Bjørnaskallen and Heggjeitlane on the Hardangervidda 1450 m. Down to 480 m in central Norway (Solhall, south of Ramstadsjøen, Midtre Gauldal).

Habitat: Prefers rich fens, and irrigated and moist cliffs. Basiphilous.

K. simpliciuscula is a circumpolar, arctic-montane species, but has large disjunctions in Eurasia. There is a gap

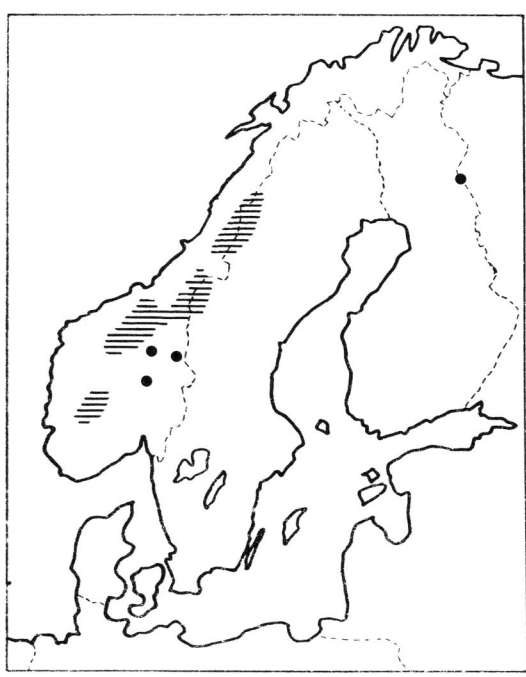

Fig. 54 The Fennoscandian distribution of *Kobresia simpliciuscula*

between Kuusamo in Finland and the Urals (where it is very rare), and further gaps of approximately 1500 km between the Urals and stations in the Altai mountains to the southeast and Khatanga in northern Siberia. In central Europe, *K. simpliciuscula* is found from the Pyrenees to the Carpathians. There are several localities in northern England and Scotland. Only a few are known in Svalbard. The disjunct distribution indicates that *K. simpliciuscula* survived at least the last glaciation in several places.

In Scandinavia, *K. simpliciuscula* is mainly a southern species with a distribution that strongly resembles that of *Pedicularis oederi*, but extending somewhat further north. The Norwegian distribution area is divided into two, a small one on the Hardangervidda and a large one in central Norway. The distribution is mainly continental, only a few coastal stations being known. The Swedish distribution is restricted to Härjedalen and Jämtland, the northernmost locality being at Övre Älsvattnet, 10 km south of the Rana locality. An extremely isolated locality is found in Kuusamo, Finland, an area with several interesting alpine species.

K. simpliciuscula usually grows in rich fens transitional to damp meadows or rich heaths. To illustrate the sociological and geographical differences, Table 10 shows an extract of synedria analyses from three parts of the Norwegian segment of the total distribution area. Three communities come from Rindal in the northwest corner of the area, group 1 from rich fens, group 2 from crevices and group 3 from meadows (Wilmann 1983). Toftaker (1969) carried out 15 analyses in the central, more continental, mountains in Oppdal. The 10 analyses (all of vascular plants) derived from Lid (1959), characterise a *K. simpliciuscula* community on the Hardangervidda. As the table shows, the analyses made by Toftaker and Lid cover a wide range of habitats, namely, rich fens, meadows and a community transitional to *Dryas octopetala* heaths.

Table 10

	Wilmann 1983			Toftaker 1969		Lid 1959	
Group no.	1	2	3	1	2	1	2
Number of sites (1 m²)	9	4	4	5	10	7	3
Kobresia simpliciuscula	341111331	5315	1553	54535	5545555455	2353113	242
Arctostaphylos alpina	....	....	....	1.121	......11.1.	....	...
Betula nana	..112111.	.111	...2	11511	...1112.1.	....	...
Empetrum hermaphroditum	...1.1.2	....	....	1.111	...1.11.1.	1.....1	1.2
Salix reticulata	....	.13.	....	.11.1	111.1.....	.2.....	12.
Antennaria dioica	....	....	....	1	...:...11.	1	111
Bartsia alpina	..1......	...1	....	1..11	.........1.	111.1..	111
Campanula rotundifolia	111	....	.11.	...11	......1..1.	1.111..	.1.
Dryas octopetala	....	....	....	11131	....	4222412	...
Euphrasia frigida	111111..3	..1.	....	11..1	11....1.11	1.....1	...
Leontodon autumnalis	11.1.111.	1.11	.1.1	.....	.......1..	.1.....	1..
Pedicularis oederi	312111323	3111	2334	1.111	11.111111.	....	...
Pinguicula vulgaris	..1211111	1111	11.1	..1.1	.111.1111.	.1....1	..1
Polygonum viviparum	.1...11..	11..	1112	11111	1121111111	111.1.1	111
Saussurea alpina	....11..1	1111	31.2	1..11	..11.1111.	.11.1.1	111
Saxifraga aizoides	1.11..21.	11.1	1455	1211.	112.1..2..	.11..1.1	.1.
S. oppositifolia	....	....	....	22211	1...1....	.11..1.	111
Selaginella selaginoides	111111111	1111	1111	1...1	.111.11111	11..111	111

The cover scales used by Wilmann (1983) and Toftaker (1969) have been changed to the five degree Hult-Sernander scale used by Lid (1959).

Nordhagen (1936a) regarded *K. simpliciuscula* as a characteristic species of the *Caricion bicolori-atrofuscae* alliance.

K. simpliciuscula is a basiphilous species. In communities in which it dominates or is abundant, the pH values vary between 6.0 and 7.0, with a concentration around 6.5.

Bodil Wilmann

	Wilmann 1983			Toftaker 1969		Lid 1959	
Silene acaulis	1	....	....	121.1	1.1.......	121.211	211
Succisa pratensis	41111.13.	...1	....	.....	....	....	...
Thalictrum alpinum	241321111	1111	1342	11111	.111.12111	1211111	111
Tofieldia pusilla	111111111	1111	...1	1.111	..1.11111.	.1...1.	111
Carex atrofusca	2211	3113	1212	11111	1..21.2121	....	...
C. capillaris	1...11211	1...	1113	11111	.111.11.1.	111.1.1	111
C. panicea	..1112.11	1.11	2...	.....	4	....	1..
C. rupestris	....	....	....	1...1	....	2.12211	.11
C. vaginata	....	....	....	211..	2113.11.3.	111.1.1	1.1
Festuca ovina	....	....	....	.1.11	1.1.....1.	11.111.	1.1
Molinia caerulea	42222.325	3511	....	.....	....	....	...
Scirpus cespitosus	232323325	..3.	3.12	.....	......12.1.	1	1..
Campylium stellatum	554454551	1.11	1211	11211	2111334211	....	...
Drepanocladus revolvens	.14154433	3555	5555	141..	252341231.	....	...
Fissidens osmundoides	1.11.1111	1.11	11..	.....	....	....	...
Hylocomium pyrenaicum	..111111.	....	....	.11.1	.1..111.21	....	...
H. splendens	.......1..	....	....	1.114	...3..1...	....	...
Rhytidium rugosum	....	....	....	4..24	..11....4.	....	...
Tomenthypnum nitens	....	1...	....	.1134	51.1.22..1	....	...

Koenigia islandica L.

Maps: Hultén (1955, 1971a, no. 630), E. Dahl (1963a). Total: Hultén (1968, 1971b, no. 55), Hultén & Fries (1986, no. 643). Local: Benum (1958), Lid (1959), Toftaker (1969), Kristensen (1981).

First Norwegian record: Reported by Gunnerus (1772, no. 1073), but not from a Norwegian locality. In Topografisk Journal for Norge, vol. 7:24 (1799-1800) the species is reported from Finnmark by O. H. Sommerfelt "i Mængde nær Boepælerne paa Vardøe, saa og paa adskillige fugtige Steder i Varanger Fjorden, skjøndt mindre hyppig" (cf. also Wahlenberg 1812).

Norwegian distribution: Slightly bicentric. In the southern area, from Bjørnaskallen on the Hardangervidda to Kjølhaugan in Verdal. In the northern area, from Jomafjellet in Røyrvik to Magerøya.

Altitude limits: Low-alpine to middle-alpine. Leirtjønnkollen in Oppdal 1650 m. Down to 625 m in Soknedal. Kirkestinden in Målselv 1000 m (Engelskjøn 1986a). Down to sea level at some places in North Norway.

Habitat: Wet, open soil in snow-beds, on the banks of small lakes and streams, and in cold springs. Indifferent to basiphilous.

K. islandica belongs to the group of circumpolar, arctic plants, but has an extremely disjunct distribution. In Sweden, it is found from Härjedalen to Torne Lappmark, but with a corresponding gap to that in Norway. In Finland, it is only

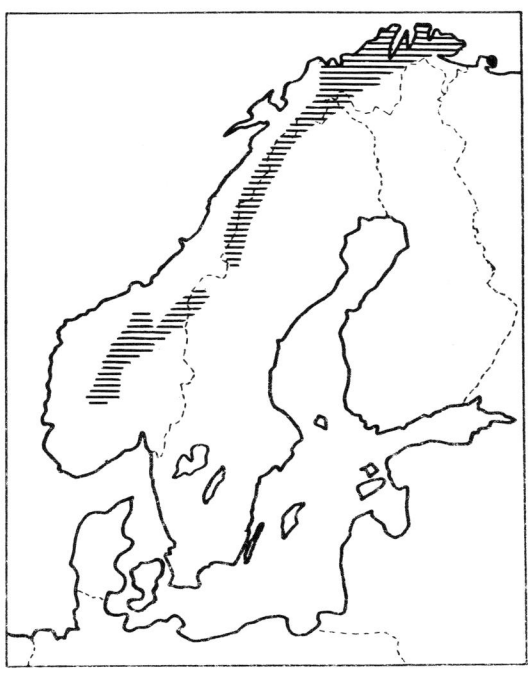

Fig. 55 The Fennoscandian distribution of *Koenigia islandica*

known from Enontekis Lappmark. Its distribution is almost continuous from Fennoscandia to the Urals. It is found on the Faroes, Iceland, Jan Mayen and Svalbard, and at a single locality in Scotland. In addition to its arctic and subarctic distribution it also grows in the mountains of central Asia. It is, moreover, a bipolar species, occurring in Tierra del Fuego.

K. islandica does not grow in the mountains of central Europe, but several finds of late-glacial and postglacial fossils have been made. Fossils are also known from Scotland, South Sweden and southernmost Norway (Hafsten 1958).

The distribution in Norway is clearly continental. Coastal localities are only found in the northernmost part of the country.

This tiny annual species has a rather narrow ecological range. It keeps to wet, open habitats where it has little competition. It prefers snow-beds, particularly areas that become snow free very late and where plants are low-growing and scattered. Nordhagen (1943) regarded it as a characteristic species of the *Salicetalia herbaceae* order. Within this order, it shows a preference for the *Saxifrago oppositifolio-Oxyrion digynae* alliance. Like *Ranunculus hyperboreus* and *Sedum villosum*, *K. islandica* occurs as a nitrophilous apophyte near upland, summer farms, in ditches and places where cattle trample creating open, muddy patches.

Leucorchis straminea (Fern.) A. Löve
(Syn. *L. albida* (L.) E. Mey. subsp. *straminea* (Fern.) A. & D. Löve, *Pseudorchis albida* (L.) Rich. subsp. *straminea* (Fern.) A. & D. Löve)

Maps: Hultén (1971a, no. 530). Total: Löve (1950), Hultén (1958, no. 90), Meusel et al. (1965), Hultén & Fries (1986, no. 561). Local: Ouren (1952), Benum (1958), Toftaker (1969), Kristensen (1981), Holten (1983, 1984).

First Norwegian record: Vest-Agder: Åseral, at the foot of Hekkfjell. Collected by Oeder in 1757, drawing in Flora Danica 2 (Oeder 1763). Oppdal, Vangsfjellet, 24.8.1764, J. E. Gunnerus (TRH).

Norwegian distribution: From Hjelmeland in Ryfylke to Magerøya, but with a lacuna in Nord-Trøndelag.

Altitude limits: Mainly low-alpine. In South Norway, rarely below 700 m. Knutsholtind in the Jotunheimen 1800 m (Jørgensen 1933). Down to sea level in North Norway, especially Finnmark. Troms: Målselv, Isdalstind 995 m.

Habitat: Meadows and damp *Dryas octopetala* heaths, solifluction soil, shallow fens and dunes. Basiphilous.

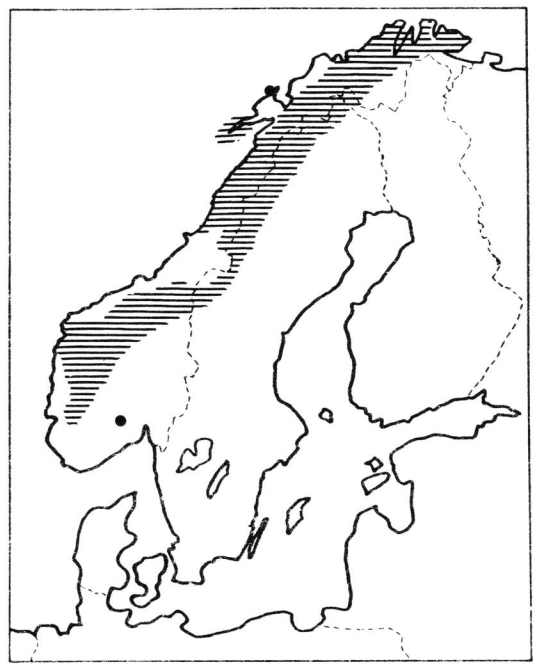

Fig. 56 The Fennoscandian distribution of *Leucorchis straminea*

L. albida s. lat. consists of two taxa, *L. albida* s. str. and *L. straminea* (cf. Löve 1950, Nilsson 1976). The former is a European species which, in Norway, occurs mainly along the southern and western coasts and in inland subalpine habitats. There is still some uncertainty about its geographical area.

L. straminea is a northern amphi-Atlantic, arctic-montane species. In Sweden, it is known from Härjedalen to Torne Lappmark, and in Finland, from Enontekis Lappmark and Enare Lappmark. Its distribution area otherwise comprises the Faroes, Iceland, southern Greenland and two small areas in Canada, one in Newfoundland and the other at Hudson Bay. This is a similar distribution pattern to that of several other amphi-Atlantic species.

L. straminea rarely occurs in abundance, but is found in a variety of plant communities. It seems to grow most frequently in sheltered, damp, north-facing heaths. Nordhagen (1943) regarded it as an indicator species of the fen alliance, *Caricion atrofuscae-saxatilis.*[x)] It particularly occurs in fens affected by frost action. In North Norway, it is frequently found in grass heaths on sand dunes along the coast.

L. straminea is a distinctly basiphilous species. Ouren (1952) reported pH = 5.9 as the average result of 5 tests from Budal.

Luzula arctica Blytt

Maps: Hultén (1971a, no. 461). Total: Hultén (1962, no. 4, 1968) Hultén & Fries (1986, no. 170). Local: Benum (1958), Gjærevoll (1963b), Nordsteien (1982), Engelskjøn (1984, 1986a).

First Norwegian record: Oppdal, Knutshø 1836. M. N. Blytt (O).

Norwegian distribution: Bicentric. The southern area is very concentrated, from Mehøa in Tynset in the east to Råtåsjøhø in the south and Blåhø in the Trollheimen in the north. The northern area extends from the mountains near Gævdnajav're in Bardu to Vassbotndalen in Alta.

Altitude limits: Mainly middle-alpine. Knutshø 1650 m, Råtåsjøhø 1170 m; rarely below 1300 m. Troms: Målselv, Kirkestinden 1330 m; down to 500 m in Dividal.

Habitat: Prefers north-facing solifluction slopes that have a more or less well-developed moss carpet. Hygrophilous and calciphilous.

L. arctica is a circumpolar, high-arctic species. In Sweden, it is known from Lule Lappmark and Torne Lappmark, but there is a considerable gap between these occurrences. Only two localities have been reported between Alta and the Urals-Novaya Zemlya, one on the Kola Peninsula, the other on Kolguyev Island. It is fairly common in Svalbard.

Its distribution in Norway is continental, no localities having been reported from coastal districts.

L. arctica prefers easily disintegrating schists, but seems to avoid pure limestone. Where solifluction takes place, it usually keeps to places where the moss carpet is well developed (*Aulacomnium turgidum*, *Drepanocladus revolvens* and *Tomenthypnum nitens*), avoiding the central, gravelly parts of solifluction lobes where frost action is most pronounced. Its occurrence on north-facing

slopes is striking. The species needs a great deal of moisture throughout the season, and may occur in large quantities in such locations.

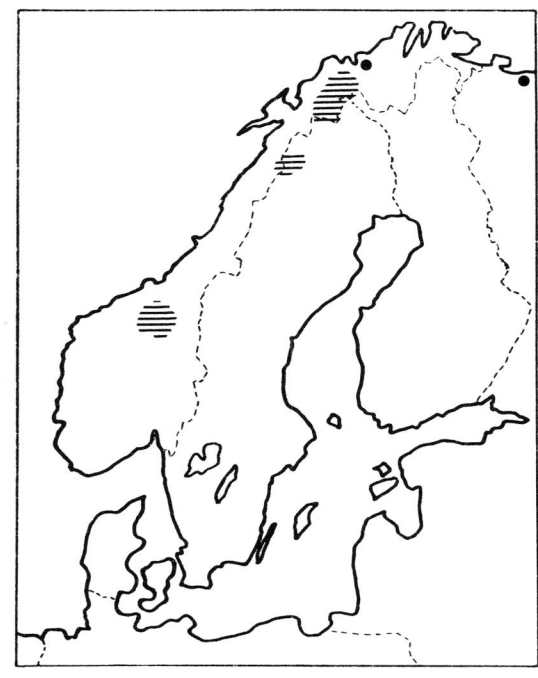

Fig. 57 The Fennoscandian distribution of *Luzula arctica*

It is most frequently met with in the middle-alpine belt. In the mountains of central Norway, the lower limit of this belt is at about 1300 m on north-facing slopes.

L. arctica is a characteristic species of the *Luzulion arcticae* alliance (Nordhagen 1936a, Gjærevoll 1956). It does not necessarily need a thick snow cover, but its occurrence at high altitudes and on north-facing slopes means that it becomes free of snow late. The community is transitional to the latest exposed, damp *Dryas octopetala* heaths. Table 11 gives an extract of 5 square analyses (1-5) from 1470 m on the north slope of Orkelhø (Gjærevoll 1956) and 3 analyses (6-8) from 1400 m on the north slope of Knutshø (Nordhagen 1954a).

Table 11

Square no. (1m²)	1	2	3	4	5	6	7	8
Salix polaris	1	1	2	.	2	1	2	1
S. reticulata	.	.	.	.	.	1	1	1
Cerastium alpinum	1	1	.	1	.	1	2	1
Draba alpina	1	1	1	1	.	2	2	1
Pedicularis oederi	.	.	.	.	.	1	1	1
Poa stricta	2	2	1	3	2	.	.	.
Polygonum viviparum	1	1	1	1	1	1	1	1

Saxifraga oppositifolia	1	1	2	1	1		2	2	2
S. tenuis	1	1	1	1	1		.	1	.
Silene acaulis	2	2	1	2	1		1	1	2
S. wahlbergella	1	.	.	1	1		1	.	.
Carex lachenalii	.	1	1	1	.		.	.	.
C. misandra	2	2	1	2	2		1	1	1
Festuca vivipara	1	1	2	1	1		.	.	.
Juncus biglumis	1	1	1	1	1		.	.	.
Luzula arctica	2	2	2	2	3		2	2	2
Aulacomnium turgidum	.	.	.	.	.		1	1	1
Blindia acuta	2	2	1	2	1		.	.	.
Calliergon sarmentosum	.	.	.	.	.		2	1	2
Campylium stellatum	.	.	.	.	.		1	1	4
Distichium capillaceum	2	2	3	2	2		1	1	1
Drepanocladus revolvens	1	.	1	1	1		5	5	3
Tortella tortuosa	.	.	.	.	.		1	2	1
Blepharostoma trichophyllum	.	.	.	.	.		2	3	2
Aneura pinguis	2	1	2	2	.		.	.	.
Stones and bare soil	5	5	5	5	5		3	2	2

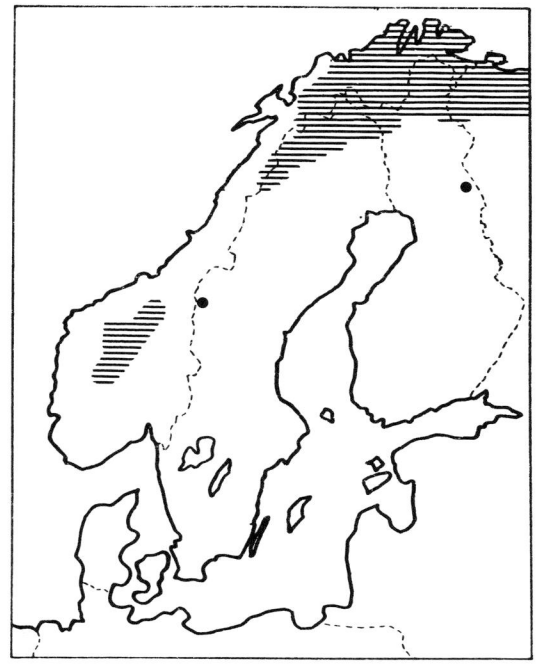

Fig. 58 The Fennoscandian distribution of *Luzula parviflora*

The disjunct distribution of *L. arctica*, and its narrow ecological amplitude, point towards glacial survival in different areas. Immigration from the east seems rather unlikely as the species is very rare between Alta and Novaya Zemlya.

Luzula parviflora (Ehrh.) Desv.

Maps: Björkman (1939), Hultén (1971a, no. 469), Böcher (1951b). Total: Hultén (1962, no. 50, 1968), Hultén & Fries (1986, no. 166). Local: Benum (1958), Bråthen (1973), Hagen (1976), Kristensen (1981), Schumacher & Løkken (1981).

First Norwegian record: "Kåbdavanka" 1828 (most probably in Storfjord) L. L. Læstadius (S).

Norwegian distribution: Bicentric. In the south, from Skarvåna, Hol in Hallingdal to the Trollheimen and Røros. In the north, Saltdal, and Skjomen to Magerøya.

Altitude limits: Low-alpine to northern boreal. Down to about 700 m in South Norway; Glittertind 1750 m (Jørgensen 1933). Down to sea level in Finnmark. Troms: Balsfjord, Tamokfjell 1030 m.

Habitat: Birch forests and willow thickets, on north-facing slopes. Basiphilous.

L. parviflora is a circumpolar species, but is lacking in Svalbard, Iceland and East Greenland. In Sweden, it is known in Härjedalen and from Lycksele Lappmark to Torne Lappmark. It is found in the northern districts of Finland.

In South Norway, it is distinctly continental with a high frequency in the Jotunheimen and Drivdalen mountains. There are some western outposts in Geiranger. Only two localities are known in the Sunndalen mountains, and only one in the Trollheimen. It is also mainly continental in North Norway, but there are many localities in coastal parts of Finnmark.

L. parviflora mainly occurs in moist birch forests and willow thickets, preferring calcareous soils. It is also frequently found, often with *L. arctica*, on cliff ledges and moist north-facing slopes with a well-developed moss cover, up to the lower part of the middle-alpine belt. It may also occur on fresh gravel and river bars, e.g. in deltas in Finnmark. Usually it grows dispersed.

The postglacial history of *L. parviflora* has been discussed by several authors, especially because, although it grows at fairly high altitudes, it is mainly a low-alpine to subalpine species. As its distribution is continuous eastwards to western Siberia, the northern area may be explainable by invoking postglacial immigration from the east. Björkman (1939), however, thought the species had survived on refugia along the coast from Nordland to Varanger and later joined up with plants immigrating from the east. This view was opposed by Samuelsson (1943). He argued that Lule Lappmark should have received most alpine species from the north because they were lacking in the part of Nordland situated west of Lule Lappmark. Arwidsson (1943) disagreed with Samuelsson, and pointed out that the area west of Lule Lappmark was not well investigated, and, furthermore, that it ought not to be more difficult to explain the situation in this area than in southern Norway where localities are lacking in the coastal mountains.

After this discussion took place, *L. parviflora* has been

found west of Lule Lappmark (Nord-Saulo and Solvåg-tind). The latter locality is of great interest in this connection since Holmsen (1913) believed that Solvågtind was a nunatak during the last glaciation (see *Carex scirpoidea*).

Returning to the situation in the northernmost part of the area, *L. parviflora* is very rare in Siberia, and there is a pronounced decline in its frequency eastwards from northern Norway. If it is accepted that the species survived the Ice Age in northern Norway, immigration from the west into northern Russia seems more likely than immigration from the east. If refugia existed, Norway has not only received incoming species, but also been a centre for dispersal. Björkman (1939) regarded *L. parviflora* as indifferent, but most observations suggest it is basiphilous, growing mainly on mica schists and dolomite.

Luzula wahlenbergii Rupr.

Maps: Hultén (1971a, no. 474). Total: Hultén (1962, no. 24, 1968), Hultén & Fries (1986, no. 167). Local: Benum (1958), Kristensen (1981), Engelskjøn (1986a).

First Norwegian record: Finnmark: Gamvik, Hopseidet, 10.7.1802 ("Juncus spadiceus"), G. Wahlenberg (UPS).

Excluded or doubtful stations: The localities in central Norway: There are some old reports of *L. wahlenbergii* in central Norway. They were not accepted by Hultén (1952) or Hultén & Fries (1986). One collection was made by C. Boeck "in Alpes Valdrisiae". There are three specimens on the sheet (O). One is *L. parviflora*, one probably *L. arcuata*, and the third is *L. wahlenbergii* (rev. Sigmund Sivertsen). Another collection was made by M. N. Blytt on Bitihorn in 1839 and consists of two specimens, one of which is *L. arcuata*.

In his flora, M. N. Blytt (1861) reported *L. glabrata* Hoppe on the basis of material collected on Lomseggen by J. Norman (undated). Three specimens are mounted together with another collection (a single specimen) made by F. Hoch in 1863 at "Lomseggen ved Andvord". Two of the specimens collected by Norman, along with that of Hoch, are *L. parviflora*, but the third, collected by Norman, seems to be *L. wahlenbergii*.

Herb. BG contains a collection (undated) made by H. Guldberg at Dovre (ex. herb. Lynge). According to other collections, Guldberg was at Kongsvoll in 1801 and 1803. He also collected *L. wahlenbergii* in Saltdal (Balvatn and Solvågtind in 1804). Finally, herb. C contains a collection by H. Mortensen "S for Jerkin 8.8.1871". Sigmund Sivertsen, who registered and revised the material in Copenhagen, added "ad int." on the card.

I have drawn special attention to these central Norwegian localities because it is amazing that *L. wahlenbergii* has not been reported from central Norway later than the beginning of this century. Comprehensive, detailed investigations have been carried out in this region since then, and this factor, along with the partially dubious material,

make it questionable whether the species occurs in central Norway.

Norwegian distribution: If we leave aside the localities dealt with above, *L. wahlenbergii* is to be found from Tydal (Sylane mountains) to North Cape and Grense Jakobselv. It is rare in the southern part of this area, but common from Hattfjelldal northwards.

Altitude limits: Mainly low-alpine. Troms: Målselv, Brøran 1250 m. Nissontjårro in Torne Lappmark 1370 m. Down to sea level in Troms and Finnmark.

Habitat: May be abundant in moist snow-beds, on sandy alluvial soil, along streams, on the banks of lakes, and even on bogs. Mainly on acid soils, but also reported from dolomite.

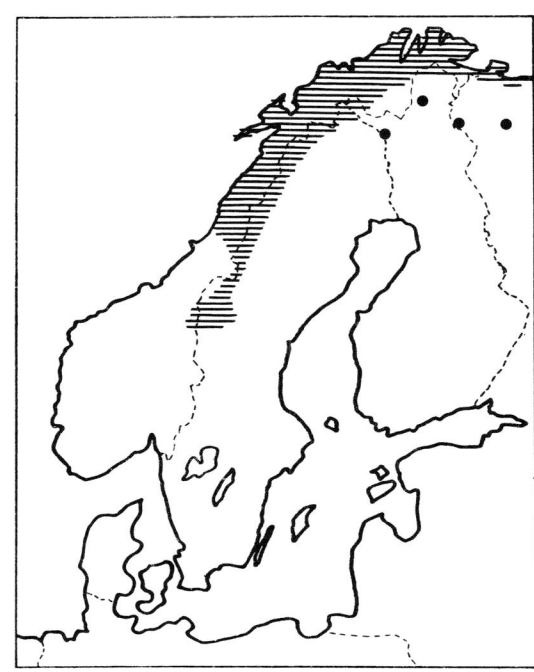

Fig. 59 The Fennoscandian distribution of *Luzula wahlenbergii*

L. wahlenbergii is a circumpolar, arctic species. The pattern is somewhat unusual as it is very rare between Scandinavia and eastern Canada. It is unknown on Iceland and extremely rare in Svalbard. Only a single locality is known in Greenland (Zackenberg, northeast Greenland). There is continuous distribution from Finnmark eastwards to Siberia. In Sweden, *L. wahlenbergii* is found from Härjedalen to Torne Lappmark, and in Finland, it occurs in Enontekis Lappmark and Enare Lappmark.

The total distribution of the species greatly resembles that of *L. parviflora*. The population in North America is regarded as a separate race, subsp. *piperi* (Cov.) Hult., a parallel to a similar racial difference in *L. parviflora* (subsp. *divaricata* (Wats.) Hult.). The distribution in the North Atlantic region is interesting. The Greenland locality, Zackenberg on Wollaston Foreland, discovered in 1947

(Holmen & Mathiesen 1953), is situated in an area thought not to have been glaciated during the Pleistocene glaciation (see also Funder 1982). A number of other extremely isolated species are found in the same area.

L. wahlenbergii may have immigrated from the east into Norway. However, as the species becomes less frequent from northern Fennoscandia eastwards the immigration route is just as likely to have been in the opposite direction. When the isolated localities in Greenland and Svalbard are taken into account, it seems evident that the species has an old history in the North Atlantic region.

L. wahlenbergii is most frequently met with in moist snow-beds closely related to the communities characterised by, above all, *Carex bigelowii*. It prefers localities that are irrigated for a short time, and depressions containing stagnant water (Gjærevoll 1956). It may also be abundant in open willow thickets in the low-alpine belt.

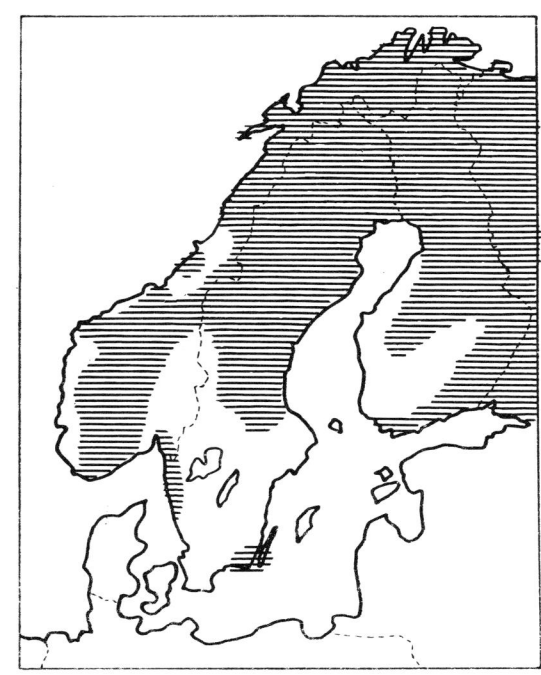

Fig. 60 The Fennoscandian distribution of *Lychnis alpina*

Lychnis alpina L.

Maps: Lid & Zachau (1928), Hultén (1971a, no. 736). Total: Hultén (1958, no. 49), Lid & Zachau (1928), Meusel et al. (1965), Hultén & Fries (1986, no. 778). Local: Benum (1958), Ouren (1966), Kristensen (1981).

First Norwegian record: Oeder (1763) − "Slidre Præstesetter i Valders", "Dovrefjeld lige ved Fogstuen". There are 14 collections in the Gunnerus herbarium, but only one has a label: "Talwig d. 20 Juni 1767".

Norwegian distribution: Throughout the mountain range from the south coast to Magerøya and Sør-Varanger.

Altitude limits: Mainly low-alpine, but also frequent in the northern boreal belt. Down to sea level in both South and North Norway. Kyrkja in the Jotunheimen 1900 m (Jørgensen 1933). NAS reported finds up to 1900 m from several mountains in the Jotunheimen. Troms: Balsfjord, Tamokfjell 1030 m.

Habitat: The ecological range is very wide. The species occurs in *Dryas octopetala* heaths as well as grass heaths on acid soils. It also grows on cliff ledges, scree slopes and shallow, rich fens. It has specialised ecological niches in that it occurs on substrate containing heavy metals, e.g. copper, and on serpentinite.

L. alpina is an amphi-Atlantic, arctic-montane species. It is widely distributed in Fennoscandia. Outside the Kola Peninsula, there are only a couple of localities in northern Russia, and a single one in western Asia, just east of the Urals. It is otherwise known in northern Spain, the Pyrenees, the Alps, the Apennines and Scotland. The amphi-Atlantic distribution also embraces Iceland, the southern half of Greenland, and an area in Canada from Newfoundland to the Hudson Bay. The plant found in Greenland and Canada is regarded as a separate race, var. *americana* Fern.

L. alpina is found throughout the mountain range in Norway, but avoids the outermost coast of the southern part of West Norway. In North Norway, there are numerous localities in coastal mountains, e.g. Lofoten.

There are some peculiarities in the distribution of *L. alpina*, obviously due to racial differences (Turesson 1927). Contrary to almost all other alpine species it occurs on the coast of South Norway. A strange occurrence is on some islets in Lake Vansjø in Østfold. Similar localities are found on the west coast of Sweden in Bohuslän and Halland, and on the southeast coast in Blekinge and on Öland. The same situation is met with on the south coast of Finland.

There is no fossil evidence of immigration from the south after the last glaciation. The coastal localities in Norway and Sweden have been regarded as glacial relics. This means that *L. alpina* was widely distributed during early postglacial time, but was ousted by competitors except in special niches such as coastal cliffs, the alpine-like alvar of Öland, and steep inland cliffs. *L. alpina* is easily dispersed by running water and is frequently found on gravel bars. This is most striking in Sweden where numerous localities are known along the northern rivers down to the Gulf of Bothnia.

L. alpina has a basiphilous tendency and is common on *Dryas* heaths. However, as Nordhagen (1943) emphasised, in Sikilsdalen, *L. alpina* shows a clear preference for grass heath communities in the *Juncion trifidi* alliance which occurs on acid soils (pH = 4.0-4.5).

Much attention has been paid to its preference for bedrock containing heavy metals, serpentine-rich bedrock (Rune 1953) and slag heaps. A number of localities are related to such sites.

Minuartia rubella (Wahlenb.) Hiern.

Maps: Hultén (1955, 1971a, no. 712). Total: Hultén (1968, 1971b, no. 43), Hultén & Fries (1986, no. 716), Local: Benum (1958), Gjærevoll (1963b), Bråthen (1973), Nordsteien (1982), Schumacher & Løkken (1981).

First Norwegian record: Troms: Lyngentind, G. Wahlenberg, 11. juli 1800 (UPS).

Excluded or doubtful stations: Røros, 11. juli 1882, E. Ryan (BG, TRH). More precise information is not given (cf. Elven 1979).

Norwegian distribution: Bicentric. In the southern area, there is a single, isolated locality on Hårteigen, Hardangervidda. It was reported by Per Størmer following an excursion to Finsehøgda in 1951 (see Blyttia 1952), but no material exists. It is otherwise found from Grinningdalshø in Vågå to the Trollheimen, westwards to Blåhø in Høydalen (Jotunheimen) and Grøvudalen, and eastwards to Mehø in Tynset. In the northern area, it is found on Kvernhatten in Velfjord, and then from Simskarfjellet in Hattfjelldal to Båtsfjord and Grense Jakobselv.

Altitude limits: Mainly low-alpine; occasionally at lower levels, on gravel bars. Lom, Smådalshø to 1900 m (NAS). In the northern area, it occurs down to sea level. Troms: Målselv, Kirkestinden 1380 m.

Habitat: Ridges and heaths exposed to deflation, screes and cliff ledges. Basiphilous.

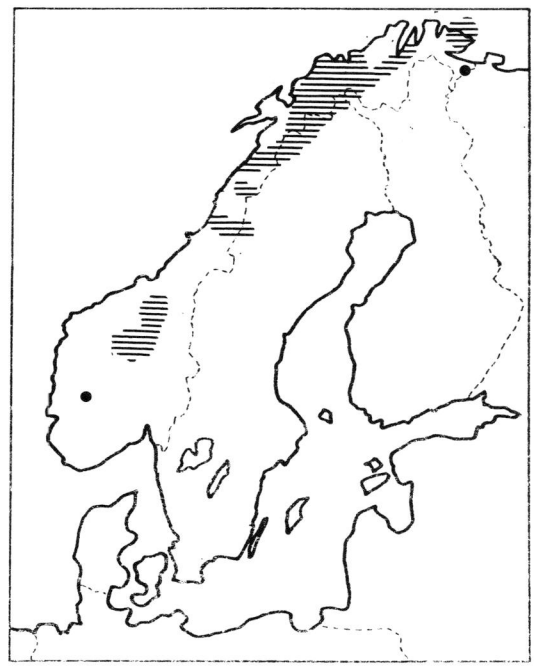

Fig. 61 The Fennoscandian distribution of *Minuartia rubella*

M. rubella is a circumpolar, arctic species. In Sweden, it is known from Åsele Lappmark to Torne Lappmark, and in Finland, in Enontekis Lappmark. The Fennoscandian area is isolated as there is a gap of about 700 km east to Kolguyev Island. The plant is fairly common in Svalbard.

The distribution in western Europe is very strange. The species is known from a few localities on the mainland of northern Scotland, Shetland and the Faroes, but is common on Iceland. This is very similar to the distribution of *Arenaria norvegica*. In South Norway, *M. rubella* is continental, like most other bicentric species. In the northern area, most localities are situated inland.

M. rubella is a characteristic species of the most exposed part of the *Kobresio-Dryadion* alliance, often growing on gravelly patches eroded by the wind. It usually grows dispersed, and is quantitatively of minor importance. Common companions are, among others, *Kobresia myosuroides*, *Potentilla nivea*, *Carex nardina*, *C. glacialis*, *Draba fladnizensis* and *D. nivalis*. It is also frequently found on screes in company with *Papaver radicatum*, *Arenaria norvegica* and *Braya linearis*, and occurs as a pioneer plant on fresh gravel, e.g. roadsides.

M. rubella is a basiphilous species growing on calcareous schists, marble and dolomite, but strangely lacking on the large area of dolomite in Porsanger.

Minuartia stricta (Sw.) Hiern.

Maps: Hultén (1971a, no. 713). Total: Hultén (1968, 1971b, no. 24), Hultén & Fries (1986, no. 717), Local: Benum (1958), Ouren (1966), Nordsteien (1982).

First Norwegian record: According to Hornemann (1821), Wahlenberg made the first record in Finnmark, but no evidence has been found. Several Swedish localities are mentioned by Wahlenberg (1812), but none from Finnmark. The oldest record known refers to C. Smith, Litlos on the Hardangervidda in 1812 (O).

Norwegian distribution: Slightly bicentric. Buardalen in Røldal (Odda) to Bjørdalsfjell in Meråker, and Lierne to Tana and Nesseby.

Altitude limits: Mainly low-alpine. Knutshø 1600 m. On gravel bars down to 480 m (Alvdal). Up to 1000 m in inner Troms (Engelskjøn 1984). Målselv, Frihetsli, on river banks at 160 m.

Habitat: Damp heaths, shallow rich fens, solifluction soil, screes and river bars. Basiphilous.

M. stricta is a circumpolar, arctic-montane species, but has a rather disjunct distribution. In Europe, a few localities have been reported from the Jura mountains and southern Germany (according to Hegi (1962) it is now probably extinct there). It is also known from Teesdale in England, Iceland and Svalbard. From Finnmark, there is a gap of more than 500 km to the easternmost part of the Kola Peninsula. In Sweden, *M. stricta* is known from Härjedalen to Torne Lappmark, displaying the same gap as in Norway. In Finland, it occurs in Enontekis Lappmark.

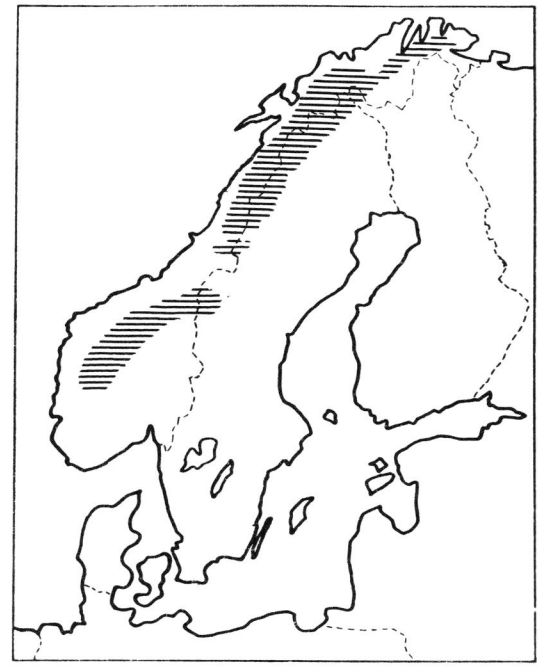

Fig. 62 The Fennoscandian distribution of *Minuartia stricta*

In South Norway, *M. stricta* has a distinctly continental distribution, further emphasised by numerous stations in Härjedalen. (Hultén (1971a) gave a locality in Nordfjord, but there is no herbarium evidence for this.) There are some coastal localities in the northern area, but the main distribution is continental.

M. stricta is a basiphilous species. Nordhagen (1943) took it as a characteristic species of his *Caricion atrofuscae-saxatilis* (*Caricion bicolori-atrofuscae*) alliance. In addition to mica schists and similar, easily disintegrating schists, it grows on marble, dolomite and serpentinite gravel.

Nigritella nigra (L.) Rchb. fil.

Maps: Holmboe (1936), Hultén (1971a, no. 531). Total: Tralau (1961), Meusel et al. (1965), Gjærevoll (1973), Hultén & Fries (1986, no. 557). Local: Benum (1958), Toftaker (1969), Moen (1976), Elven (1984).

First Norwegian record: Oppdal, probably at Håker, near the present railway station (cf. O. Dahl 1894), 1.8.1764, J. E. Gunnerus (TRH). Surprisingly, Gunnerus (1772) did not mention Oppdal, but instead wrote "In alpibus værdalicis & snaasensibus versus Jemtiam".

Excluded or doubtful stations: Hedmark: Vang, Vendkvern. Oppland: Vestre Slidre. Sør-Trøndelag: Orkdalen. Nord-Trøndelag: Verdal and Snåsa. Nordland: Velfjord. Comments on all five localities are given by Holmboe (1936). A very doubtful collection from Aust-Agder,

Dybvåg, leg. Åselius (S) has also been excluded.

Norwegian distribution: From Toten, west of Mjøsa, to the Trollheimen and Holtålen, except for a single locality in Nordreisa, Troms.

Altitude limits: Most common in the northern boreal belt, but also occurs in the lowlands (Toten) and the low-alpine belt. Ascends to 1270 m in the Trollheimen. In Nordreisa, up to 630 m.

Habitat: Grows in the rich vegetation of open grasslands and heaths, and in communities that are transitional between rich fens, grasslands and heaths. Calciphilous.

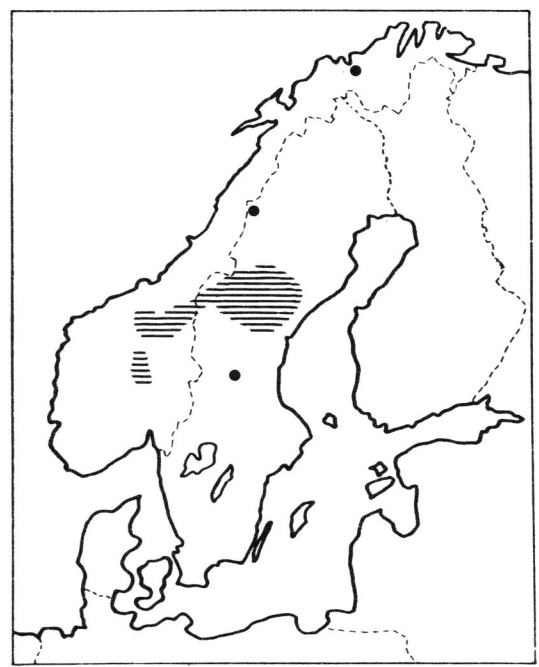

Fig. 63 The Fennoscandian distribution of *Nigritella nigra*

N. nigra is a European, boreal-alpine species. In addition to Scandinavia, it occurs in the mountains of central and southern Europe. It has a disjunct distribution in Scandinavia. In Sweden, it has been recorded from a large number of localities in Jämtland and Härjedalen as well as some other provinces in central Sweden, even on the Baltic coast. The northernmost locality is in Lycksele Lappmark. According to Afzelius (1932, 1943), *N. nigra* is apomictic in the boreal areas, in contrast to its character in southern areas.

Numerous stations in Sweden as well as Norway have been on outlying land harvested for hay for many centuries, but decreasingly so during the present century. This change in agricultural practice has resulted in a sharp decline in *N. nigra*. Both the number of localities and the number of specimens in each locality have fallen dramatically in Jämtland since the beginning of the present century (Björkbäck & Lundqvist 1982, Björkbäck et al. 1976, 1986). In 1940, there were 200-300 localities in Jämtland, but in 1975 this figure was reduced to 85. The traditional use of outlying land lasted longer in Norway than Jämtland, so the process of overgrowing and crowding out of N. nigra is less advanced. Nevertheless, the plant has also disappeared from a number of localities in Norway in this way. In addition, some localities have been destroyed by, for example, ditching and the construction of reservoirs for hydroelectric power plants (Moen 1976).

The distribution map therefore gives a false depiction of the number of localities existing now. No N. nigra localities seem to survive in Oppland county, the last collection (made in Gausdal) dates from 1902. There are now actually about 30 localities in northern Hedmark and southern Trøndelag, and the one in Nordreisa. The highest locality in Scandinavia, on the Gjevilvasskammene mountains, was first recorded by Gjærevoll & Sørensen in 1948 (Sørensen 1949). It was subsequently visited by Gjærevoll on a number of occasions, plants being last seen in 1957. Despite extensive searching, they have not been recorded since then (see Nordsteien 1982).

More than 2000 specimens, representing one of the largest populations of N. nigra in Scandinavia today, are to be found in the Sølendet Nature Reserve, east of Røros. The frequency of their flowering fluctuates greatly. This was registered at four selected localities in 1978-88. There were only 34 flowering specimens in 1988, whereas 700 individuals were counted in 1979, which was also a good year for N. nigra in Jämtland (Björkbäck et al. 1986). The same specimens usually do not flower two years in succession. Similar variations are known in other orchid populations, e.g. Gymnadenia conopsea.

The discovery of N. nigra on Balgesoai've in Nordreisa in 1934 made the distribution bicentric, and still more interesting from a phytogeographical viewpoint. Holmboe (1936) investigated the history of N. nigra. He held that its isolation in Scandinavia, and the racial differences it shows, indicated that it had survived the Ice Age. This view was supported by Tralau (1961). N. nigra is not a pioneer plant, and always occurs in closed plant communities. Ecologically, it does not seem particularly well adapted to refugium conditions.

The plant communities in which N. nigra occurs are always rich in species. Table 12 combines a number of analyses from Sølendet, Røros (A and B), Innerdalen, Tynset (C), Nordreisa (D, Engelskjøn & Skifte (1984)) and Ålbu, Oppdal (E, a single quadrat). Frequency values and degree of cover (Hult-Sernander-Du Rietz scale) are shown for 0.25-1 m² squares representing a selection of the most important species. Localities A-C are situated in areas transitional between the middle and northern boreal regions, D is low-alpine, and E is middle boreal. As might be expected, the analyses display heterogeneity, partly depending on the influence of past and present grazing and, in part, scything.

Table 12

Locality no. No. of squares (0.25-1 m²)	A 5	B 5	C 4	D 6	E 1
Nigritella nigra	100-1	100-1	100-1	100-1	1
Betula nana	.	80-1	50-2	.	.
Dryas octopetala	.	.	.	50-1	.
Salix phylicifolia	80-1	60-1	.	.	.
Vaccinium vitis-idaea	80-1	40-1	25-1	17-1	.
Achillea millefolium	80-3	.	.	.	2
Antennaria dioica	100-1	60-1	50-1	83-1	1
Bartsia alpina	.	30-1	100-1	33-1	.
Botrychium lunaria	80-1	.	.	67-1	.
Campanula rotundifolia	100-1	.	25-1	83-1	1
Erigeron boreale	100-1	.	50-1	17-1	.
Gentiana nivalis	80-1	.	.	83-1	.
Geranium sylvaticum	100-2	100-1	50-2	.	1
Parnassia palustris	20-1	60-1	25-1	50-1	.
Pedicularis oederi	20-2	100-1	75-1	.	.
Polygonum viviparum	100-2	100-1	25-1	100-2	2
Potentilla crantzii	80-1	.	50-1	100-1	1
Saussurea alpina	100-2	100-2	75-1	83-1	1
Selaginella selaginoides	100-2	80-1	25-1	.	.
Solidago virgaurea	80-2	60-1	.	100-2	.
Succisa pratensis	80-2	100-3	.	.	.
Thalictrum alpinum	100-3	60-4	75-4	100-2	3
Viola biflora	.	.	25-1	100-1	.
Agrostis capillaris	80-1	60-1	.	67-2	3
Anthoxanthum odoratum	100-1	100-1	50-1	33-1	2
Carex capillaris	80-1	60-2	100-1	33-1	1
Carex rupestris	.	.	25-1	83-3	.
Carex vaginata	100-1	100-1	100-1	50-2	1
Deschampsia cespitosa	60-2	60-1	50-2	67-1	2
Festuca ovina	100-1	100-2	25-1	17-1	1
Nardus stricta	80-2	100-3	75-2	.	3
Aulacomnium palustre	60-1	100-3	25-2	.	.
Homalothecium nitens	.	80-2	.	.	.
Hylocomium splendens	100-2	100-2	75-4	.	.
Pleurozium schreberi	80-2	60-3	50-2	.	.
Sphagnum warnstorfii	.	80-3	.	.	.
Tortula ruralis	100-3	.	.	.	.

Asbjørn Moen

Oxyria digyna (L.) Hill

Maps: Hultén (1955, 1971a, no. 629). Total: Meusel et al. (1965), Hultén (1968, 1971b, no. 65), Hultén & Fries (1986, no. 657). Local: Kristensen (1981).

First Norwegian record: Mentioned by Otto Sperling from Valdres in 1628. He took it from Valdres and planted it in Bergen (O. Dahl 1888-1911). Drawing in Oeder (1761, no. 14). Oppdal, Vangsfjellet, 3.8.1764, J. E. Gunnerus (TRH).

Norwegian distribution: Ubiquitous, Very common throughout the mountain range.

Altitude limits: Mainly low-alpine to middle-alpine. Jotunheimen, Raudhamrin 2160 m (NAS). Down to the lowlands and seashores in western and northern Norway.

Troms: Målselv, Maddagai'si 1450 m (Jørgensen 1937).

Habitat: Irrigated snow-beds that become snow-free late. Indifferent.

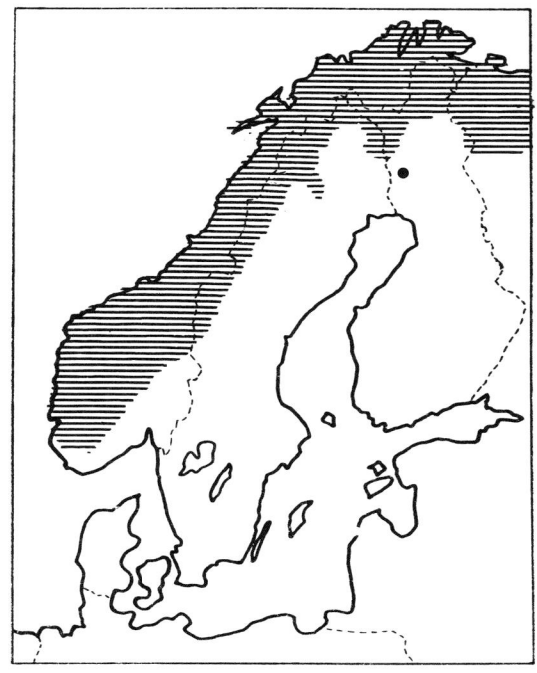

Fig. 64 The Fennoscandian distribution of *Oxyria digyna*

O. digyna is a widely distributed circumpolar, arctic-montane species. In addition to the continuous area in the north, it is common in the central European and central Asian mountains, and the Rocky Mountains. The species is very homogenous. In Sweden, it is found south to Dalarna, and in Finland, south to Kuusamo. It is fairly rare on the Kola Peninsula.

Though mainly an alpine species, *O. digyna* is common on river bars down to the lowlands, especially in western and northern parts of its distribution area; it is rarer along rivers in southeast Norway. On bars in the River Lågen, it is found as far south as Fåberg. With a few exceptions, it is surprisingly not found along rivers in Sweden, indicating that the river-bar occurrences are associated with a cool, moist summer climate. In West Norway, *O. digyna* may also grow on cliffs down to sea level. It is still more commonly found in ravines.

O. digyna is a hygrophilous species that is indifferent to soil conditions. It inhabits gravelly, irrigated snow-beds that become free of snow late. Under oligotrophic conditions, it forms a characteristic community with, among others, *Saxifraga stellaris*, *Carex lachenalii*, *Deschampsia alpina*, *Eriophorum scheuchzeri* and some hygrophilous mosses (*Saxifrago stellaris-Oxyrion digynae* Gjærevoll 1956). This community is poor in species. When conditions are eutrophic, *O. digyna* is accompanied by numerous other species, many of which have a disjunct distribution. Important companions are *Saxifraga oppositifolia*, *S. cernua*, *S. rivularis*, *S. tenuis*, *Phippsia algida*, *Ranunculus nivalis*,

R. sulphureus, *Cerastium arcticum* and *Sagina intermedia* (*Saxifrago oppositifolio-Oxyrion digynae* Gjærevoll 1956).

O. digyna is often a pioneer plant on fresh moraines (Elven 1975, 1978) and may grow as an apophyte on gravelly roadsides. It was abundant on Jensen's Nunataks in Greenland (Gjærevoll & Ryvarden 1977). Late-glacial fossil finds in Denmark show that it apparently grew near the ice margin during the last glaciation.

Oxytropis lapponica (Wahlenb.) Gay

Maps: Hultén (1971a, no. 1145). Total: Hultén & Fries (1986, no. 1194). Local: Benum (1958), Hagen (1976), Halvorsen & Salvesen (1983).

First Norwegian record: According to Hornemann (1821), the species was first recorded from Finnmark by Wahlenberg, but no proof is to be found. Hornemann also mentioned finding *O. lapponica* at Tofte in Dovre. Since he was there in 1808, this may be the first reliable record.

Norwegian distribution: From Ramnaberg near Litlos, Ullensvang, to Kåfjord, Alta, and Sil'bacåkka in Porsanger.

Altitude limits: Ulvik, Sanddalsnut 1555 m. Down to 470 m in Alvdal. Troms: Målselv, Maddagai'si 1170 m (Engelskjøn 1986a). Down to sea level in Nordland and Troms.

Habitat: Mainly low-alpine. In various heath communities, on gravel derived by weathering of the local bedrock, and on scree slopes. Carbonate rocks.

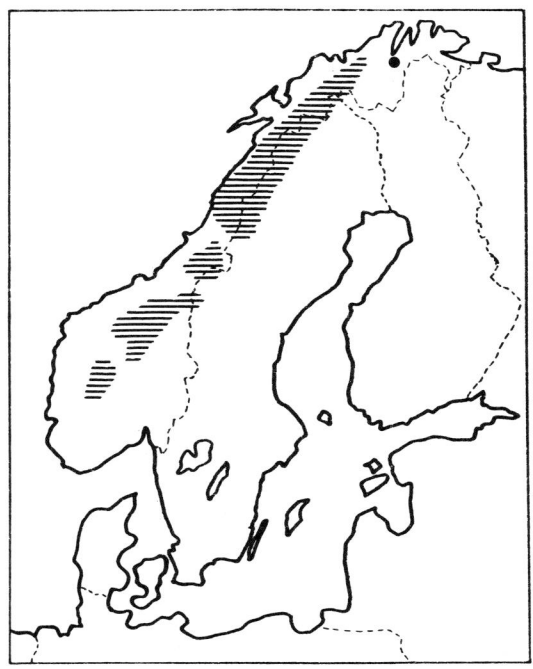

Fig. 65 The Fennoscandian distribution of *Oxytropis lapponica*

O. lapponica is a Eurasian, arctic-montane species with a very disjunct distribution. The Fennoscandian occurrence is extremely isolated. Its total range otherwise comprises the Pyrenees, the Alps, Albania, the Caucasus and the mountains of central Asia. It is not known from the Urals, or from Siberia north of 50° N, and therefore shows a very different distribution pattern from that of *Astragalus norvegicus*.

O. lapponica has a continuous distribution in Norway, but is rare between the Sylane mountains and Nordland. The same applies to Sweden, where it is known from Härjedalen to Torne Lappmark, although very rarely in Jämtland. In Finland, it has only been reported from Enontekis Lappmark.

In South Norway, *O. lapponica* is a highly continental species with many stations on inland mountains as far east as Tron (Alvdal). It avoids the coastal mountains of South Norway, even when geological conditions seem favourable. It should also be characterised as a continental species in North Norway, even though there are some coastal localities, particularly in Nordland. Like *Astragalus norvegicus*, it is rare in Finnmark.

As Selander (1950a) suggested, its overall disjunct distribution and its isolation in Fennoscandia indicate that *O. lapponica* survived the Ice Age. This view was supported by Nordhagen (1964) who pointed out that immigration from the east seemed very unlikely. He found it more plausible that the species had survived in coastal refugia in Nordland.

O. lapponica is frequently accompanied by *Astragalus alpinus*, *A. norvegicus* and *A. frigidus*, but can withstand more open conditions, e.g. deflated areas lacking humus. E. Dahl (1987a) took it as a preferential species of the *Kobresio-Dryadion* alliance. It may just as well serve as a preferential species of the scree-slope alliance, *Arenarion norvegici* (Nordhagen 1935, 1936a). It usually grows dispersed, but may be abundant under specific conditions. It is partly dominant on the steep south-facing slope between Hemre and Midtre Gjevilvasskammene in the Trollheimen. This slope may be characterised as a *Dryas octopetala* heath, and was originally a scree which is now well stabilised by dense plant cover. There is no doubt that *O. lapponica* shows great vitality on calcareous, sandy soils. In Nordland, it is found on active as well as stabilised dunes.

O. lapponica may also be a pioneer plant, often inhabiting gravelly roadsides. In Drivdalen, it grows beside the road for long distances, together with *Astragalus alpinus*.

It is a basiphilous species. Lunde (1962) made 136 pH tests in North Norway obtaining a total amplitude of 5.5-7.6 and a mean of 6.7. Ouren (1952) reported a value of 6.0 from Budal, whilst E. Dahl (1956) gave one from Grimsdalen of 6.4.

Papaver dahlianum Nordh.

Maps: Nordhagen (1931a, 1933, 1936b, 1963a), Johansen (1979), Hultén (1971a, no. 841). Total: Hultén (1958, no. 12).

Norwegian distribution: Confined to a few stretches of valley and estuaries on the Varanger Peninsula, Finnmark, that are separated from each other. One population (Østerelva in Syltefjord) has yellow flowers, the remainder, at Gulgofjord, Leirpollen, Julelva and Reindalen are creamy white (Nordhagen 1931a, 1933, 1936b, 1963a, Knaben 1959a, Johansen 1979).

Altitude limits: Although it is an extremely hardy species in the Arctic (Sunding 1962, Engelskjøn 1986b), *P. dahlianum* is confined to foothills in Norway, reaching 140 m in Gulgofjord (O. Dahl 1934) and 160 m in Reindalen (Johansen 1979).

Habitat: Open, gravelly communities such as river banks and shaley screes. Some populations seem to be ephemeral, being poor in individuals and in danger of extinction (Høiland 1986a, b). In order to flower, *P. dahlianum* requires the long-daylight regime of the Arctic (Knaben 1959a).

This species was first identified as a separate taxon by Tolmatchev (1927), who named it *P. radicatum* subsp. *brachyphyllum*. Nordhagen (1931a) pointed out that two characters of the eastern Finnmark strains, a pyriform capsule and a curved flower scape, distinguished these strains from *P. radicatum*, and he defined *P. dahlianum* as a distinct species within the Scapiflora papavers.

The total range map of Hultén (1958) is incomplete, indubitable specimens having been identified from Ellesmere Land (Knaben 1959a), Novaya Zemlya (Lynge (O) and many collections in herb. LE), and also further eastwards, e.g. Ajon Island (Sverdrup (O)). The typification and distribution of *P. polare* sensu Tolmatchev (1975) suggest that this taxon is heterogeneous, the type collection being a stunted *P. lapponicum* biotype from the coast of Vaygach Island (LE), (cf. also Tolmatchev 1923).

P. dahlianum is a member of the northern Arctic element (Engelskjøn 1986b). The extrazonal occurrences in Norway were discussed by Knaben (1959b) and Nordhagen (1963a) who thought they indicated localised survival through the Ice Age of discrete populations in coastal areas of Finnmark.

Torstein Engelskjøn

Papaver laestadianum (Nordh.) Nordh.

Maps: Nordhagen (1931a, 1933, 1936a), Benum (1958), Hultén (1971a, no. 843). Total: These maps also comprise the entire area of the species, which is considered endemic to North Scandinavia.

Excluded stations: The station named 'Blåbærfjell' was wrongly interpreted by Benum (1958, map 276) as the mountain of Sarrevarre, and the same map needs other minor corrections of stations.

Norwegian distribution: Restricted to a few mountain areas in Balsfjord and Storfjord. In Balsfjord, it occurs on summit 1371, southeast of Tamokfjell ("Blåbærfjellet" in the sense of Devold, cf. Jørgensen (1937)). In Storfjord, it has been found on Cew'cesgai'si, Paras and innermost Sørdalen (Kitdalen) (Schilling & Pollard 1964). This small area borders onto the well-documented localities on the Pältsan massif in northernmost Sweden (Nordhagen 1939, Hedberg et al. 1952, Nilsson & Gustafsson 1979).

Altitude limits: Most stations in Sørdalen are at about 1100 m, beneath the eastern cliffs of Markusfjellet (Schilling & Pollard 1964). In 1961, it ranged from 900 to 1160 m a.s.l. (D. W. F. Pollard, personal communication). However, it has now retreated from the lower periglacial areas mapped by Schilling and Pollard, probably due to a change in the vegetational succession (K. Engelskjøn, observations in 1972, K. Høiland, observations in 1984 (personal communications)). The Pältsan stations range from 760 to 1300 m a.s.l. (Hedberg et al. 1952, and own observations). A *Papaver* species, most probably *P. laestadianum*, has also been observed on river banks in Signaldalen (S. Spjelkavik, personal communication).

Habitat: Some ephemeral growths of this weakly competitive species have been seen on alluvial and morainic gravel, but permanent populations are confined to middle-alpine, mica schist lithosol and solifluction lobes (Nordhagen 1939, Hedberg et al. 1952, and own observations on Pältsan). *P. laestadianum* associates with the *Luzulion arcticae* alliance (Hedberg et al. 1952, Gjærevoll 1956) and has been seen as a companion of the extremely rare *Stellaria crassipes* (Nordhagen 1939).

Originally named *P. radicatum* subsp. *polare* (Tolmatchev 1923, 1927), or *P. radicatum* subsp. *laestadianum* (Nordhagen 1931a), the species was raised to the specific level following the observation by Horn (1938) of an octoploid chromosome number, 2n = 56, in contrast to 2n = 70 in other *P. radicatum* subspecies (Nordhagen 1939).

In North Scandinavia, *P. laestadianum* is replaced by *P. radicatum* subsp. *hyperboreum* in adjoining areas to the south and north, in corresponding alpine belts and usually in similar species-rich communities referred to *Luzulion arcticae*. This example of an isolated enclave is chorologically important, especially in relation to survival during periglacial conditions in the Quaternary and to areal restrictions during late-glacial and Holocene times.

Torstein Engelskjøn

Papaver lapponicum (Tolm.) Nordh.

Maps: Nordhagen (1931a, 1933, 1936a, 1963a), Benum (1958), Knaben (1959a), Hultén (1971a, no. 844). Total: Knaben (1985). – North Atlantic only.

Norwegian distribution: Limited to an area at Burfjord in Kvænangen, Troms, where two of four populations have recently become extinct, and another at Talvik in Alta, Finnmark, where there are at least two intact populations.

Altitude limits: Restricted to the subalpine and low-alpine belts, ascending to 400-450 m a.s.l. on mountains between Burfjord and Kvænangen (O. Larsen, O) and the northern slopes of Talviktoppen in Alta (Høiland 1986b).

Habitats: Unstable moraines and shaley scree; less permanently on alluvial outwash fans below 50 m a.s.l. Secondary invasion was observed in roadcuts at Talvik. The number of individuals is usually small, seldom more than 100 at each station.

This widespread and polymorphous, presumably circumpolar, species (Knaben 1959a) was first distinguished as *P. radicatum* subsp. *lapponicum* by Tolmatchev (1927), and raised to specific rank by Nordhagen (1931a). It is characterised by a pyriform capsule, distantly pinnatifid leaf lobes and the octoploid chromosome number of 2n = 56.

Specimens on file in herb. LE leave no doubt as to the occurrence of *P. lapponicum* in the Chibinskiy and Lovozerosk mountains in the central part of the Kola Peninsula, where it even spreads to railway embankments.

Torstein Engelskjøn

Papaver radicatum Rottb.

Maps: Nordhagen (1931a, 1933, 1936b, 1963a), Knaben (1959a), Gjærevoll (1963b), Hultén (1971a, nos. 842a, 845-849). Total: Knaben (1985). Local: Nordsteien (1982), Hagen (1976), Engelskjøn (1986a, 1987).

Norwegian distribution: The subspecies occurring in South Norway are subsp. *relictum* (1), *intermedium* (2), *ovatilobum* (3), *groevudalense* (4), *oeksendalense* (5), *gjærevollii* (6), and in North Norway, subsp. *subglobosum* (7), *hyperboreum* (8), *avkoënse* (9) and *macrostigma* (10); the numbers correspond with those on the map. All these taxa, each with a defined local distribution, have been distinguished by experimental taxonomical work. In some instances, their morphological distinctness may seem poorly defined unless cultivation experiments are performed under uniform conditions. The subspecies display partial sterility barriers ascribed to the different structural composition of chromosome sets (Knaben 1959a, b, 1970, 1979).

Altitude limits: The ranges of several subspecies of *P. radicatum* are considerable, and nearly all may migrate downwards or be permanently established on subalpine river banks and screes. The upper limits are distinctly high-alpine, subsp. *intermedium* reaching 1820 m (Smådalshø in the Jotunheimen, NAS), subsp. *ovatilobum* 1750 m (Oppdal, Tjernglupen west of Drivdalen, K. Klaveness (O)), and subsp. *hyperboreum* 1675 m (Målselv, Kirkestinden, Engelskjøn 1986a). The species is found at ca. 1800 m on Kebnekaise, Sweden (Åberg in Hedberg 1947).

Habitats: Like other Scapiflora papavers, their poor ability to compete is common to all Norwegian strains of *P. radicatum*. Preferred habitats are open lithosol, scree

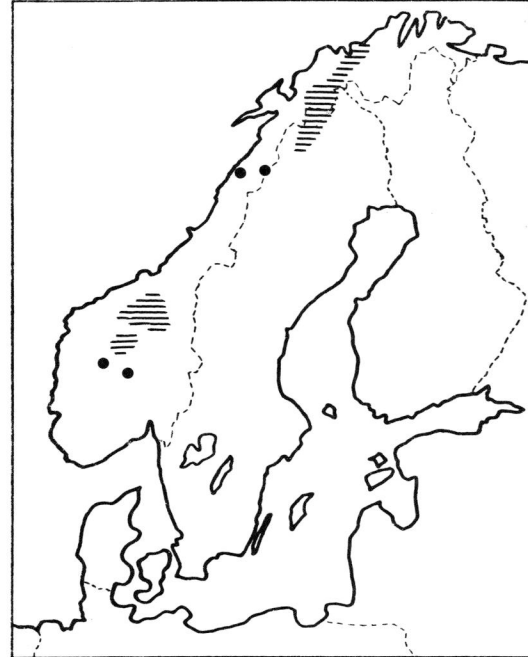

Fig. 66 The Fennoscandian distribution of *Papaver radicatum* coll.

soil and river banks on a substratum of mica schist, amphibolite or gabbroid rocks. None of the subspecies have been observed on calcite marble or dolomite. Subsp. *gjærevollii* and subsp. *hyperboreum* show a preference for the middle-alpine altitudinal belt (Gjærevoll & Sørensen 1954, Edvardsen et al. 1983, Engelskjøn 1985, Gabler & Elvestad 1988). Judging by its representation in herbarium collections, subsp. *ovatilobum* was more abundant at higher altitudes in the Dovre mountains prior to indiscriminate collecting around the turn of the century.

The middle-alpine habitats usually have a species assemblage resembling that of the *Luzulion arcticae* alliance, whereas the low-alpine ones are located on scree soil and contain several species of the *Arenarion norvegici* alliance, but not the most calcicolous ones.

P. radicatum was described by Rottböll (1770). There has been some dispute about the geographical origin of the type material, Iceland or Greenland (Löve 1962), but evidence has been provided (Knaben & Hylander 1970) that *P. radicatum* should be typified on Icelandic material and defined by its barrel-shaped capsule, coarsely pinnate leaves and 2n = 70 chromosomes. Other Arctic papavers in the North Atlantic region are referred to either *P. dahlianum* or *P. lapponicum* (Knaben 1985).

Nordhagen (1933, 1936b, 1963a) suggested that *Papaver* taxa survived the last glaciation in parts of the present Scandinavian landmass or on emerged continental shelf areas to the west or north. Sørensen (1949), Gjærevoll & Sørensen (1954), Gjærevoll (1959, 1963b) and Gjærevoll & Ryvarden (1977) pointed out indications for plant life having persisted further inland, for instance on summits left unglaciated due to topographical and glacidynamic conditions. *P. radicatum* subsp. *gjærevollii* was suggested

as an example of *in situ* survival in the Trollheimen mountains (Gjærevoll 1963b).

The phytogeographical implications of the microevolution of *P. radicatum* have been much debated (see review articles by Löve & Löve (1963) and Knaben (1985)). Three alternative models for migration, which may be combined, are briefly outlined here.

1 The 10 biotypes defined, persisted as discrete populations in some West Norwegian coastal, mountain and shelf areas during the culmination of the last glacial period. Whether precursors of, or identical with, the present-day taxa, they underwent independent migration histories relating to their present sites.

2 A less differentiated pool of biotypes survived the glaciation. *P. radicatum* populations surged and perhaps interbred under the new environmental conditions created in Younger Dryas and Pre-Boreal times, and were finally differentiated and isolated during later Holocene times.

3 Acknowledging the possibility of large-scale expulsion of the vascular flora from the Scandinavian landmass, *P. radicatum* would have survived on ground left dry in the North Sea and in unglaciated parts of the British Isles.

The first alternative is compatible with the general results obtained by Knaben (1985). However, the course of glacial events is still disputed and in need of clarification (Andersen 1981). The second model implies a shorter time span for the microevolution of *P. radicatum*. The third alternative approaches the *tabula rasa* model advocated by some glacial geologists, and seems mainly applicable to the history of hemi-arctic and temperate floral elements.

Torstein Engelskjøn

Pedicularis flammea L.

Maps: Fries (1913), Hultén (1971a, no. 1586), Engelskjøn (1986a). Total: Hultén (1958, no. 163, 1968), Hultén & Fries (1986, no. 1691). Local: Benum (1958), Engelskjøn (1984).

First Norwegian record: According to Hornemann (1821), Vahl reported it from North Norway, but no voucher specimens exist. Troms: Storfjord, Čacca in 1839, L. L. Læstadius (UPS).

Norwegian distribution: Northern unicentric. From Tausafjell in Junkerdalen to Potkavarre in Nordreisa.

Altitude limits: Mainly low-alpine. Målselv: Kirkestinden 1200 m (Engelskjøn 1986a). 1320 m in Lule Lappmark. Down to 325 m in Nordreisa (Norman 1900). Most localities in inner Troms are between 800 and 900 m.

Habitat: Prefers moist localities affected by solifluction, irrigated ledges and shallow fens, but also found in *Dryas* and *Empetrum* heaths. Calciphilous.

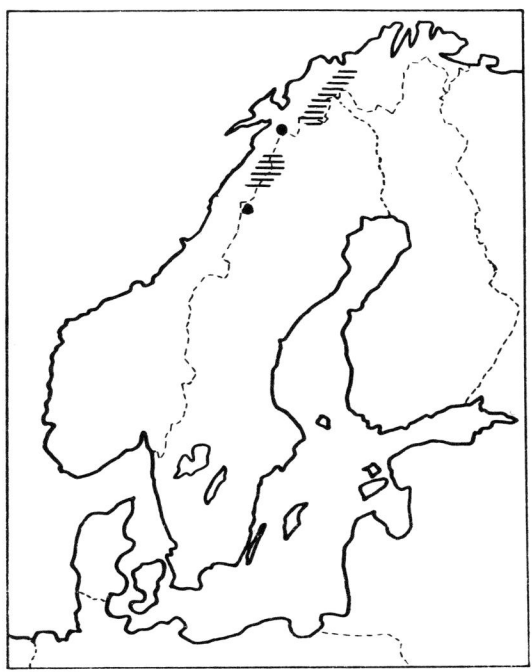

Fig. 67 The Fennoscandian distribution of *Pedicularis flammea*

P. flammea is one of the amphi-Atlantic species occurring in a very limited area on the European mainland where, apart from Norway, it is only found from Lycksele Lappmark to Torne Lappmark in Sweden. It otherwise occurs in Iceland, East and West Greenland (except northernmost parts) and on through central northern Canada to the Great Bear Lake, thus showing a low-arctic distribution.

The Scandinavian distribution is disjunct. The recently discovered locality in Lycksele Lappmark is very isolated, with a gap of about 120 km from there to Tausafjell. Lule Lappmark has numerous localities. The Tysfjord locality is also isolated, with a gap of 80 km in each direction. Then there is continuous distribution from Pesisvarre, north of Torneträsk, to Nordreisa. The localities are very concentrated within the two main areas.

The species is clearly continental, with no coastal localities. Ryvarden (1974) pointed out that the highly specialised pollination biology exhibited by *P. flammea* should be taken into account since the insect(s) involved avoid coastal areas. Warming (1890) and Nordhagen (1957) thought that *P. flammea* may be partially autogamous. The distribution in Scandinavia closely conforms with that of numerous other species.

P. flammea usually avoids the most exposed heaths, but has been reported from *Carex nardina* and *Kobresia myosuroides* heaths. It needs good snow protection, but fairly early exposure from snow, and hence avoids snowbeds. It quite often grows in shallow fens.

Generally, *P. flammea* is regarded as a calciphilous plant, preferring areas of mica schist, marble and dolomite. It may, however, also occur in *Empetrum* and *Juncus trifidus* heaths lacking distinctive calciphilous species.

Pedicularis hirsuta L.

Maps: Hultén (1955, 1971a, no. 1587), Engelskjøn (1986a). Total: Hultén (1958, no. 9), Hultén & Fries (1986, no. 1689). Local: Benum (1958), Aune & Kjærem (1978), Engelskjøn (1984).

First Norwegian record: Reported by Wahlenberg from Tsatsekaise. The oldest collection is from there, leg. L. L. Laestadius 1832 (UPS).

Excluded or doubtful stations: The plant referred to as *P. hirsuta* by Gunnerus (1772, no. 469) ("ex alpibus snaasensibus") is not this species. Vahl reported *P. hirsuta* from Beito in Valdres (C), (cf. O. Dahl 1895), but this is apparently a mistake.

Norwegian distribution: Northern unicentric. From Gilatinden in Rana to Kårhamnstind on Seiland.

Altitude limits: Low-alpine to middle-alpine. Troms: Målselv, Kirkestinden 1395 m (Engelskjøn 1986a). 1610 m in Lule Lappmark. Down to 250 m at Bergsfjord, Alta.

Habitat: Ground affected by solifluction and irrigation, gravelly snow-beds and cliff ledges; may also occur in damp heaths with *Dryas octopetala* and *Cassiope tetragona*. In the middle-alpine belt, it is found on flat, stony or gravelly ground. Prefers calcareous schists, but may also be frequent on more acid soils, especially in the middle-alpine belt.

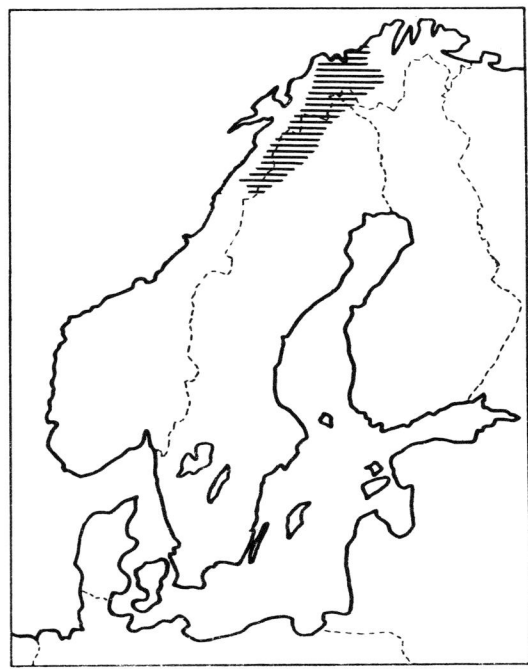

Fig. 68 The Fennoscandian distribution of *Pedicularis hirsuta*

P. hirsuta is an amphi-Atlantic, arctic plant. In North America, it is found as far west as the Keewatin district, and in Siberia, as far east as the New Siberian Islands. It is common in Svalbard. In Sweden, it occurs from southernmost Pite Lappmark to Torne Lappmark, but is limited to Enontekis Lappmark in Finland. From the easternmost locality at Čærro in Karasjok there is a large gap eastwards to Kolguyev Island. Where it occurs in Fennoscandia it is fairly common and has continuous distribution. Most localities are inland, but there are a number of coastal ones, especially in the northernmost part, e.g. Nord Fugløy in Troms. It occurs further east on the Finnmarksvidda than *P. flammea*. Whereas *P. flammea* is mainly a low-alpine species, *P. hirsuta* is also frequent in the middle-alpine belt. This agrees with its more northerly, arctic distribution. Its ecological range is fairly wide, from snow-beds to damp heaths. As several authors have emphasised (e.g. Selander 1950b), it is most frequent on north-facing slopes, where it may descend to the subalpine belt. It may also be a pioneer plant on new soil.

On high altitude plateaus, it may occur, usually dispersed, in a variety of plant communities, chiefly *Cassiope tetragona* and *Poa arctica* heaths, and *Saxifraga oppositifolia* and *Salix polaris* snow-beds, but also in others which become snow-free late.

Generally, *P. hirsuta* should be characterised as calciphilous, and is undoubtedly most frequent on calcareous mica schist and similar substrate. However, it also occurs on acid soils (cf. Engelskjøn 1984), especially at high altitudes.

Pedicularis oederi Vahl

Maps: Tengwall (1913), Nordhagen (1931b), Hultén (1955, 1971a, no. 1589), E. Dahl (1957). Total: Hultén (1968), Meusel et al. (1978), Hultén & Fries (1986, no. 1690). Local: Lid (1955), Ouren (1961, 1964, 1966), Toftaker (1969), Skogen (1971), Elven (1979), Sæther et al. (1980), Moen & Kjelvik (1981), Wilmann (1983), Holten (1984).

First Norwegian record: A drawing of the species is found in Flora Danica (Oeder 1761), the location being given as "Paa Dovrefield i Norge, paa veien mellem Tofte og Fogstuen". Oppdal, Vangsfjellet, 3.8.1764, J. E. Gunnerus (TRH). Both floras refer to it as *P. flammea*.

Excluded or doubtful stations: Oppland: Ringebu, Stulsbroen (Norman 1851). This locality has been visited by numerous botanists since, but the species has never been found. Nord-Trøndelag: Snåsa, Skjækerfjella (Tengwall 1913); no collections are known. "Vid Tanaelven i Østfinmarken" (undated), P. V. Deinboll (O), (cf. O. Dahl 1934).

Norwegian distribution: From Rauhellaren on the Hardangervidda and the upper Hallingdal mountains to Kjølhaugan in Meråker.

Altitude limits: Northern boreal to middle-alpine. Trollsteinhøin in the Jotunheimen 2050 m (NAS). Down to 270 m in Kvenndalen in the northwesternmost Trollheimen.

Habitat: Prefers moist or wet, species-rich communities (fens, willow thickets and meadows). The ecological range of the species is, however, very wide, and it may be found in a variety of communities from exposed ridges to extreme snow-beds, both on rich and fairly poor soils. In its southernmost area, it is found only in rich communities at high elevations.

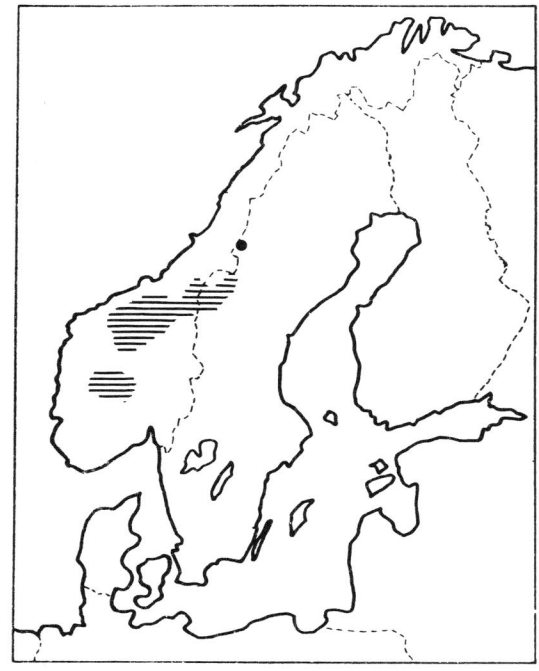

Fig. 69 The Fennoscandian distribution of *Pedicularis oederi*

P. oederi has an unusual total distribution. In central Europe, it occurs from the western Alps to the Carpathians. The Scandinavian area is very isolated as there is a gap eastwards to Kolguyev Island and the Urals. It is widely distributed in northern Siberia and the mountains of central Asia. In North America, it is known only from the Mackenzie District to Alaska and some localities in the Rockies. There is accordingly a huge gap between Scandinavia and northwest America. Apart from Scandinavia, *P. oederi* is absent from the area occupied by amphi-Atlantic species. The Scandinavian occurrence is difficult to explain by postglacial immigration from south or east.

The distribution of *P. oederi* in Scandinavia closely resembles that of *Kobresia simpliciuscula*. The Norwegian area is divided into two, a small one embracing the Hardangervidda and the mountains of upper Hallingdal, and a large one extending from the Jotunheimen to Meråker. The Swedish distribution is restricted to Härjedalen and Jämtland, the northernmost station being Ornäsfjäll in Frostviken.

Bodil Wilmann

Petasites frigidus (L.) Fr.

Maps: Hultén (1971a, no. 1730). Total: Hultén (1968), Hultén & Fries (1986, no. 1830). Local: Benum (1958), Lid (1959), Kristensen (1981), Mølster (1981), Holten (1983), Elven (1984).

First Norwegian record: "Habitat ad rivulum in pascius sylvaticus prædii æstivi Engan ad Røraas, 1/4 milliare circiter ab ædibus prædii memorati per declivia meridiem versus eundo, ubi cum seminibus maturis d. 26. Julii 1764 a me lecta est". Gunnerus (1766, no. 81) (TRH).

Norwegian distribution: From Grønafjell in Suldal and Bykle to Magerøya and Sør-Varanger.

Altitude limits: Northern boreal to low-alpine. Tverrbotnhorn in the Jotunheimen 1750 m (Jørgensen 1933); NAS reported 1960 m in the Jotunheimen, but details are lacking. Down to 300 m in Stor-Elvdal. Troms: Målselv, Rostafjell 1380 m (Jørgensen 1937). Down to sea level in North Norway.

Habitat: Irrigated snow-beds, margins of intermediate and rich fens, springs, stream banks and willow thickets. Neutral to calciphilous.

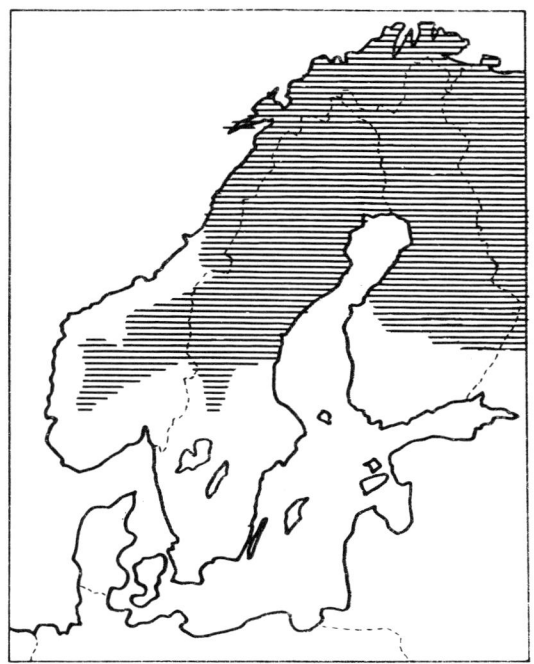

Fig. 70 The Fennoscandian distribution of *Petasites frigidus*

P. frigidus is a northern Eurasian-American species, but is not circumpolar. It is not known in Iceland, Greenland or the eastern Canadian Arctic, but occurs in Svalbard and on the Kola Peninsula. It is widely distributed in the middle boreal to low-alpine belts in Finland and Sweden. It occurs at fairly low altitudes in southeastern parts of Norway, e.g. in southern Gudbrandsdalen and Trysil. Elsewhere in Norway, it is restricted to the northern boreal and low-alpine belts. The distribution in Norway shows an eastern tendency, the species being absent from mountain massifs in the west. The westernmost localities are at Gloppen.

P. frigidus is a hygrophilous species, growing in habitats where the groundwater level is high for most of the year, but not stagnant. It occurs as a dominant species on gently sloping snow-beds irrigated by meltwater for a long time in summer, in communities belonging to the *Ranunculo acris-Poion alpinae* alliance (Nordhagen 1943, Gjærevoll 1956).

P. frigidus is also common in the vegetation of springs, and mainly belongs to the upland types of the *Cardamino-Montion* and *Cratoneurion commutati* alliances. Lowland occurrences are found in shady localities in or near springs. In upland areas, it also occurs in fen margin communities (often with willow scrub), mainly in the *Caricion canescenti-nigrae* and *Caricion bicolori-atrofuscae* alliances.

Even though *P. frigidus* can be found growing on soils poor in lime, it is far more common on lime-rich soils. The pH in 6 stands analysed on Knutshø (Gjærevoll 1956) was 5.4-6.1.

Asbjørn Moen

Pinguicula alpina L.

Maps: Fægri (1940), Hultén (1971a, no. 1603). Total: Meusel et al. (1978), Hultén & Fries (1986, no. 1715). Local: Ouren (1952), Benum (1958).

First Norwegian record: Troms: Skånland, Astafjord, 9.6.1767, J. E. Gunnerus (TRH), (Gunnerus 1772, no. 640).

Norwegian distribution: Bicentric. It has an easterly distribution in the southern area, the westernmost localities being found on Finnvasspiken, Rennebu, and in the Langvella valley in Oppdal. From there, the species has a continuous distribution towards the Swedish border. In the northern area, it is found from Skarmodalen in Hattfjelldal to North Cape.

Altitude limits: Northern boreal to low-alpine. In the southern area, between 800 and 1150 m. Troms: Bardu, Råkkunbårri 1050 m (Engelskjøn 1986a). Down to sea level in many places in North Norway.

Habitat: Shallow rich fens and snow-beds. Calciphilous and hygrophilous.

P. alpina is a Eurasian, arctic-alpine species distributed eastwards from Iceland to eastern Siberia. It is common in the central European mountains from the Pyrenees to the Carpathians. Most localities in Asia are in Siberia, but it has been reported from several stations in the Himalayas. In Sweden, it is known from Härjedalen close to the southern Norwegian area and from Åsele Lappmark to Torne Lappmark. It also has a peculiar occurrence on Gotland where numerous localities are found in rich fens. In Finland, it has been recorded from the northern provinces southwards to Nordbotten and Kuusamo. From Fennoscandia, there is continuous distribution eastwards to Siberia.

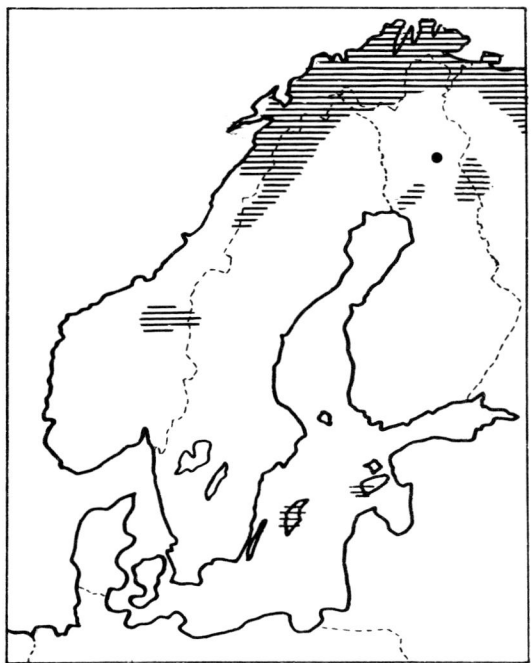

Fig. 71 The Fennoscandian distribution of *Pinguicula alpina*

The distribution in the southern area is not congruent with the typical area for bicentric species. The majority of localities are found east of the Dovre-Trollheimen area. Already Th. C. E. Fries (1913) drew attention to this special distribution (the *Pinguicula alpina* group). Berg (1963) classified it among his "aberrant disjuncts". It does not show the same continental tendency in North Norway.

The disjunct distribution in Europe makes the postglacial immigration history of *P. alpina* intricate. The occurrence of alpine plants in southern Sweden, including *P. alpina*, has been regarded as due to immigration from northern Finland via Karelia and the Baltic States (Erlandsson 1942), or from somewhere in central Europe east of the ice margin (Sterner 1944). There is no fossil evidence between the Alps and Scandinavia. The continuous distribution eastwards from Scandinavia makes migration from the east probable. However, the puzzling occurrence on Iceland remains. The species has also been reported from northern Scotland, but is now believed to be extinct (Clapham et al. 1962).

In the southern area, the ecological and altitudinal ranges are narrow. Normally, *P. alpina* grows in rich fens with low or fairly low vegetation. The ranges are wider in North Norway where, in addition to rich fens and springs, it may be abundant in snow-beds downslope from *Dryas octopetala* heaths, together with dominating *Salix reticulata* (Gjærevoll 1956). It is also frequently found down to sea level in a variety of plant communities, especially in places affected by calcareous spring water. It is often seen along rivers and streams, on the banks of small lakes and

on solifluction soils.

In addition to schistose soils, it has been reported from marble and serpentinite. Lunde (1962) reported 107 pH tests from North Norway which ranged from 5.0 to 7.8 and had a mean of 6.55. Three from Budal (Ouren 1952) gave a pH of 5.3, 5.7 and 5.9. A pH of 6.6 was found in a *Salix reticulata* snow-bed locality with abundant *P. alpina*, near Guolasjav'ri, Kåfjord, Troms (Gjærevoll 1956).

Platanthera obtusata (Pursh.) Lindl. subsp. *oligantha* (Turcz.) Hultén

Maps: Alm (1929), Hultén (1943, 1971a, no. 538). Total: Nordhagen (1933), Hultén (1962, no. 150, 1968), Hultén & Fries (1986, no. 565). Local: Benum (1958).

First Norwegian record: Alta, Sak'kubadni, M. N. Blytt 1841.

Norwegian distribution: Northern unicentric. From Evenstadskaret in Målselv to Sak'kubadni in Alta and Sil'bacåkka in Porsanger.

Altitude limits: M. N. Blytt collected the species at the foot of Sak'kubadni, and T. M. Fries also reported it from this station, which is near sea level. At this classical locality, the plant occurs from the foot of the hill to the top (358 m) (Zetterstedt 1854). The highest locality is at 700 m on Jav'reoai'vit in Nordreisa.

Habitat: Northern boreal to low-alpine. Fairly dry or damp calcareous soils.

P. obtusata s. lat. is a circumpolar species. Subsp. *obtusata* is widely distributed in North America from Newfoundland through central Canada to westernmost Alaska.

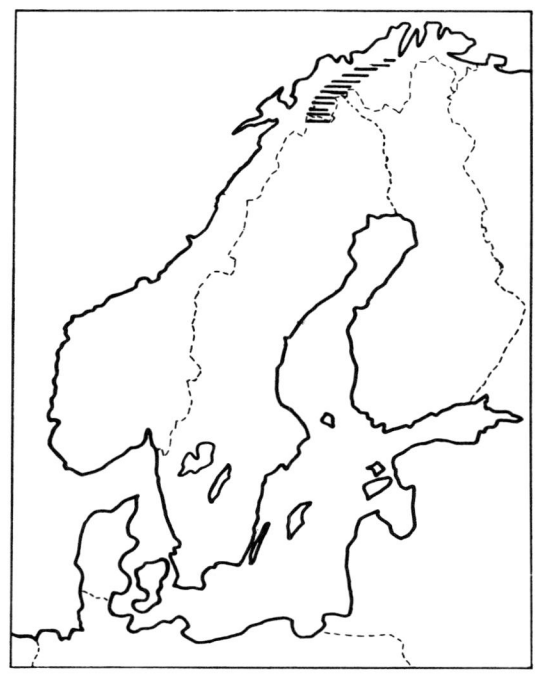

Fig. 72 The Fennoscandian distribution of *Platanthera obtusata* subsp. *oligantha*

Subsp. *oligantha* (*P. oligantha* Turcz.) is a Eurasian race with an extremely disjunct distribution. The Scandinavian localities are very isolated. The nearest locality is at the River Yenisey, and in Asia there are a few widely separated localities from there to Kamchatka. There is a single Swedish locality, on the mountain of Nuolja in Torne Lappmark, which is therefore the southernmost one. Hence, the distribution of *P. *oligantha* conforms with that of several other northern unicentric species.

Much attention has been paid to the disjunct distribution of this species (cf. Nordhagen 1933, 1935, 1936a). Nordhagen thought it difficult to accept postglacial immigration from Siberia to northern Scandinavia. As he emphasised (Nordhagen 1935), *P. *oligantha* should be viewed in connection with plants having similar disjunct and isolated occurrences, such as *Scirpus pumilus*, *Oxytropis deflexa* and *Crepis multicaulis* (the last-mentioned is now extinct in Norway).

P. *oligantha is mainly found a short distance above the birch timber line, mostly on heaths with a solid moss cover providing at least some damp or moist conditions. At one of its best localities, on Jav'reoai'vit, it grows in a *Cassiope tetragona* heath together with, among others, *Rhododendron lapponicum*, *Tofieldia pusilla*, *Pedicularis flammea* and *Carex misandra*, in a well-developed moss carpet of *Hylocomium splendens*.

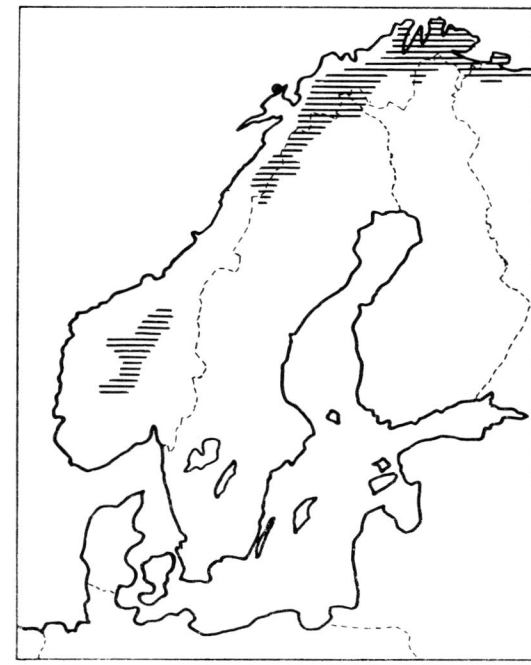

Fig. 73 The Fennoscandian distribution of *Poa arctica*

Poa arctica R. Br.

Maps: Nannfeldt (1940), Hultén (1971a, no. 210), Meusel et al. (1965). Total: Hultén (1958, nos. 5 and 6, 1968), Hultén & Fries (1986, nos. 218 and 219). Local: Benum (1958), Toftaker (1969), Elven (1979), Schumacher & Løkken (1981).

Norwegian distribution: Bicentric. In the southern area, from Bjøberg in Hemsedal to Geithetta in the Trollheimen. In the northern area, from Krokki on Saltfjellet to Magerøya and Grense Jakobselv.

Altitude limits: Mainly middle-alpine. Trollsteinhøin in the Jotunheimen 2050 m (NAS). Såtbakkollen in the Trollheimen and Nordstølen, Lesja, about 1000 m. Troms: Målselv, Kirkestinden 1530 m (Engelskjøn 1986a). Almost down to sea level on river banks, especially in Finnmark.

Habitat: *Dryas octopetala* heaths, grass heaths, *Polygonum viviparum-Potentilla crantzii* heaths, and solifluction lobes. Calciphilous.

P. arctica is a highly polymorphic, circumpolar, arctic plant. The taxonomy of the Scandinavian forms was studied by Nannfeldt (1940), (cf. also Nygren 1950). They are generally apomictic. Some forms are well defined, others display an element of continuous variation. To a large extent, the taxonomical situation is still not clarified, particularly in North Norway. The map shows the distribution of *P. arctica* coll.

According to Nannfeldt, five fairly distinct races occur in Norway (excluding *P. stricta* which is treated as a separate species in this atlas):

1 Subsp. *depauperata* (Fr.) Nannf. is restricted in Norway to the mountains of central southern Norway from Dovre and Lesja to the Trollheimen, but is also known from Iceland. 2n = 75-79.

2 Subsp. *elongata* (Blytt) Nannf. is endemic to the mountains of South Norway from Hemsedal to the Trollheimen and Røros. 2n = 68-76.

3 Subsp. *caespitans* (Simm.) Nannf. is known in Scandinavia from a few localities in Troms (Gratangen-Bardu) and Torne Lappmark. The taxon is circumpolar, but has a gap eastwards to Novaya Zemlya. This is the dominating race of *P. arctica* in many parts of Svalbard. 2n = 56.

4 Subsp. *microglumis* Nannf. is only known from Polvartind in Storfjord, Norway, and from Torne Lappmark and Lule Lappmark in Sweden. 2n = 68, 82.

5 Subsp. *tromsensis* Nannf. is only known from Fløyfjellet in Tromsø.

Most material from North Norway and Sweden has so far been reported as *P. arctica* coll. In Sweden, *P. arctica* is known from Pite Lappmark to Torne Lappmark, and in Finland, from Enontekis Lappmark and Enare Lappmark.

Nannfeldt (1940) looked upon the southern Norwegian races as the "last remnants of a very polymorphous *P. arctica* population that once inhabited the whole of the Scandinavian mountains". He thought they had survived the last glaciation somewhere in central Norway. Subsp.

caespitans must have survived the last, and probably also the last-but-one glaciation in northern Scandinavia.

As far as I can see, the various races display a fairly homogeneous ecological behaviour. *P. arctica* is preferably a heath species in the middle-alpine belt, but may also be frequent in the upper part of the low-alpine belt. As Nordhagen (1954a) pointed out, it avoids the exposed *Kobresia myosuroides* heaths, is common in the more sheltered *Dryas* heaths and abundant in the *Polygonum viviparum-Potentilla crantzii-Silene acaulis-Hylocomium splendens* communities (*Potentilleto-Polygonion vivipari*).

Table 13 gives an extract of 5 square analyses carried out by Nordhagen (1954a) on Knutshø, at 1500-1600 m a.s.l.

Table 13

Square no. (4 m²)	1	2	3	4	5
Poa arctica (exl. stricta)	4	4	4	4	3
P. alpina	3	4	3	2	1
P. alpigena	.	1	1	.	.
Festuca vivipara	2	3	3	3	3
Trisetum spicatum	2	1	1	1	1
Carex lachenalii	.	.	1	.	.
C. misandra	1	1	.	.	.
Luzula spicata	1	1	1	2	1
Cerastium alpinum	1	2	2	2	1
Draba alpina	2	1	1	1	2
Erigeron uniflorus	1	1	.	2	2
Silene wahlbergella	1	.	.	.	.
Minuartia biflora	.	.	1	1	1
Petasites frigidus	1	1	1	1	.
Polygonum viviparum	4	3	3	3	3
Potentilla crantzii	2	1	1	1	3
Pedicularis oederi	3	2	1	1	3
Salix polaris	2	3	2	3	2
Saxifraga cespitosa	1	.	.	.	.
S. oppositifolia	1	1	2	1	1
Saussurea alpina	1	1	2	2	2
Sedum rosea	1	1	2	2	2
Silene acaulis	4	4	4	3	4
Thalictrum alpinum	2	2	2	3	2
Aulacomnium turgidum	3	2	1	2	3
Blindia acuta	.	.	1	.	2
Campylium stellatum	.	2	1	.	.
Climacium dendroides	.	.	.	.	2
Dicranum sp.	3	2	3	1	2
Drepanocladus uncinatus	.	2	.1	1	.
Hylocomium splendens	4	2	2	5	5
Hypnum callichroum	1	1	1	1	2
Oncophorus virens	1	1	2	1	.
Polytrichum alpinum	1	2	2	1	1
Rhytidium rugosum	1	.	.	1	2
Tomenthypnum nitens	2	1	1	2	2
Ptilidium ciliare	2	1	.	1	1
Cetraria cucullata	1	.	.	1	1
C. nivalis	.	.	.	1	1
Peltigera aphthosa	1	1	1	1	2
P. scabrosa	.	.	.	1	.
Thamnolia vermicularis	1	.	.	.	1
Nostoc cfr. commune	1	1	1	.	.

Poa flexuosa Sm.

Maps: Nannfeldt (1935), Hultén (1971a, no. 214). Total: Hultén (1958, no. 48), Meusel et al. (1965), Hultén & Fries (1986, no. 222). Local: Nordsteien (1982), Holten (1984).

First Norwegian record: Saltdal, Solvågtind and Båtfjellet 1826, S. C. Sommerfelt (UPS). Telemark: Gausta, Sept. 1826, M. N. Blytt (O). Aust-Agder: Bykle, Sept. 1826, M. N. Blytt (O).

Norwegian distribution: From Svarteheii in Rogaland to Fonnfjellet in Meråker, and from the Børgefjell mountains to Lappfjellet in Sørfold.

Altitude limits: Mainly middle-alpine, but frequently also high-alpine. Store Memurutind 2350 m (NAS). Hyen in Nordfjord 400 m. Rana 320 m.

Habitat: Flat ground, stony and gravelly places, summits, moraines and snow-beds. Neutral to basiphilous.

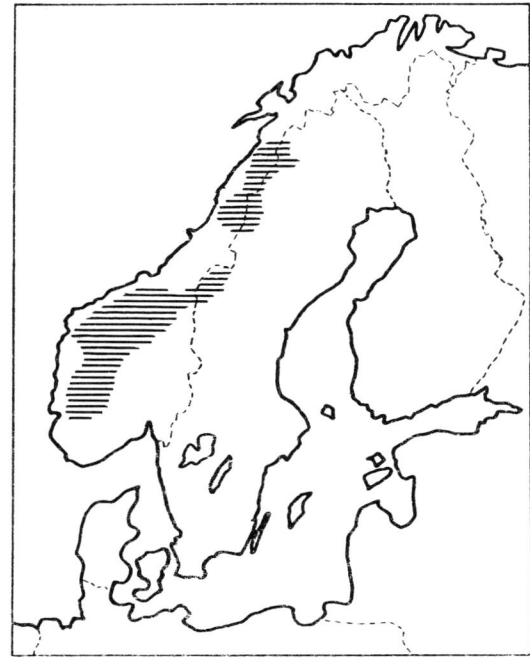

Fig. 74 The Fennoscandian distribution of *Poa flexuosa*

P. flexuosa is an amphi-Atlantic taxon. In addition to Norway and Sweden, its distribution comprises Scotland, Iceland and Greenland (Jensen's Nunataks). It is closely related to *P. laxa* in the central European mountains and *P. fernaldiana* in eastern North America. Its taxonomy has been thoroughly dealt with by Nannfeldt (1935).

P. flexuosa has a distinctive distribution that is not

easily comparable with that of other species. It occurs widely in South Norway. In North Norway, it is known from numerous localities in Hattfjelldal, as are many slightly bicentric species. The most remarkable feature is its northern limit in Sørfold, which is not far from the southern limit in North Norway of a number of centric species. In Sweden, *P. flexuosa* is known from Härjedalen to Lule Lappmark, but has a considerable gap in Jämtland.

Nannfeldt (1935), who studied the *P. laxa* group, compared the distribution of *P. flexuosa* with that of *Carex rufina*. Both show a strong westerly tendency. However, *C. rufina* has a distribution that is comparable with many ubiquitous species, and its ecological behaviour is very different from that of *P. flexuosa*.

As Nannfeldt pointed out, *P. flexuosa* is a rather old species since identical forms occur in Scandinavia, Scotland, Iceland and Greenland. In his opinion, it must have reached its present range before the last glaciation and survived in refugia in western parts of South Norway. The special northern limit was regarded as a postglacial migration limit. It should, however, be borne in mind that refugia may have existed in the Saltdal area. The species was common on Jensen's Nunataks in Greenland (Gjærevoll & Ryvarden 1977), and therefore seems well adapted to nunatak conditions.

P. flexuosa is first and foremost found at high altitudes in similar locations to *Ranunculus glacialis* and is a characteristic species of the *Luzulo-Ranunculetum glacialis* association (Nordhagen 1943), a typical middle-alpine community. It accompanies *R. glacialis* almost to the altitude record for vascular plants in Norway. According to NAS, it is frequently met with above 2000 m in the Jotunheimen. It is also often found at high altitudes in the western mountains, e.g. the Romsdalen mountains. *P. flexuosa* is also a pioneer plant on moraines.

Poa stricta Lindeb.

Maps: Nannfeldt (1940), Nordhagen (1954a), Gjærevoll & Sørensen (1954), Hultén (1971a, no. 220a). Local: Toftaker (1969).

First Norwegian record: Oppdal, Knutshø 1854, C. J. Lindeberg (O).

Norwegian distribution: Endemic to central Norway, being found between Grønhø in Tynset in the east and Store Åmotshytten in the west, and Søndre Knutshø in the south and the Finnpiggene mountains in the north, all these last three limits being in Oppdal.

Altitude limits: Mainly middle-alpine. Knutshø 1655 m. Not reported below 1300 m.

Habitat: Moist, gravelly solifluction soil.

P. stricta was discovered by C. J. Lindeberg on Knutshø in 1854 and described shortly afterwards (Lindeberg 1855). There has been much discussion about its taxonomical position. Nannfeldt (1940) regarded it as a subspecies of *P. arctica*, whereas Hylander (1955) reduced it to a variety of that species. Nordhagen (1954a), however, strongly advocated that it should have species rank. The morphological characters are obvious, and above all it displays an ecological behaviour that differs greatly from the non-viviparous races of *P. arctica* (subsp. *depauperata* and subsp. *elongata*) in the same area. I share this view.

P. stricta mainly grows at an altitude of 1400-1600 m. Within its restricted area it is frequently found in abundance. The typical locality is slightly sloping ground affected by irrigation and solifluction. Even when irrigation has ceased, the gravelly soil remains fairly moist. *P. stricta* may be characterised as a hygrophilous and chionophilous species, whereas *P. arctica* is a characteristic species of lichen-rich, xerophilous and mesophilous heath communities.

P. stricta is a characteristic species of the *Luzulion arcticae* alliance (Nordhagen 1936, 1954a, Gjærevoll 1956). Table 14 gives an extract of 10 square analyses carried out on Orkelhø, stand I at 1400 m, stand II at 1490 m (Gjærevoll 1956). The pH values from the two stands were 6.9 and 6.7, respectively. My experience from numerous localities is that *P. stricta* should be classified as a calciphilous species.

Table 14

Stand no.	I						II			
Square no.(1 m²)	1	2	3	4	5	6	7	8	9	10
Salix polaris	.	1	1	.	2	1	1	2	1	1
S. reticulata	.	1	1	1	.	.	.	.	.	.
Cardamine bellidifolia	.	1	.	1	.	1	1	1	1	1
Cerastium alpinum	1	.	1	1	.	1	1	1	1	1
Draba alpina	1	1	1	1	1	1	1	1	1	1
Koenigia islandica	1	1	1	1	1	1	1	1	1	1
Minuartia biflora	.	.	.	.	1	1	1	1	1	1
Polygonum viviparum	1	1	1	1	1	1	1	1	1	1
Sagina intermedia	1	1	1	.	.	.	.	.	.	.
Saxifraga cernua	1	1	1	1	1	1	1	1	1	1
S. oppositifolia	1	1	1	.	.	.	.	.	.	.
Silene acaulis	.	2	.	2	.	2	2	1	2	2
Carex misandra	1	.	1	.	1	.	.	.	.	.
Juncus biglumis	1	2	2	1	2	1	1	1	1	1
Luzula confusa	1	1	1	1	1	.	.	.	.	.
Poa stricta	3	3	3	2	3	3	2	2	3	3
Blindia acuta	1	1	1	1	1	1	1	2	1	1
Distichium capillaceum	1	.	1	.	.	1	1	1	1	1
Drepanocladus revolvens	2	2	3	1	2	1	2	1	3	1
Philonotis tomentella	1	1	1	1	1	.	.	.	.	.
Aneura pinguis	.	.	.	.	.	2	1	2	2	1
Stones and bare soil	5	5	5	5	5	5	5	5	5	5

Potentilla chamissonis Hultén

Maps: Hultén (1945, 1971a, no. 1013). Total: Hultén (1945, 1958, no. 8, 1968, 1971b, no. 20), Hultén & Fries (1986, no. 1105). Local: Benum (1958).

First Norwegian record: Finnmark: Alta, Kåfjord 1841, M. N. Blytt (O), Alta 1841, N. Lund (S). As Blytt and Lund were together and Lund (1846a) reported *P. nivea* from "Sakkavarre ved Kaafjorden" the locality is Sak'kubadni. Both collections were determined as *P. chamissonis* by Hultén (1945).

Norwegian distribution: Northern unicentric. From Øverbygd in Troms to Porsa in Kvalsund and Gjelhaugan at Vadsø, in Finnmark.

Altitude limits: Mainly low-alpine. Garanasgai'si in Øverbygd 1020 m. Down to 100 m in eastern Finnmark.

Habitat: Cliff ledges, scree slopes and rocks. Calciphilous.

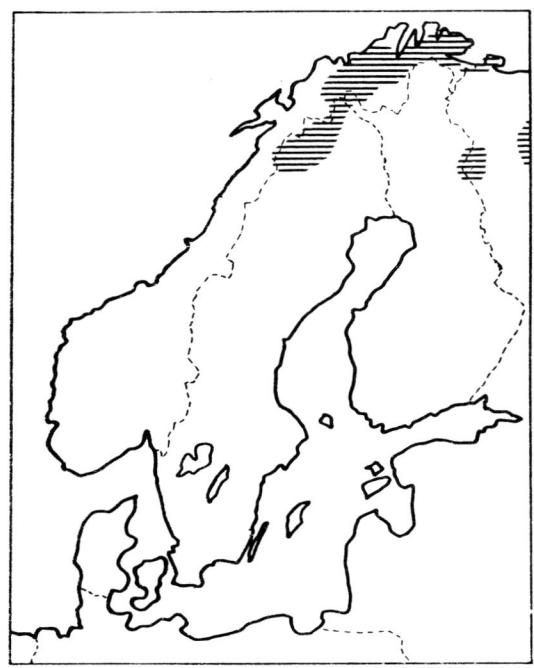

Fig. 75 The Fennoscandian distribution of *Potentilla chamissonis*

P. chamissonis was identified and described when Hultén (1945) revised the *Potentilla nivea* group. It was found to be a northern species in Fennoscandia and amphi-Atlantic in its full distribution. In Sweden, it is known from Lule Lappmark and Torne Lappmark. In Finland, it occurs in Enontekis Lappmark, Enare Lappmark and Kuusamo. In Russia, it is found in northern Karelia and central Kola.

Hultén (1971b) reduced it to a subspecies of *P. hookeriana*, and this view was repeated by Hultén & Fries (1986). Together, *P. hookeriana* s. str. and *P. chamissonis* are circumpolar, but the same gap exists between *P. chamissonis* in the Kola Peninsula and *P. hookeriana* in northern Russia as is displayed in the distribution of *P. nivea*.

P. chamissonis and *P. nivea* may grow together, but the former prefers cliff ledges and scree slopes and occurs more rarely in the windswept localities favoured by *P. nivea*. It is a distinctly calciphilous species, growing on calcareous schists and dolomite. Because of its critical position and the possibility of confusion with *P. nivea*, the geographical distribution of *P. chamissonis* may be somewhat incompletely known. It seems clear, however, that it is a continental species.

Potentilla nivea L.

Maps: Hultén (1971a, nos. 1023 and 1024). Total: Hultén (1945, 1968, 1971b, no. 63), Hultén & Fries (1986, no. 1104). Local: Ouren (1952), Benum (1958), Lid (1959), Ryvarden (1967), Toftaker (1969), Bråthen (1973), Hagen (1976), Aune & Kjærem (1978), Sæther et al. (1980), Nordsteien (1982).

First Norwegian record: Gunnerus (1772, no. 512) reported *P. nivea* from Varanger. The drawing (Plate III, Fig. 1) is of *Fragaria vesca*, which agrees with the herbarium specimen. No collection of *P. nivea* is found in the Gunnerus herbarium. Vahl reported *P. nivea* from Lom in 1787. The material was used in Flora Danica 18 in 1792 (drawing no. 1035) – "Paa græsrige steder ved Biergfødderne i Loms Præstegield i Gudbrandsdalen".

Norwegian distribution: Slightly bicentric. In the southern area, from Skardheii in Hjelmeland to Steinfjellet in Meråker. In the northern area, from Strømsfjellet on the border between Bindal and Velfjord to Båtsfjord and Grense Jakobselv.

Altitude limits: Mainly low-alpine. Ryggehø and Raudhamrin in the Jotunheimen 1700 m (NAS). Down to about 500 m in Sunndalen. Troms: Målselv, Dolpasvarre 1200 m. Down to about 90 m in Persfjord.

Habitat: Windswept ridges and heaths, and rock ledges, on calcareous soils.

P. nivea is a circumpolar, arctic-montane species. It is variable and partly apomictic. Hultén (1945) carried out a thorough study of the *P. nivea* complex, and his opinion is followed here. Hence, subsp. *subquinata* (Lange) Hultén is included on the map. It is frequent in the Båtsfjord area and is probably hybridogene in origin (Flora Europaea 2). With very few exceptions, the Norwegian distribution shows a continental tendency.

In Sweden, *P. nivea* is very rare in Härjedalen and Jämtland, and then more common from Pite Lappmark to Torne Lappmark. In Finland, it is known from Enontekis Lappmark, Enare Lappmark and Kuusamo. The localities in Kuusamo are continuous with occurrences in neighbouring Russian Karelia. The species is also found in central parts of the Kola Peninsula from where a large gap of about 1000 km ensues eastwards to northern Russia. Occurrences in the Alps and the Caucasus contribute to the circumpolar distribution.

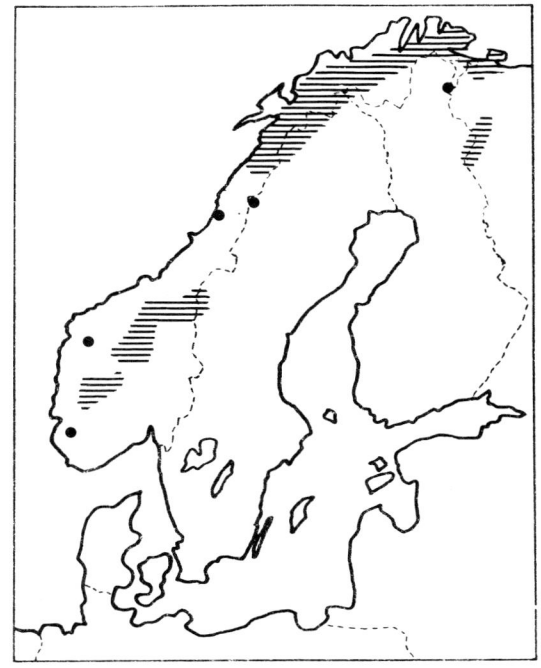

Fig. 76 The Fennoscandian distribution of *Potentilla nivea*

P. nivea is most frequently met with in the low-alpine belt and the lower part of the middle-alpine belt. It is also quite common in subalpine sites with open landscape, e.g. around upland, summer farms. These localities are probably secondary. When the birch forest was cleared, *P. nivea* was given new opportunities on rocks and ridges within the opened-up landscape. It may even grow as an apophyte on turf roofs.

The ecological range of *P. nivea* is fairly narrow. It is one of the preferential species of the *Kobresio-Dryadion* alliance, growing on the most windswept heaths and on ridges and peaks. It also shows preference for scree slopes. Nordhagen (1943) took it as a characteristic species of his *Veronico-Poion glaucae* alliance, found in both the low-alpine and subalpine belts. *P. nivea* also occurs occasionally on gravel bars.

In addition to calcareous schists, *P. nivea* grows on dolomite and marble. Lunde (1962) gave results of 153 pH tests from the northern area, which had a total amplitude of 5.2-7.9 and a mean of 6.64. Ouren (1952) reported pH = 7.7 at Budal.

Primula scandinavica H. Bruun

Maps: Bruun (1938), Hultén (1971a, no. 1390), Gjærevoll (1973), Meusel et al. (1978), Hultén & Fries (1986, no. 1470). Local: Lid (1959), Toftaker (1969), Hagen (1976), Nordsteien (1982), Halvorsen & Salvesen (1983), Moe (1985).

First Norwegian record: Reported from Oppdal (Gunnerus 1766, no. 26) as *P. farinosa* "Habitat in pratis

& pascuis uliginosis subalpinus e.g. Opdaliæ, mense Majo & Junio florens". There are numerous collections of *P. scandinavica* in the Gunnerus herbarium, but none labelled Oppdal. It contains a collection from Tromsø, 23.7.1767, and several from Græsholmen in Troms (Harstad), 9.6.1770.

Excluded or doubtful stations: Vardø: O. Dahl (1934) judged this locality to be dubious. The same seems to apply to Alta. There is no reliable material from either locality. Hitterdal: "på Taarefjeld nær Gaustad, 5.7.1912, frk. Berglund" (O); the locality cannot be identified.

Norwegian distribution: Slightly bicentric. From Skardheii in Hjelmeland to Tromsø.

Altitude limits: Mainly low-alpine to northern boreal. Hardangervidda 1450 m (Lid 1959). Troms: Målselv, Vakkerfjell 680 m. Down to sea level in Trøndelag and North Norway.

Habitat: Heaths, meadows and screes. Calciphilous.

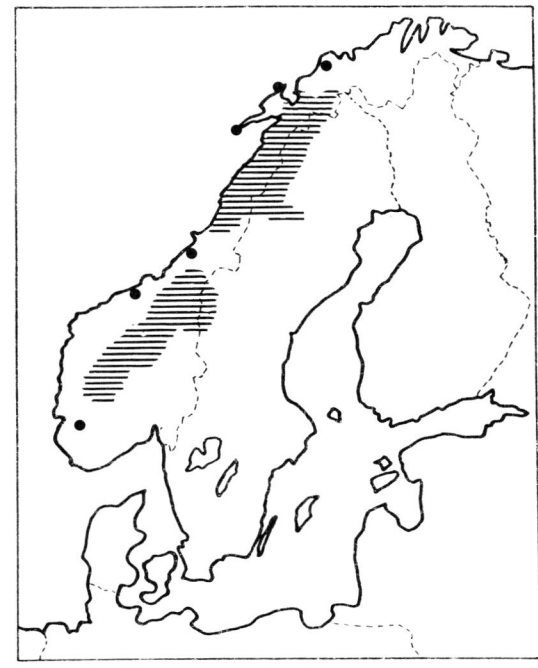

Fig. 77 The Fennoscandian distribution of *Primula scandinavica*

P. scandinavica is endemic to Scandinavia. Previous reports of occurrences on Vaygach Island and Novaya Zemlya have not been verified; the species is not mentioned in Flora USSR. It was originally taken to be identical with *P. farinosa*, and later with *P. scotica*, until Bruun (1938) showed that it was an endemic taxon with a chromosome number of 2n = 72, whereas *P. farinosa* has 2n = 18,36 and *P. scotica* has 2n = 54. In Sweden, it is known from Dalarna to Jämtland and Lycksele Lappmark to Torne Lappmark.

The distribution shows many strange features. In 1966, it was discovered by Ryvarden & Kaland (1968) at Skardheii in Hjelmeland, far south of its previously known southern limit and in the same area as they had made sensational discoveries of *Artemisia norvegica* and several other rare alpine species. In South Norway, it has its main concentration from Dovre and the Trollheimen, east to Røros.

Its distribution, as regards mountain occurrences, is slightly bicentric and closely resembles that of *Oxytropis lapponica*. But, whereas *O. lapponica* is restricted to the mountains, *P. scandinavica* also occurs at low altitudes, e.g. on the coast from Osen northwards. Its northern boundary, apart from the somewhat isolated occurrence at Tromsø, is almost identical with that of *Saxifraga adscendens*.

The ecological behaviour of *P. scandinavica* also differs in some respects from most alpine species. In the mountains, it prefers *Dryas* heaths and various meadow communities in the low-alpine belt where it is fairly frequent, but usually scattered. It is also met with on scree slopes, shallow rich fens and moist cliffs. In many places (e.g. Oppdal, Budal, Røros and Nord-Østerdal), it is frequently found as an apophyte in hayfields and previously cultivated fields. Along the coast, the favourite habitat is meadows on shell sand, but these must be regarded as secondary localities. The species is clearly basiphilous, growing on all kinds of calcareous mica schists and marble, and also on serpentinite.

P. scandinavica and *P. scotica* (which is endemic to Scotland) must have had a common pre-glacial history. *P. scotica* is restricted to the northernmost part of Scotland where several other interesting species are found (see *Artemisia norvegica* and *Arenaria norvegica*).

Ranunculus glacialis L.

Maps: Hultén (1971a, no. 809). Total: Böcher (1938), Hultén (1958, no. 73), Meusel et al. (1965), Hultén & Fries (1986, no. 864). Local: Ouren (1952), Benum (1958), Holten (1984), Engelskjøn (1986a).

First Norwegian record: According to Oeder (1761, Table XIX), *R. glacialis* occurs "I stor Mængde i Norge allevegne paa de høieste Toppe af Fieldene, tæt ved hvor den stedsevarende Snee ligger, og er tilligemed den safranrøde Steenmosse nesten den sidste Indbygger i den øvre Luft-Kreds. I størst Mængde paa Søndenfield ikke langt fra Præstesætter i Slidre samt paa Sulutinde den høieste Fieldtop af Fillefield". During a visit to Røros on 26.7.1764, Gunnerus received *R. glacialis* from a mining counsellor named Schindel and labelled it "Habitat in alpibus Selboensibus" (TRH).

Norwegian distribution: Slightly bicentric. In the southern area, from the border between Suldal and Vinje to the mountain of Fongen in Meråker. In the northern area, from Heilhornet in Bindal and Hattfjelldal to North Cape.

Altitude limits: Mainly middle-alpine, but also conspicuous in the high-alpine belt, ascending to 2370 m on Galdhøpiggen (Jørgensen 1933), the highest station of any vascular plant in Scandinavia. Down to 400-500 m in western Norway. Found at an altitude of 1900 m in the Okstindene mountains. Troms: Målselv, Njunis 1700 m (Engelskjøn 1986a). In northern Sweden 2055 m. In North Norway, it may descend almost to sea level (O. Dahl 1934), but is then usually near glaciers.

Habitat: Preferably in snow-beds. Indifferent.

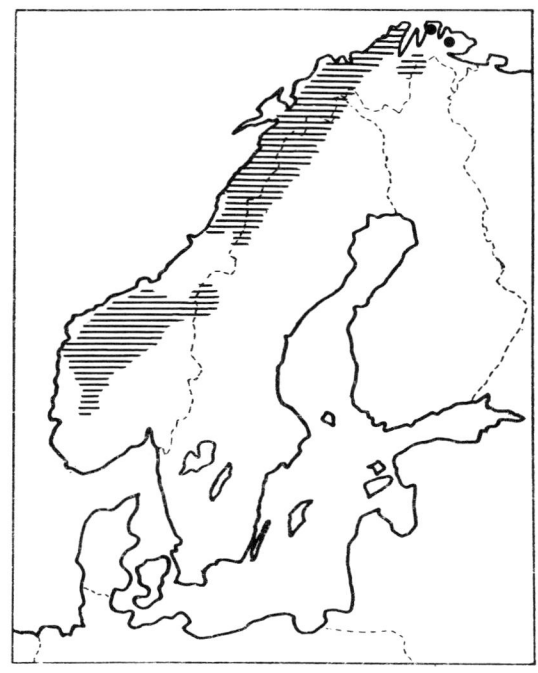

Fig. 78 The Fennoscandian distribution of *Ranunculus glacialis*

R. glacialis belongs to the amphi-Atlantic species that are also found in the central European mountains, from the Sierra Nevada and Pyrenees to the Carpathians. In Sweden, it is found from Härjedalen to Torne Lappmark, and in Finland it occurs in Enontekis Lappmark. It is also known from an isolated area in the central part of the Kola Peninsula, and occurs in Svalbard (very rare), on Jan Mayen, the Faroes, Iceland and in East Greenland. There is a closely related taxon, usually regarded as a subspecies (subsp. *chamissonis* Hultén), on both sides of the Bering Strait.

Unlike most typical bicentric species, *R. glacialis* is not limited to calcareous rocks. It occurs abundantly on acid soils (Nordhagen 1943, Gjærevoll 1956), and Nordhagen considered it an acidophilous species. It is an important constituent of an easily recognisable community, *Luzulo-Ranunculetum glacialis* (Samuelsson 1917, Nordhagen 1943, Gjærevoll 1956), in which pH values are usually close to 5.0.

The gap in Trøndelag (and the corresponding one in Sweden) may be due to lack of suitable high localities, as is also the case in central and eastern Finnmark.

R. glacialis is mainly a snow-bed species, but displays a wide range of tolerance as regards snow cover. At high altitudes, it may be abundant in lichen heaths, flowering already in early July. A common habitat is flat summits, where it is accompanied by *Luzula confusa*. The snow cover on *R. glacialis* communities is therefore not always deep, but the summer season is short because of the high altitudes. In the upper part of the middle-alpine belt and in the high-alpine belt the entire vegetation has the character of snow-bed communities at lower altitudes.

R. glacialis prefers more or less open soil and is common in areas with solifluction and frost action. The bottom layer is often composed of small liverworts (*Anthelia* and *Gymnomitrium*). It is a common pioneer plant on loose, wet gravel exposed by the retreat of perennial snowfields and glaciers. In ravines and along streams, *R. glacialis* may descend to lower altitudes.

Ranunculus hyperboreus Rottb.

Maps: Tralau (1963), Hultén (1971a, no. 811). Total: Hultén (1968, 1971b, no. 16), Hultén & Fries (1986, no. 856). Local: Benum (1958), Kristensen (1981), Alm et al. (1987).

First Norwegian record: Vadsø, undated, leg. C. Weldingh (TRH), (Gunnerus 1772, no. 627).

Norwegian distribution: Bicentric. In the southern area, from Gol in Hallingdal to Oppdal and Røros. In the northern area, from Dønna to Magerøya.

Altitude limits: Blåhø in Dovre, 1594 m; down to at least 400 m in the southern area. In the northern area, Narvik, Skjomen, Cunojav'ri 1040 m; often down to sea level.

Habitat: In springs, shallow pools, muddy places, and manured habitats at upland, summer farms and along the coast. Neutral or slightly basiphilous; nitrophilous.

R. hyperboreus is a circumpolar, arctic species with continuous distribution from North Norway eastwards. Its distribution in Norway differs from that of most alpine species, clearly because of its special ecological behaviour. In South Norway, it is found in continental areas, mainly as an alpine species, but also at some low stations. The natural habitats seem to be shallow pools in the low-alpine and middle-alpine belts, pools which frequently dry up during summer so that *R. hyperboreus* creeps on the mud. It may also occur in cold springs, sandy, muddy and clayey places along streams and on open soil in shallow mires. It is very often found as a nitrophilous apophyte at upland, summer farms and on wet cattle tracks where open mud is available. Nordhagen (1943) described a *Ranunculus hyperboreus-Agrostis stolonifera-Montia* sociation from localities at Lom.

There are relatively few alpine localities in North Norway. Instead, it is mostly found at low altitudes on the coast, particularly on shores with decaying seaweed and in small, shallow pools strongly manured by seabirds (Alm et al. 1987), thus emphasising its nitrophilous character. It may also grow in brackish water.

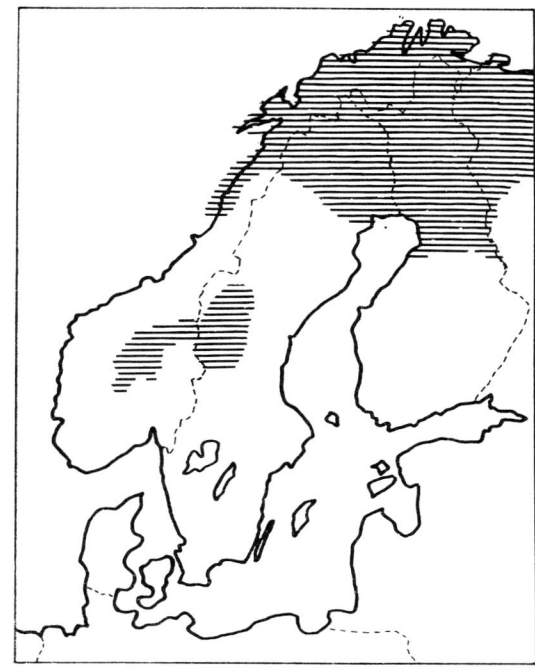

Fig. 79 The Fennoscandian distribution of *Ranunculus hyperboreus*

In Sweden, *R. hyperboreus* is known from Dalarna to Jämtland and from Lycksele Lappmark to Torne Lappmark. In Finland, it is found in the northern provinces south to Kajanaland. It does not occur in the central European mountains.

During the last glaciation, the species grew south of the Scandinavian ice sheet (Tralau 1963). Björkman (1939) maintained that it might have had the same late-glacial and postglacial history as, for example, *Luzula parviflora*, but had had insufficient time to close the gap between the two areas.

Ranunculus nivalis L.

Maps: Hultén (1971a, no. 817). Total: Hultén (1968, 1971b, no. 15), Hultén & Fries (1986, no. 854). Local: Ouren (1952), Benum (1958), Bråthen (1973).

First Norwegian record: The plant published as *R. nivalis* in Flora Norvegica (Gunnerus 1772, no. 627) is *R. hyperboreus*. The oldest collection dates back to 27.7.1802 – Finnmark, Rastigaissa, G. Wahlenberg (UPS).

Excluded stations: (1) Resvoll-Holmsen (1932) mentioned that roots of *R. nivalis* from Hallingskarvet were brought to the Botanical Gardens in Oslo in 1810 by Christen Smith and P. V. Deinboll. No herbarium specimens exist and the locality was not noted by A. Blytt (1874). (2) Alvdal, Tronfjeld, Smith 1807. O. Dahl (1895) published a list of plants registered by Christen Smith from various localities in South Norway. Smith visited Tron in 1807 together with Hornemann and recorded finding, among others, *R. nivalis*. The following year, a pharmacist named Kolstad also visited Tron. *R. nivalis* is not mentioned in his list (O. Dahl 1895). Numerous botanists have made investigations on Tron without collecting or mentioning *R. nivalis*. I feel quite sure that Smith's reported find is erroneous, as also is the locality "Snehætten" in the same list.

Norwegian distribution: Bicentric. In the southern area, from Berakupen in Hardangervidda to the Trollheimen and Sandfjellet in Soknedal. In the northern area, from Røyrvik (the Børgefjell mountains) to Magerøya and Grense Jakobselv.

Altitude limits: Middle-alpine to low-alpine. In the southern area, from 1050 to 1550 m on Midtre Knutshø. Troms: Målselv, Maddagai'si 1450 m (Jørgensen 1937). Down to sea level in the northern area.

Habitat: Snow-beds, preferably irrigated ones. Calciphilous to neutral.

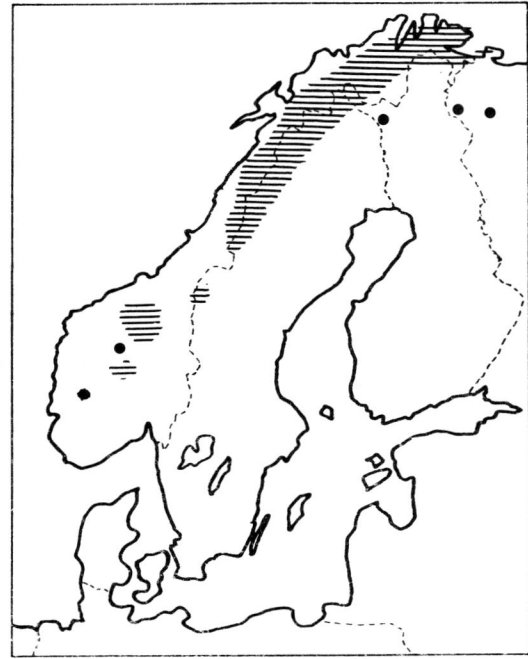

Fig. 80 The Fennoscandian distribution of *Ranunculus nivalis*

R. nivalis is a circumpolar, arctic species. The Fennoscandian distribution area is somewhat isolated as there is a gap of 600 km between Pechenga-Enare Lappmark and the Kanin Peninsula. The species occurs in Svalbard. According to Hultén (1971a) and Hultén & Fries (1986), it grows on Bear Island, but this is incorrect (Rønning 1959).

In Sweden, *R. nivalis* is known from Jämtland to Torne Lappmark, but there is a considerable gap in northern Jämtland. In Finland, it occurs from Kittilä Lappmark to Enare Lappmark. Apart from the localities in Sør-Varanger, the distribution is continuous in the northern area and shows a continental tendency, although there are some coastal localities.

A different situation is met with in South Norway. There, *R. nivalis* is very common and abundant in the central Dovre area, above all in the mountains on the east side of Drivdalen where it occurs continuously from

Knutshø to Ålmenberget. It is extremely rare west of Drivdalen. The occurrences in Hardangervidda and Hemsedal-Valdres are isolated. A single locality is known in each of the three mountain areas, the Jotunheimen, the Trollheimen and the Sunndal district, which is strange considering that suitable snow-beds are found everywhere there.

R. nivalis is a definite snow-bed species, preferring habitats that are irrigated for at least part of the season, or north-facing snow-beds where the soil remains moist during the whole summer. It is a characteristic species of the *Saxifrago oppositifolio-Oxyrion digynae* alliance (Gjærevoll 1956), which is characterised by becoming snow-free late, but not extremely late, and being rich in calciphiles. Within this alliance, *R. nivalis* forms the *Ranunculetum nivalis* association (Gjærevoll 1956).

Table 15 shows the results of analyses carried out in Dunderlandsdalen and Saltdalen in Nordland (A), the Knutshø mountains at Oppdal (B & C), and Spikaloabme in Bardu (D). A is situated in the low-alpine belt, and B-D in the middle-alpine belt. A selection of the most important species is given.

Table 15

Locality	A	B	C	D
Salix herbacea	70-2-	100-2-	100-1	.
S. polaris	55-1	.	.	100-2
Arabis alpina	55-1	30-1	100-1	60-1
Cerastium cerastoides	90-1	100-1	100-1	60-1
Gnaphalium supinum	55-1	40-1	100-1	.
Minuartia biflora	20-1	70-1	80-1	.
Oxyria digyna	100-1+	100-2	100-2	100-2
Polygonum viviparum	95-1	100-1	100-1	100-1
Ranunculus nivalis	100-2	100-2	100-2	100-2
R. pygmaeus	50-1	100-1	100-1	80-1
Saxifraga cernua	55-1	80-1	100-1	.
S. oppositifolia	20-1	70-1	20-1	100-2
S. rivularis	20-1	100-1	100-1	60-1
S. tenuis	95-1	30-1	50-1	100-1
Sibbaldia procumbens	45-1	80-1	40-1	.
Taraxacum croceum	80-1	100-2	100-2	40-1
Veronica alpina	40-1	60-1	100-1	80-1
Carex lachenalii	60-1	90-1+	100-2	100-1
Juncus biglumis	75-1	90-1	100-1	.
Poa alpina	90-1	100-1	100-1+	100-1
Drepanocladus uncinatus	65-2-	90-1	100-1-	.
Philonotis fontana	50-2	40-1	100-5	.
P. tomentella	20-2-	50-1	.	100-1
Pohlia wahlenbergii	70-2	.	.	100-5
P. drummondii	.	100-3	100-2	.
Anthelia sp.	100-3	40-1	40-1	100-2

19 measurements gave a pH range of 5.6-7.2.

Ranunculus sulphureus C. J. Phipps

Maps: Hultén (1971a, no. 830). Total: Hultén (1968, 1971b, no. 25), Hultén & Fries (1986, no. 855). Local: Benum (1958), Ryvarden (1969), Kristensen (1981).

First Norwegian record: Finnmark: Rastigaissa, 29.7.1802, G. Wahlenberg (S).

Norwegian distribution: Northern unicentric. From Malla in Storfjord to Magerøy and Persfjord in Sør-Varanger.

Altitude limits: Mainly low-alpine. Troms: Storfjord, Favresvarri 1240 m. Down to sea level in Tana.

Habitat: Irrigated snow-beds and cold springs. Mainly calciphilous.

R. sulphureus is a circumpolar, arctic species. The area it occupies in northern Fennoscandia is somewhat isolated. There is a gap between East Finnmark and the central part of the Kola Peninsula, and another east of there to the Urals and Novaya Zemlya. The species occurs at only one locality in Sweden, on Jebrentjåkko, north of Torneträsk. In Finland, it is known from Enontekis Lappmark.

The distribution in Troms is continental. There is a rather strange disjunction in Finnmark with a gap between Bogn-elvdalen in Alta and the "gaissa" area, even though many ecologically suitable localities should be available there. It is abundant in the mountains between Lebesby and Tana, and on the Varanger Peninsula, growing at low altitudes in some places, even down to sea level.

The ecology of *R. sulphureus* corresponds closely with that of *R. nivalis*, the two species often growing together. Nordhagen (1943) published analyses from Rastigai'si and Arisondalen which show *R. sulphureus* to be the most conspicuous species. Important constituents are *Salix herbacea*, *Ranunculus acris*, *R. nivalis*, *Rumex acetosa*, *Saxifraga cernua*, *Veronica alpina*, *Carex lachenalii* and *Poa alpina*.

A *Ranunculus sulphureus* sociation has also been described by Gjærevoll (1956) from inner Troms and the Varanger Peninsula, and an extract is given in Table 16.

Table 16

Locality	A	B
Salix polaris	53-1	60-1
Cerastium cerastoides	70-1	100-1
Polygonum viviparum	94-1	60-1
Ranunculus nivalis	70-1	60-1
R. sulphureus	100-2	100-2
Saxifraga cernua	65-1	40-1
Carex lachenalii	82-1	80-1
Juncus biglumis	41-1	80-1
Poa alpina	41-1	100-1
Philonotis tomentella	53-1	.
Pohlia wahlenbergii	59-1	100-3
Stones and bare soil	100-5	100-5

A comprises 15 analyses (1 m²) from the low-alpine belt in the Kåfjord-Nordreisa mountains (710-740 m), and B represents 5 analyses from Båtsfjorddalen (130 m).

The pH amplitude is similar to that of the *R. nivalis* communities, 12 determinations varying from 5.7 to 6.6.

Rhododendron lapponicum
(L.) Wahlenb.

Maps: Th. C. E. Fries (1913), Hultén (1955, 1971a, no. 1368). Total: Hultén (1958, no. 181, 1968), Hultén & Fries (1986, no. 1450). Local: Benum (1958), Nordhagen (1965b), Aune & Kjærem (1978), Engelskjøn (1984), Holten (1984).

First Norwegian record: Ramus (1719) published a description of Norway which included a list of plants. One of these "Myrtillus semper virens sine fructu foliis majoribus. Joh. Bauh. Chamæbuxus" was found by O. Dahl (1888-1911) to be *Azalea lapponica* (= *Rhododendron lapponicum*). The first reliable discovery was made by Vahl on Lomseggen in 1787. This collection was used for drawing no. 966 in Flora Danica 17 in 1790.

Norwegian distribution: Bicentric. Restricted in the southern area to Lom, Vågå, Lesja and Rauma. The northern area extends from Beveråmyrene in Rana to Garssumvarri and Buoidijav'rit in Porsanger.

Altitude limits: Most localities in the southern area are situated between 1000 and 1200 m. Ryggehø in the Jotunheimen between 1500 and 1600 m (NAS). Troms: Målselv, Njunis 1170 m (Engelskjøn 1986a). Most localities in Troms are between 500 and 1000 m; down to 50 m in Lyngen.

Habitat: Exposed heaths on calcareous soils.

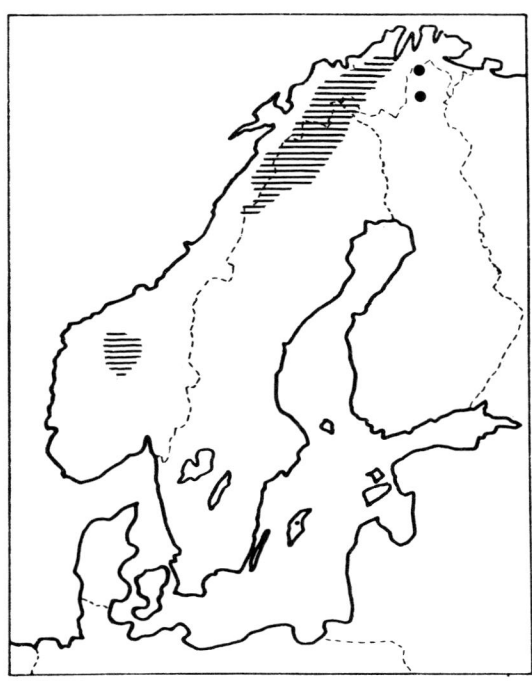

Fig. 81 The Fennoscandian distribution of *Rhododendron lapponicum*

R. lapponicum belongs to the amphi-Atlantic, arctic species. It has a large distribution area from Greenland to Alaska. In Siberia, it is known from the Bering Strait westwards to the Sayanskiy mountains. Russian botanists consider the Siberian population to be a separate species, *R. parvifolium*, but this view was not accepted by Hultén (1958). An old report from Bear Island was due to incorrect information (Rønning 1959). In Sweden, *R. lapponicum* occurs from Mieskattjåkka in Lycksele Lappmark to Torne Lappmark. In Finland, it is found at a few localities in Enontekis Lappmark and Enare Lappmark. The Fennoscandian area is therefore very isolated.

R. lapponicum has a continental distribution in South Norway. The same applies to its northern area. There are a few localities in coastal mountains, but frequency and abundance are far greater in inland mountains.

Since Th. C. E. Fries (1913) pointed out its bicentricity, this species has played a key role in discussions on the history of the Scandinavian mountain flora. In the northern area, *R. lapponicum* follows the normal pattern of bicentric plants. However, the southern area differs, as the species is not known in the Dovre area or the Sunndalen and Trollheimen mountains. Nordhagen (1965b) drew attention to the different distributions of *R. lapponicum* and *Artemisia norvegica*, both of which are of special interest in relation to the refugium theory. These two species never meet despite having very similar ecological demands. "This fact suggests that during the last glacial epoch there existed separate refuges along the coast of Møre, with a different flora" (Nordhagen 1965b, p. 33).

R. lapponicum may be a dominant species in low-alpine heaths, especially in inner Troms, often growing so abundantly that the heaths are coloured in the early flowering season. This community is closely related to *Dryas octopetala* communities. In most places, *R. lapponicum* is a characteristic species of the *Kobresio-Dryadion* alliance. Although it reaches its maximum vitality in species-rich *Dryas* heaths, *R. lapponicum* also grows in open, gravelly places affected by solifluction, and is seen in shallow, stony fens. It usually grows on calcareous soils of all kinds, but is also found in oligotrophic localities. It occurs in *Loiseleuria procumbens-Arctostaphylos alpina* heaths, especially in its southern area (Nordhagen 1965b).

Lunde (1962) reported 156 pH determinations from the northern area which showed a total amplitude of 4.8-7.7, most values lying between 5.6 and 6.9; the mean was 6.21.

R. lapponicum may also occur in subalpine localities. The southern area has several localities in pine forest. Nordhagen (1965b) paid special attention to this ecological behaviour. "Some Scandinavian geologists suppose that *R. lapponicum* in late-glacial time had a rather continuous distribution from northern to southern Scandinavia, but that the Hypsithermal destroyed it because the pine-woods during this long period went higher up in the mountains than in subatlantic and recent times, and because the mountains between South Trøndelag and southern Nordland county are lower than in northern Nordland and

in South Trøndelag respectively. With other words: they suppose that the recent gap which the distribution of *R. lapponicum* shows in Scandinavia, is due to the prehistoric pine-woods, which by their shade killed *R. lapponicum*, a light-growing plant". Referring to the pine forest occurrence, Nordhagen concluded that this argument did not hold true.

Roegneria borealis (Turcz.) Nevski
(Syn. *Agropyron latiglume* auct. scand., *A. boreale* (Turcz.) Drob., *Elymus alaskanus* (Scribn. & Merr.) A. Löve subsp. *scandicus* (Nevski) Melderis.)

Maps: Hultén (1971a, no. 269). Total: Hultén (1968), Hultén & Fries (1986, no. 289). Local: Benum (1958), Halvorsen & Salvesen (1983), Holten (1983).
First Norwegian record: Oppdal, Kongsvold 1836, M. N. Blytt (O).
Norwegian distribution: From Hjelmeland in Rogaland to Porsanger and Grense Jakobselv.
Altitude limits: Chiefly low-alpine. Hordaland: Eidfjord, Tverrgavlen 1457 m. Down to 350 m in Lærdal. Troms: Balsfjord, Tamokfjell 1030 m (Jørgensen 1937). Down to sea level in some places in North Norway.
Habitat: Exposed heaths, scree slopes, cliff ledges and gravel bars. Basiphilous.

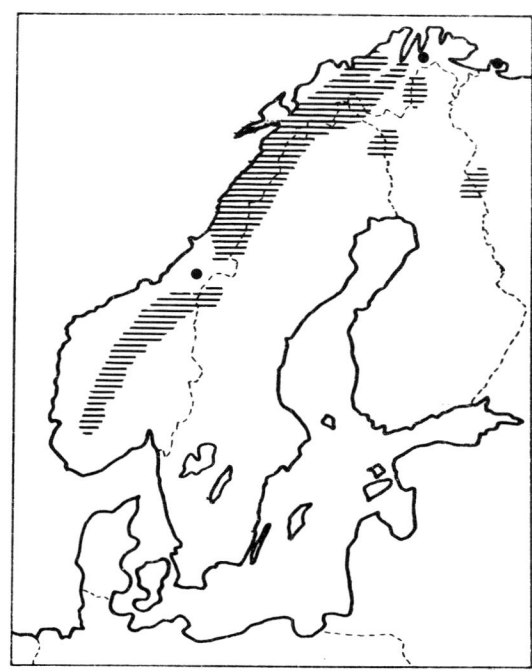

Fig. 82 The Fennoscandian distribution of *Roegneria borealis*

R. borealis belongs to a critical arctic-montane taxon that is usually divided into several subspecies. The species is fairly rare in South Norway, where it has a distinctly continental tendency. It is more common in North Norway, where it has several stations on coastal mountains. In Finnmark, it is very rare north and east of Alta. In Sweden, *R. borealis* occurs from Härjedalen to Torne Lappmark. There is a distinct gap in Nord-Trøndelag and Jämtland. In Finland, it is known from Kuusamo to Enare Lappmark.

Nordhagen (1943) took *R. borealis* as a characteristic species of his *Veronico-Poion glaucae* alliance, occurring on screes on calcareous substrate. It is most frequently found in the low-alpine and subalpine belts. Like, for example, *Kobresia myosuroides*, it is frequent and abundant on hills and ridges in the open landscape around upland, summer farms.

Sagina caespitosa (J. Vahl) Lge

Maps: Nannfeldt (1941), Hultén (1971a, no. 702). Total: Hultén (1958, no. 161), Hultén & Fries (1986, no. 759). Local: Benum (1958), Gjærevoll (1963b), Engelskjøn et al. (1968), Engelskjøn (1984, 1986a), Schumacher & Løkken (1981), Nordsteien (1982).

First Norwegian record: The oldest collection dates back to 1828, Boeck, Dovre (O).

Norwegian distribution: Bicentric. Found in South Norway from the central Jotunheimen to the Trollheimen. The westernmost locality is Krosshø near Grotli, and the easternmost is Rødalshø in Tynset. In North Norway, it is known from Tausa in Saltdal to North Cape.

Altitude limits: Mainly middle-alpine. Kvitingshø in the Jotunheimen 1850 m (NAS). Havyrja in Lom ca. 1000 m. Troms: Målselv, Kirkestinden 1330 m (Engelskjøn 1986a). Nissontjårro 1600 m in Torne Lappmark. North Cape 60 m.

Habitat: Most frequently found in gravelly places that become snow-free late and are strongly affected by solifluction. Hygrophilous and basiphilous.

S. caespitosa belongs to the group of northern amphi-Atlantic species. In Sweden, it is known from Lule Lappmark and Torne Lappmark, but with a considerable gap. It is also known from Svalbard (very rare), Jan Mayen, Iceland, East Greenland, and from West Greenland to Keewatin in central northern Canada. *S. caespitosa* is one of the species with a distribution that strongly indicates that they have an old history in the North Atlantic region. The centricity and isolated occurrence favour the refugium theory. The distribution in South Norway is fairly concentrated, but that in North Norway is more disjunct, and the station at North Cape is isolated.

S. caespitosa has a distinct continental tendency in both areas. The two coastal localities in Troms are both limited to outcrops of olivine-bearing rock. At one of them, Bø in Vesterålen, the species has a mass occurrence on a low-alpine outcrop of olivine pyroxenite, at an elevation of 400 m (Engelskjøn 1986a). Engelskjøn believes that *S. caespitosa* is a relict species in these areas due to

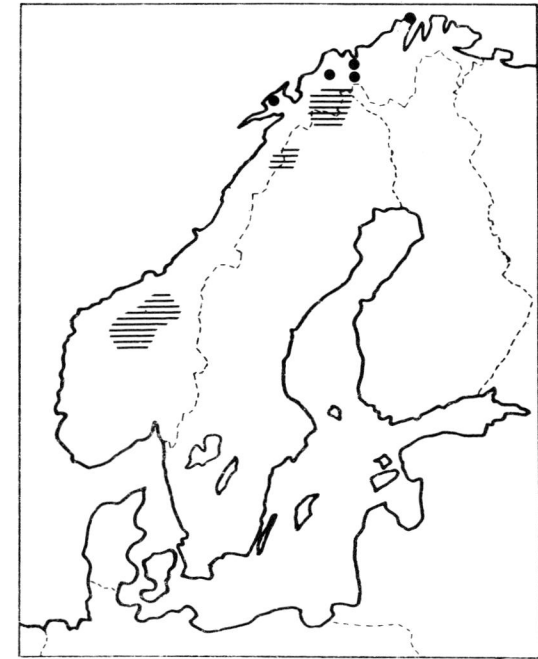

Fig. 83 The Fennoscandian distribution of *Sagina caespitosa*

the special edaphic conditions on olivine-rich bedrock.

Although the species displays considerable variation in altitude range, the great majority of localities are found in the middle-alpine belt. It most often grows on central parts of solifluction lobes where vegetation is open. It may also occur as a pioneer plant on fresh moraines consisting of schistose rocks. Usually it keeps to mica schists and similar substrate having a pH of about 6.5.

The middle-alpine localities become snow-free late, and *S. caespitosa* should be characterised as a snow-bed species. Nordhagen (1954a) looked upon the *S. caespitosa* communities as an association in their own right, *Saginetum caespitosae*. The snow-bed communities in which species like *S. caespitosa*, *Luzula arctica*, *Stellaria crassipes* and *Poa stricta* have their main occurrence are somewhat heterogeneous — as the middle-alpine occurrences often are — but they are so closely related sociologically that it is most correct to unite them in one alliance, *Luzulion arcticae* (Nordhagen 1954a, Gjærevoll 1956).

The result of 5 square analyses from 1460 m on Orkelhø in Oppdal is given in Table 17.

Table 17

Square no. (1m²)	1	2	3	4	5
Salix herbacea	.	1	.	1	1
S. polaris	1	2	2	2	1
Cerastium alpinum	1	1	1	1	1
Draba alpina	1	1	1	1	1
Minuartia biflora	1	1	.	1	.
Polygonum viviparum	1	1	1	1	1

100

Ranunculus glacialis	1	1	1	.	.
Sagina caespitosa	2	2	1	2	2
Silene acaulis	.	.	2	2	2
Carex misandra	1	1	2	1	2
Festuca vivipara	1	2	1	1	1
Juncus biglumis	1	1	1	1	1
Luzula arctica	1	.	.	1	1
L. confusa	2	1	3	2	1
Poa alpina	1	1	.	1	.
Trisetum spicatum	.	1	.	1	1
Blindia acuta	2	1	1	2	1
Distichium capillaceum	1	.	2	2	1
Drepanocladus revolvens	2	3	2	1	3
Polytrichum alpinum	1	.	.	1	1
Pohlia drummondii	1	.	.	1	1
Blepharostoma trichophyllum	1	2	.	2	2
Lophozia sudetica	1	1	.	1	1
Stones, gravel, bare soil	5	5	5	5	5

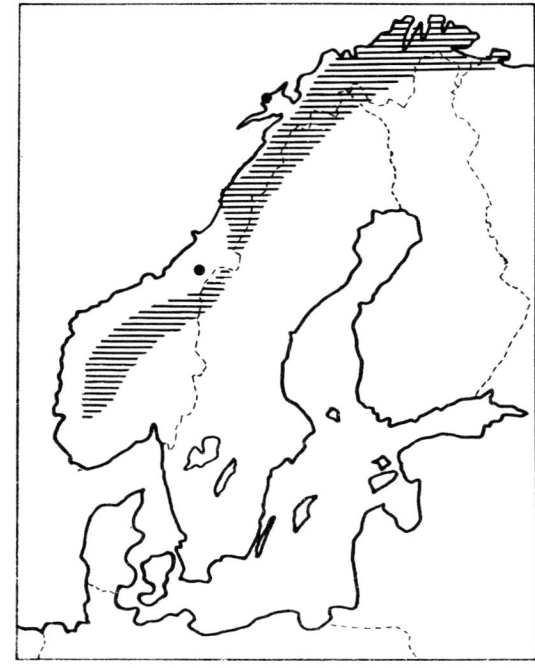

Fig. 84 The Fennoscandian distribution of *Sagina intermedia*

Sagina intermedia Fenzl

Maps: Hultén (1955, 1971a, no. 704). Total: Hultén (1968, 1971b, no. 23), Hultén & Fries (1986, no. 758). Local: Benum (1958), Lid (1959), Ryvarden (1969), Bråthen (1973), Kristensen (1981), Schumacher & Løkken (1981).

First Norwegian record: Oppdal, Kongsvold, "vid öfre delen af Sprenbäcken på Knudshö", 24.9.1837, A. E. Lindblom (LD).

Norwegian distribution: From Meien, a mountain in Aust-Agder, to Magerøya.

Altitude limits: Low-alpine to middle-alpine. To above 1800 m on Bukkeholstindane in the Jotunheimen (NAS). Down to 750 m in South Norway. Troms: Målselv, Kirkestinden 1395 m (Engelskjøn 1986a). Down to sea level in Sør-Varanger.

Habitat: Wet or irrigated snow-beds. Basiphilous.

S. intermedia is a circumpolar, arctic species. In Sweden, it is found from Härjedalen to Torne Lappmark, and in Finland, from Enontekis Lappmark to Enare Lappmark.

There is a lacuna in the distribution in the mountains of inner Sogn, and the species is rare between the Sylane and Børgefjell mountains. Its absence from inner Finnmark may be due to lack of suitable localities. *S. intermedia* is clearly continental in South Norway, and the same tendency is apparent in the northern area despite many coastal stations, particularly in Finnmark.

The species is mainly found in the upper part of the low-alpine belt and in the middle-alpine belt, and shows preference for irrigated snow-beds.

S. intermedia is a preferential species of the *Saxifrago oppositifolio-Oxyrion digynae* alliance (Gjærevoll 1956). It may also occur in gravelly places, on flat-topped summits and in places influenced by frost action and solifluction. At lower altitudes, it may occur more or less occasionally on gravel bars. It shows a clear preference for mica schists and other basicolous substrate, and may also grow on serpentinite.

Saxifraga adscendens L.

Maps: Hultén (1971a, no. 956), Knaben (1954). Total: Hultén (1958, no. 229, 1968), Hultén & Fries (1986, no. 1023). Local: Benum (1958), Gjærevoll (1963b), Toftaker (1969), Nordsteien (1982), Holten (1983), Engelskjøn (1984).

First Norwegian record: Gunnerus (1772, no. 427) named it *S. petraea* – "A me lecta in Gilleskaal norl. Maasöe vestfinm." The drawing clearly shows that it is *S. adscendens*. There are numerous collections named *S. tridactylites* in the Gunnerus herbarium, a number of which are *S. adscendens*. The oldest labelled ones are from Nordland: Sortland, Trones Præstegaard 2.7.1770 (TRH).

Norwegian distribution: In the mountains from Bykle in Aust-Agder to Rennebu, and from Grong to Salangen in Troms. There are also several localities at lower altitudes in southeast Norway.

Altitude limits: In the mountains, mainly low-alpine to northern boreal. There are no records in herbaria of localities above 1400 m (Trollheimen). Troms: Bardu, Sanjavarri 800 m.

Habitat: Screes, stream beds, rocks and ledges. Basiphilous.

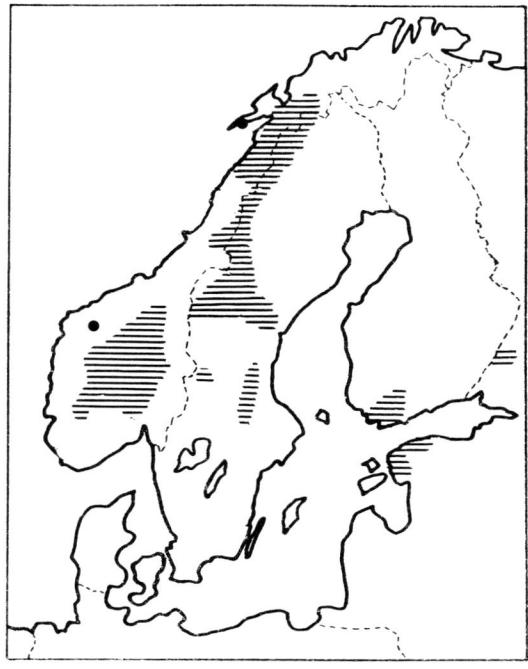

Fig. 85 The Fennoscandian distribution of *Saxifraga adscendens*

The total distribution of *S. adscendens* is very strange and disjunct. Outside Scandinavia, it occurs in the lowlands in southern Finland, Estonia and Karelia, and also in the mountains of central and southern Europe and the Caucasus. A second area is found in the Rocky Mountains, the taxon found there usually being taken as a subspecies, subsp. *oregonensis* (Raf.) Bacigalupi. The Scandinavian distribution is also unusual. Although mainly an alpine species in South Norway, it has several localities at low altitudes in the southeastern part of the country. In Sweden, it is found in the mountains from Härjedalen to Torne Lappmark. In addition, there are lowland stations south to Västmanland and an almost continuous distribution from Jämtland through Medelpad east to the Baltic Sea.

In South Norway, *S. adscendens* is continental, avoiding the westernmost mountains. In southeastern Norway, there are localities almost down to sea level. It is found down to sea level in North Norway, too, and the distribution is peculiar with an unusual northern limit in southern Troms.

The strange distribution pattern in South Norway is clearly a consequence of its ecology. It is frequently found on cliff ledges and scree slopes, and such habitats are also available in the lowlands; north-facing ravines seem to be a particularly favourite habitat for *S. adscendens*.

Nordhagen (1943) regarded *S. adscendens* as a preferential species of his *Veronico-Poion glaucae* alliance, a community found on unstable substrate and cliff ledges.

The species is calciphilous, growing on calcareous mica schists, limestone and marble. The extreme disjunction between the European *S. adscendens* and the American subsp. *oregonensis* must be older than postglacial time. Knaben (1954) viewed *S. adscendens* as belonging "to the **Arctic** group of plants that survived the last glacial period

somewhere on or near the Norwegian coast". The northern limit was explained by Engelskjøn (1984) as a postglacial migration boundary inasmuch as northern Troms and western Finnmark offer plenty of suitable habitats.

Saxifraga cernua L.

Maps: Hultén (1971a, no. 959). Total: Hultén (1968, 1971b, no. 60), Hultén & Fries (1986, no. 1032). Local: Benum (1958), Toftaker (1969), Kristensen (1981), Holten (1983, 1984).

First Norwegian record: Oeder (1761) reported it being "Alminnelig paa de norske Fielde ved smaa Bække og paa steenagtige fuktige Steder". Gunnerus (1772, no. 1006), using the name Saxifraga bulbifera, stated: "Habitat in Tromsöen norlandiæ & alibi a. 1762 & 1767, a me lecta" (TRH).

Norwegian distribution: Slightly bicentric. From Suldal in Rogaland to North Cape and Sør-Varanger.

Altitude limits: Chiefly middle-alpine, but also frequent in the high-alpine belt. Store Memurutind in the Jotunheimen 2350 m (NAS). In South Norway, occasionally in the subalpine belt along streams. The bulbils are easily transported by water. Troms: Målselv, Kirkestinden 1675 m (Engelskjøn 1986a). Down to sea level in East Finnmark.

Habitat: Snow-beds, moist, stony plateaus and slopes, and along streams. Calciphilous.

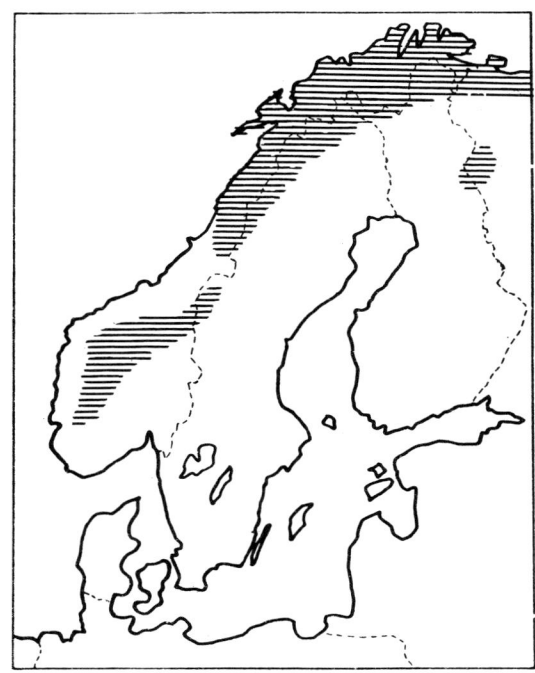

Fig. 86 The Fennoscandian distribution of *Saxifraga cernua*

S. cernua is a circumpolar, arctic-montane species. It is widely distributed in all Arctic areas, as well as Scotland, the central European and central Asian mountains, and the Rocky Mountains. In Sweden, it is known from Härjedalen to Torne Lappmark, but has a gap corresponding to that in Norway. The Finnish distribution comprises Kittilä Lappmark, Enontekis Lappmark, Enare Lappmark and Kuusamo. There is continuous distribution eastwards from Fennoscandia.

In South Norway, *S. cernua* mainly occurs in the central mountains; there are few coastal localities. In the central mountains, it often grows together with *Ranunculus glacialis*, but unlike *S. cernua* the former is frequent in the coastal mountains of Fjordane and Møre og Romsdal. The explanation for this may be that *S. cernua* is at least somewhat calciphilous. In North Norway, the distribution of the two species is more alike, apart from eastern Finnmark where *R. glacialis* is very rare.

Sociologically, *S. cernua* is a fairly characteristic species of hygrophilous and calciphilous communities, generally being found in the alliance, *Saxifrago oppositifolio-Oxyrion digynae* (Nordhagen 1943, Gjærevoll 1956). Common companions are, among others, *Saxifraga rivularis*, *S. oppositifolia*, *Oxyria digyna*, *Ranunculus nivalis*, *R. sulphureus* (North Norway), *Cerastium arcticum* and *Phippsia algida*.

Saxifraga cotyledon L.

Maps: Holmboe (1937), Hultén (1971a, no. 960). Total: Hultén & Fries (1986, no. 1035), Meusel et al. (1965), Gjærevoll (1973). Local: Benum (1958), Ouren (1966), Toftaker (1969), Holten (1981), Schumacher & Løkken (1981).

First Norwegian record: *Saxifraga cotyledon* was mentioned by Otto Sperling in connection with his journey to Valdres in 1628 ("Cotyledon"). It was also reported by Ramus (1719, cf. O. Dahl 1888-1911) "in saxis Nordmöriae". Gunnerus (1761, no. 13) recorded *S. cotyledon* from numerous localities, including a very conspicuous one between Hovin and Støren. The Gunnerus herbarium contains a labelled, undated collection enumerating several localities some of which are the same as in Flora Norvegica, but there are some additional ones (TRH). The species is illustrated in Flora Danica 5 (Oeder 1766, no. 241).

Norwegian distribution: From Rusknuten in Lund, Rogaland, to Porsa in Kvalsund, Finnmark.

Altitude limits: Mainly low-alpine to subalpine in the mountains. Reineskarvet in Ål 1450 m. Frequent in the lowlands in both South and North Norway, down to sea level.

Habitat: Mainly a chasmophyte, often on very steep cliffs. Also frequently on seashore rocks.

S. cotyledon is a European species, chiefly a Norwegian one, and has a very peculiar and disjunct distribution. By far the greatest number of localities are found in Norway, where it may be regarded as a common species in the area

where it occurs. It is then found in southeastern Iceland, the Pyrenees and the Alps.

It has a westerly distribution in Scandinavia, and is more frequent in western parts of Norway than the map perhaps gives the impression of. It is very rare in Sweden.

Sociologically, it is a characteristic species of the *Saxifragion cotyledonis* alliance (Nordhagen 1936a, 1943). On steep cliffs, *S. cotyledon* seems to be dependent on high humidity or a rich supply of moisture, either as seeping water or spray from waterfalls. Some of the most beautiful concentrations of *S. cotyledon* are found close to waterfalls. It also thrives in saltwater spray.

In central Europe, *S. cotyledon* is looked upon as an acidophilous-neutrophilous species, but in Scandinavia it would be more correct to regard it as a neutrophilous-basiphilous one. It is often found on gabbro and greenstone, especially in coastal areas. In the mountains it may grow in exposed heaths, usually in the *Kobresio-Dryadion* alliance. In the eastern part of its area, it is generally met with in ravines.

S. cotyledon may also grow on gravel bars and moraines, often close to glaciers, and thus occurs as a pioneer plant. The most famous specimen reported from Norway was found on a gravel bar in Litledalen, Sunndalen. It measured 75 cm and carried 1525 flowers.

Holmboe (1937), when discussing the distribution of *S. cotyledon* and *S. paniculata*, was of the opinion that both species survived the last glaciation near the western coast of Norway. Whereas *S. paniculata* is dependent upon calcareous rocks, *S. cotyledon* has a much wider range and was therefore able to spread more effectively.

Taking into consideration the special ecological behaviour of *S. cotyledon*, the species might have had possibilities for survival on steep rocks on the west coast. It is difficult to imagine postglacial migration to Norway from the southern occurrences.

Saxifraga foliolosa R. Br.

Maps: Hultén (1971a, no. 961). Total: Hultén (1958, no. 92, 1968), Hultén & Fries (1986, no. 1017). Local: Benum (1958), Mølster (1981), Nordsteien (1982), Schumacher & Løkken (1981), Holten (1984).

First Norwegian record: Nordland: Fauske, "I en brant midt emot Skönstugan d. 10 Juli 1807", G. Wahlenberg (UPS).

Excluded or doubtful stations: Røros 8.8.1895, A. Saxe (BG). Found in a mixed collection labelled *Saxifraga nivalis*.

Norwegian distribution: Slightly bicentric. From Såleggi in Lom to North Cape and Vadsø.

Altitude limits: Low-alpine to middle-alpine. Blåhø in Dovre 1590 m. Raudbergi in Lesja 980 m. Troms: Målselv, Kirkestinden 1675 m (Engelskjøn 1986a). Usually at high altitudes in North Norway, but down to sea level in Finnmark.

Habitat: Irrigated snow-beds, solifluction soil, banks of streams and lakes, cliff ledges and rich fens. Indifferent.

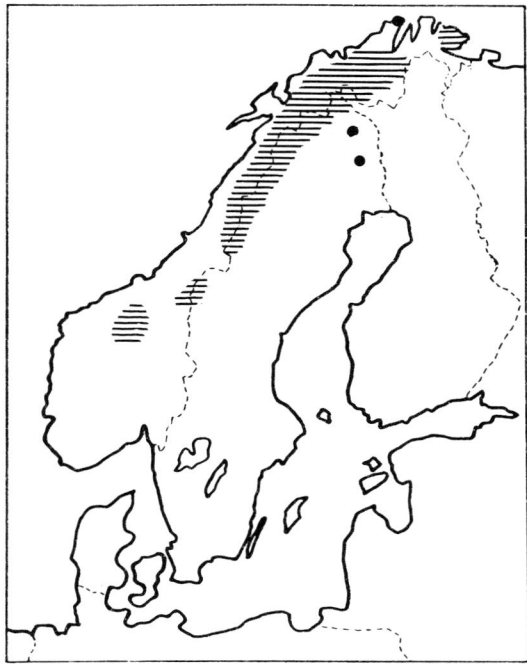

Fig. 87 The Fennoscandian distribution of *Saxifraga foliolosa*

S. foliolosa is a circumpolar, arctic plant. In Sweden, it is known from Jämtland to Torne Lappmark, and in Finland, from Enontekis Lappmark and Enare Lappmark (?). It is rare on the Kola Peninsula, but its distribution is almost continuous eastwards from Fennoscandia. It is found throughout Svalbard, but is very rare on Iceland. Its distribution in North Norway is like that of many slightly bicentric species, and there are few coastal localities. In South Norway, the picture is somewhat unusual. The species is very common in the mountains adjacent to Grimsdalen in Dovre, and also those around Lesja on the south side of the River Lågen. It is very rare or absent in areas where centric species are usually common, i.e. in the Jotunheimen, Sunndalen, Trollheimen and Drivdalen mountains.

S. foliolosa shows a preference for snow-beds and occurs in a variety of communities. It usually grows dispersed, but I have seen it in abundance in fairly open *Carex bigelowii* heaths. It is most commonly met with where moist, gravelly soil is affected by solifluction, in *Ranunculus glacialis* associations, and in various communities in the *Saxifrago oppositifolio-Oxyrion digynae* and *Luzulion arcticae* alliances.

Saxifraga hieracifolia Waldst. & Kit.

Maps: Hultén (1971a, no. 964). Total: Hultén (1968, 1971b, no. 40), Hultén & Fries (1986, no. 1014). Local:

Nordhagen (1931b, 1935, 1944), Benum (1958), Hagen (1976), Schumacher & Løkken (1981).

First Norwegian record: "Grasvigseter i Lom", C. Smith 1813 (O).

Norwegian distribution: Bicentric. In the southern area, from the central Jotunheimen north to Geitådalen in the Sunndalen mountains and west to the mountain of Litleskorkja in Ørskog. In North Norway, in two separate areas, one from Skibotndalen to Nordreisa, the other from Stjernøy to Sarabyfjell in Kvalsund.

Altitude limits: Low-alpine to middle-alpine. Smådalshø in the Jotunheimen 1675 m (NAS). In ravines, down to 440 m (Gjerdingsgjelet in Lom). Troms: Storfjord, Favresvarri 830 m.

Habitat: Damp slopes and heaths with a good snow cover, and shallow rich fens. Calciphilous.

S. hieracifolia is a circumpolar, arctic-alpine species. In addition to the Norwegian localities it is only known from a single locality in Fennoscandia, in the central part of the Kola Peninsula. In central Europe, it occurs in central France, the western Alps and the Carpathians. It is widely distributed in Svalbard, occurring very abundantly at some localities. In Greenland, it is only known from the east coast between Scoresby Sound and Wollaston Foreland, and is therefore very isolated from the localities in Arctic Canada.

The Norwegian distribution presents some interesting features. In southern Norway, the great majority of localities are found in the Jotunheimen. In occurring in this continental area, it greatly resembles numerous centric species. However, unlike those it does not occur in the Drivdalen mountains or the Trollheimen, which is surprising since ecological conditions seem favourable, especially in the Drivdalen area. Contrary to most centric species, it also has some localities closer to the coast.

Nordhagen (1931b, 1935, 1944) paid great attention to the coastal localities in connection with the possible survival of *S. hieracifolia* and other species on coastal nunataks. The geologist, Anders M. Heltzen, who discovered *S. hieracifolia* on Litleskorkja in the Lauparen area between Romsdal and Sunnmøre, concluded (Heltzen 1949) that the group of mountains to which Litleskorkja belongs had been a nunatak area: "The highest upper limit of the glacier surface during the last ice age cannot have exceeded some 1250 m a.s.l." In Svalbard, *S. hieracifolia* grows very abundantly on Ossian Sars, a semi-nunatak in Kongsfjorden.

The distribution in northern Norway is also peculiar, consisting of two areas, one fairly continental, the other coastal.

S. hieracifolia shows a preference for damp slopes, often where there is a well-developed moss carpet. It needs a substantial snow cover, but is not a snow-bed species. In the Sunndalen mountains, *S. hieracifolia* is only known from Geitådalen where it grows abundantly in north-facing *Dryas* heaths. The same is found at several localities in the Jotunheimen. Otherwise, it grows on the banks of lakes and streams, and in the Jotunheimen is found in several ravines and canyons down into the boreal belt. It always keeps to calcareous schists.

Saxifraga paniculata Miller

Maps: Nordhagen (1965a), Hultén (1971a, no. 958). Total: Löve & Löve (1951), Hultén (1958, no. 25), Hultén & Fries (1986, no. 1036). Local: Nordhagen (1965a), Ryvarden (1969).

First Norwegian record: Nordland: Saltdal, Balvatnet, Lars Levi Læstadius 1825 (GB).

Norwegian distribution: In three isolated, restricted areas. 1. Ryfylke: Hjelmeland-Suldal. 2. Nordland: Saltdal, Balvatnet area. 3. Troms: Nordreisa, Lulisfjell.

Altitude limits: Hjelmeland, Nordre Svartheii 1100 m; not below 950 m in Ryfylke. Between 700 and 750 m in the Balvatnet area. 330 m on Lulisfjell.

Habitat: In the Ryfylke area, it is a typical chasmophyte, growing on steep cliffs. In Saltdal, it occurs on flat heaths. On Lulisfjell, it is found on exposed, stony heaths. It occurs on calcareous soils in all three areas.

S. paniculata belongs to the amphi-Atlantic plants. It is widespread in the mountains of central and southern Europe and in the Caucasus. It is also known from Iceland, Greenland and North America east to Saskatchewan.

The Saltdal occurrence is restricted to a small area near Balvatnet, and the plant is particularly abundant on Rosnevarri. In 1989, a locality was surprisingly discovered in Nordreisa, on Lulisfjell, where the species occurs in thousands of individuals, but in a very restricted area. The southern area, where the species was first discovered by Ove Dahl in 1905, is somewhat larger and there are numerous localities in Hjelmeland and Suldal.

Both Neuman (1905) and Nordhagen (1965a) maintained that there is a racial difference between the Ryfylke and Saltdal populations. Neuman described the northern race as subsp. *laestadii*. *S. paniculata* is a variable species, and subsp. *laestadii* seems to fall within the range of variation of the central European material. The plants in Nordreisa have a typical appearance.

There is, however, a striking ecological difference between the southern and northern populations. In Ryfylke, *S. paniculata*, with few exceptions, grows on steep cliffs together with several other chasmophytes. To my knowledge, it grows on exposed ridges and rocks, and fairly often in rock crevices, in central Europe. It has been assumed that the Ryfylke population could have immigrated from the low-lying land area that occupied some of the present North Sea during the last glaciation. It is, however, difficult to see how a chasmophyte like *S. paniculata* could find suitable localities on that flat area. It should be added that *Artemisia norvegica* and *Oxytropis campestris* have isolated stations in the same part of southwest Norway, likewise some other interesting species such as *Potentilla nivea* and *Primula scandinavica*.

In the Saltdal area, the species always grows in exposed heath communities, never closed ones. It inhabits open patches, and obviously has weak competitive ability. This corresponds with its ecological behaviour in Iceland and Greenland. According to Nordhagen (1965a), the heaths containing *S. paniculata* clearly belong to the *Kobresio-Dryadion* alliance. It avoids the most windswept flats. The species cannot tolerate a long-lasting snow cover, or moist conditions. Table 18 gives an extract of 10 square analyses (1 m²) carried out by Nordhagen on Rosnevarri.

Table 18

Square no. (1 m²)	1	2	3	4	5	6	7	8	9	10
Saxifraga paniculata	3	3	2	3	3	2	1	3	3	2
Empetrum hermaphroditum	.	.	1	1	.	+	1	1	.	1
Dryas octopetala	3	2	2	3	2	1	2	2	1	3
Cassiope tetragona	.	.	1	1	.	+	1	1	.	1
Salix reticulata	.	.	1	1	.	.	.	1	.	.
Antennaria dioica	.	1	+	.	1	1	1	.	.	.
Chamorchis alpina	.	.	+	.	1	.	.	1	.	.
Polygonum viviparum	.	.	+	1	1	.	.	1	1	1
Saxifraga oppositifolia	1	1	1	1	1	1	1	2	1	1
Carex rupestris	2	2	3	2	3	3	4	4	2	2
Festuca ovina	1	1	1	1	1	1	1	1	1	+
Luzula spicata	+	1	1	1	1	+	1	2	1	1
Cetraria nivalis	2	1	2	2	3	3	2	3	2	2
C. islandica	1	1	1	1	1	1	1	1	1	1
Sphaerophorus globosus	+	.	1	.	2	1	1	.	1	.
Cornicularia divergens	1	.	.	.	1	.	1	1	1	1
Solorina saccata	1	+	1	1	1	1	.	1	.	1
Thamnolia vermicularis	1	1	1	1	1	1	1	1	2	1
Ochrolechia frigida	3	1	3	2	3	1	3	2	3	2
Cladonia coccifera	1	.	.	1	1	.	1	1	+	1
C. gracilis	.	1	.	1	1	+	1	1	1	+
Encalypta rhapdocarpa	1	1	1	1	1	1	1	1	1	1
Polytrichum juniperinum	2	+	1	1	1	1	2	1	1	1
Racomitrium lanuginosum	1	.	+	.	2	2	1	.	2	3
Gymnomitrion corallioides	.	.	.	.	1	2	1	2	3	3

The Saltdal population is situated in an area containing numerous amphi-Atlantic species, e.g. *Carex scirpoidea*, *Braya linearis*, *Draba crassifolia*, *Sagina caespitosa*, *Arenaria humifusa* and *Potentilla hyparctica*. Nordhagen (1965a) assumed that a coastal refugium was located further west. O. T. Grønlie (1927) maintained that there might have been ice-free areas further inland and that Saltfjorden served as a drainage channel for the inland glaciers, leaving ice-free areas on both sides.

At the Nordreisa locality, *S. paniculata* is accompanied by, among others, *Carex capillaris*, *C. misandra*, *Chamorchis alpina*, *Dryas octopetala*, *Rhododendron lapponicum*, *Salix reticulata*, *Saxifraga oppositifolia*, *Silene acaulis* and *Woodsia glabella*; *Braya linearis* was found nearby. The new locality at Nordreisa is very interesting. This area is rich in centric and amphi-Atlantic species. In particular, there is the extremely isolated occurrence of *Nigritella nigra* on Balgesoai've and the endemic *Papaver avkoënse* on Avvko, both these being neighbouring mountains to Lulisfjell. Bearing in mind the present situation on the Greenland coast, where glaciers fill the valleys and inner fjords, it is easy to imagine a similar situation with glaciers moving down the Nordreisa and Kvænangen valleys leaving ice-free refugia in the intervening area.

Leif Ryvarden

Scirpus pumilus Vahl

Maps: Nordhagen (1935, 1963a), Hultén (1971a, no. 302). Total: Hultén (1958, no. 227), Gjærevoll (1973), Hultén & Fries (1986, no. 407). Local: Nordhagen (1935, 1963a), Benum (1958), Gjærevoll (1963b).

First Norwegian record: Finnmark: Porsanger, Børselv, 15-17.9. 1864, J. M. Norman (O).

Norwegian distribution: Bicentric. In South Norway, from Grimsdalen in Dovre to Grønbakken in Oppdal. In North Norway, from Leirbuktfjell in Nordreisa to Børselv in Porsanger.

Altitude limits: Between 760 m (Dalholen) and 980 m (Grønbakken) in South Norway. Troms: Nordreisa, Leirbuktfjell 400 m. Most localities near sea level.

Habitat: Shallow eutrophic fens, dolomite terraces and river flats. Calciphilous.

S. pumilus was regarded by Hultén (1958) as an amphi-Atlantic species. Hultén & Fries (1986) pointed out a tendency towards a circumpolar distribution. The species has a highly disjunct distribution in boreal-montane areas in Eurasia and North America. It is known from the central European and central Asian mountains. Fernald (1943) regarded the American strains as a species of their own (*S. rollandii* Fernald). Printz (1921) held that the Siberian biotypes, named *S. pumilus* subsp. *oliganthus* (C. A. Meyer) Printz, are "distinctly divergent from the European forms". Hylander (1945) objected to splitting the species, arguing that mainly modificative characters are involved. Counts from South and North Norway show the same chromosome number, 2n = 76-78 (Engelskjøn & Knaben 1974).

Apart from the localities in the dolomite area in Porsanger (Kolvik, Reinøya and Børselv) all the stations have been discovered after 1939. The species was first discovered in South Norway in 1968. *S. pumilus* is easily overlooked and more stations will probably be found. The locality at Storvik-Kidelv in Alta has been added recently.

Although *S. pumilus* occurs at Børselv on river flats reached by tidal water, on subalpine eutrophic river flats in Grimsdalen, and in alpine and subalpine eutrophic fens at several localities, the habitats have much in common.

In North Norway, mixed assemblages of hygrophytes, mesophytes, and even xerophytes, are met with, including, *Carex capillaris*, *C. microglochin*, *Equisetum variegatum*, *Saxifraga aizoides*, *Selaginella selaginoides*, *Tofieldia pusilla*, *Festuca ovina* and *Dryas octopetala* (Engelskjøn & Skifte 1987). Nordhagen (1935) published analyses of a *Scirpus pumilus* sociation from Børselv. These show a fairly close relationship with the sociological behaviour of *S. pumilus* in South Norway where the habitats are dominated by hygrophytic and mesophytic species such as *Selaginella selaginoides*, *Carex capillaris*, *C. vaginata*, *Thalictrum alpinum* and *Saussurea alpina*. Important companions are *Kobresia simpliciuscula*, *Carex capitata*, *C. microglochin* and *Juncus arcticus*. This community seems to fall within the present concept of the *Caricion bicolori-atrofuscae* alliance.

Soil analyses from South Norway show pH between 6.4 and 6.9, and from North Norway between 6.8 and 7.5 (Engelskjøn & Skifte 1987). A test from the community analysed by Nordhagen (1935) gave 7.02.

Ola Skifte

Sedum villosum L.

Maps: Hultén (1971a, no. 952). Total: Hultén (1955, 1958, no. 94), Meusel et al. (1965), Hultén & Fries (1986, no. 1010). Local: Benum (1958), Engelskjøn (1984).

First Norwegian record: Oeder (1761, no. 24) reported it from "Præstesetter i Slidre i Valders tæt ved Sætterhusene". The oldest collection dates back to G. Wahlenberg: "Vid Saltenfjord d. 19 Juli 1807" (UPS). More information is given by Wahlenberg (1812): "A me lectum in peninsula Öneset sinus Saltensis frequenter".

Excluded or doubtful stations: Haukelifjell 15.7.1893, F. Svendsen (C). There have been no other reports from that area.

Norwegian distribution: Slightly bicentric. From Ullensvang to Vågå. Tydal. From Røyrvik to Magerøya, with disjunctions.

Altitude limits: Mainly low-alpine. Øystre Bukkeholstindane in the Jotunheimen 1880 m (NAS). Down to sea level in Sogn. Troms: Bardu, Gævdnjajaure 620 m. Mainly at low altitudes in North Norway, often down to sea level.

Habitat: Springs and wet cliff ledges. Calciphilous.

S. villosum belongs to the amphi-Atlantic, arctic-montane species. It is widely distributed in central Europe and occurs in the British Isles, Iceland, southernmost parts of East and West Greenland and at one locality in Labrador. In Sweden, a few localities are known from Jämtland to Torne Lappmark. In Finland, it is only known from Kittilä Lappmark.

The distribution of *S. villosum* has many strange features. The plant grows frequently and abundantly in the mountains of much of central Norway, particularly the

Valdres-Jotunheimen area, but is curiously absent from the Drivdalen, Sunndalen and Trollheimen mountains. Its occurrence in South Norway is mainly continental. This is not so in North Norway, and if the total northern Scandinavian distribution is considered it is found that all the Swedish localities are situated in the western part of the mountains, close to the Norwegian border.

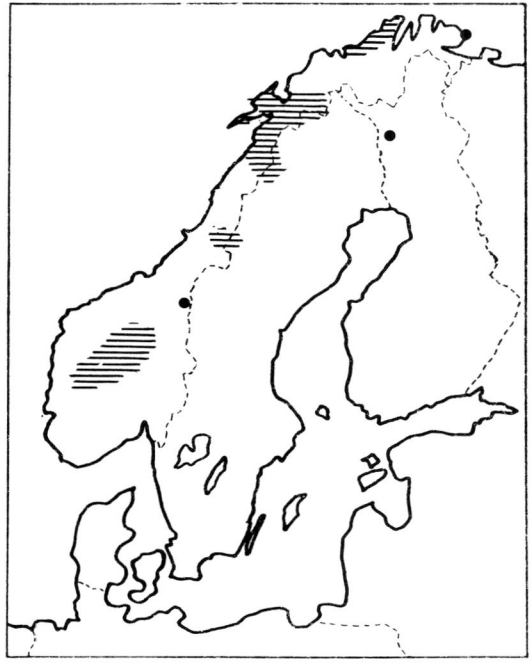

Fig. 88 The Fennoscandian distribution of *Sedum villosum*

Table 19

| Stand no. | I | | II | | III | IV | V | | VI | | VII | | VIII | | |
Square no. 1 m²)	1	2	3	4	5	6	7	8	9	10	11	12	13	14	15
Cerastium cerastoides	3	1	1	1	3	1	1	1	1	1	1	1	1	1	1
Epilobium hornemannii	.	.	1	.	1	1	1	1	.	.	1	1	.	.	.
E. palustre	1	1	.	.	2	1	.	1	.	1	.	1	1	.	1
Equisetum arvense	3	1	1	1	1	1	1	1	1	1	1	1	2	2	1
Montia fontana	3	2	1	1	5	3	2	2	2	2	1	2	3	2	1
Polygonum viviparum	+	.	.	+	.	1	1	.	2	1	1	1	1	+	+
Saxifraga stellaris	1	1	1	2	2	.	1	.	1	1	1	2	2	1	1
Sedum villosum	3	3	2	3	3	3	3	2	4	4	4	4	3	3	2
Agrostis capillaris	1	1	3	1	.	.	.	1	2	1	2	1	1	1	.
Deschampsia cespitosa	1	1	3	2	1	1	1	2	1	1	1	2	1	1	2
Phleum alpinum	2	2	1	1	2	1	2	4	2	1	1	1	1	1	1
Poa annua	2	1	1	1	.	2	1	1	1	2	.	.	1	1	1
P. alpigena	3	2	1	1	.	.	.	.	1	1	1	.	1	1	.
Carex canescens	1	1	.	.	.	1	.	.	1	.	+	.	.	1	1
C. nigra	1	4	.	3	.	4	.	.	3	4	2	4	4	4	4
C. bigelowii	.	.	2	1	.	.	.	.	1	.	1	1	.	.	.
Eriophorum angustifolium	.	.	.	.	1	1	1	2	1	.	3	2	.	.	1
Cinclidium subrotundum	.	.	1	.	.	.	.	.	1	1	1	1	1	1	1
Philonotis fontana	4	2	4	4	5	3	5	3	2	3	1	4	2	4	1
Tayloria lingulata	1	1	2	1	.	1	3	3	4	2	2	1	2	2	1
Scapania paludosa	.	.	.	.	.	1	.	1	.	2	1	1	1	1	1

Its ecological behaviour is very variable, but the plant is first and foremost found in springs. Nordhagen (1943) regarded it as a characteristic species of his *Mniobryo-Epilobion hornemannii* alliance. He carried out analyses from an *S. villosum-Philonotis fontana* sociation in this alliance, which illustrate the rich occurrence of *S. villosum* in many places. Table 19 gives an extract of this investigation carried out on 8 stands at ca. 1000 m a.s.l. Mass occurrence has also been reported by Benum (1958) from Tvibotntjønna in Bardu, which is a small lake only fed by springs. Such mass occurrence is at least partly due to efficient vegetative propagation by offsets (Nordhagen 1923, Knaben 1966a). The plant may be found far down into the forest belts along streams originating from springs.

S. villosum is a calciphilous species and is also nitrophilous. This is evident both near upland, summer farms and in the vicinity of bird biotopes in North Norway. In the latter area, it is found on coastal mountains in places fertilised by birds and also in spring-like or fen-like shore communities (Norman 1894-1901, Engelskjøn 1970). It may also occur on permanently wet cliffs both in the mountains and at low altitudes, even close to sea level in Sogn.

Silene acaulis (L.) Jacq.

Maps: Hultén (1971a, no. 738). Total: Hultén (1958, no. 180, 1968), Meusel et al. (1965), Hultén & Fries (1986, no. 791).

First Norwegian record: Oeder (1761, no. 21) reported it as "Meget alm. i Norge, i Nordlandene ned til havet". Gunnerus (1766, no. 117) stated: "Habitat ut alpes taceam, in Maasöen alibique in Finmarchiæ, ubi a 1759, initio mensis Julii florens a me lectus est". There are several collections in the Gunnerus herbarium (TRH).

Norwegian distribution: Throughout the mountain range from Bjerkreim in Rogaland to Magerøya and Sør-Varanger.

Altitude limits: Low-alpine to middle-alpine. Ymisfjell in the Jotunheimen 2210 m (NAS). Down to sea level in western and northern Norway. Troms: Målselv, Kirkestinden 1530 m (Engelskjøn 1986a).

Habitat: The ecological range is very wide, embracing *Dryas octopetala* heaths, barren, gravelly ridges and plateaus, cliff ledges, rich fens, scree slopes, grassy hillsides and river banks. Most common on calcareous soils.

S. acaulis is an arctic-montane species with an amphi-Atlantic distribution that is somewhat unusual. The species is widely distributed in North America and also occurs in easternmost Siberia and Kamchatka. There is a large gap from there to Novaya Zemlya and the Ural mountains. In Sweden, it is known from the whole mountain range, and in Finland, from Enontekis Lappmark and Sompio Lappmark. In central Europe, it is found from northern Spain to

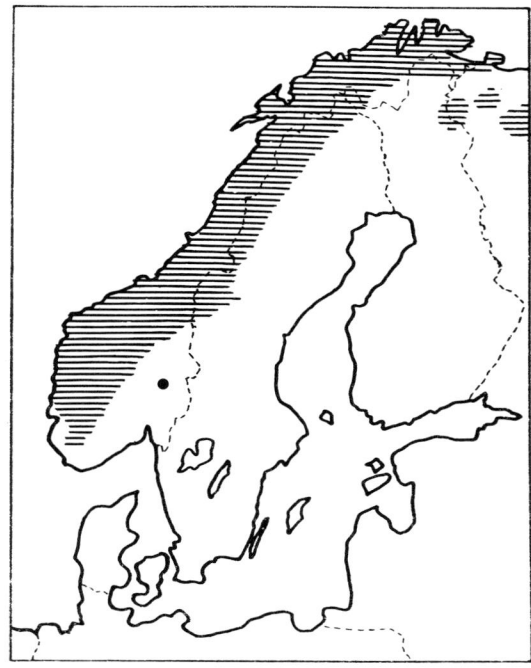

Fig. 89 The Fennoscandian distribution of *Silene acaulis*

the Carpathians. It occurs in the British Isles, and is very common in Svalbard.

S. acaulis grows in a variety of plant communities from exposed heaths to snow-beds, both in dry and wet habitats, and from the low-alpine to the high-alpine belt. It prefers calcareous soils, but also grows in acid ones with a minimum pH of about 5.0. Particularly in the middle-alpine belt on mountains consisting of calcareous schist, it is difficult to find a community in which *S. acaulis* does not occur. Contrary to most calciphilous alpine species, it is frequently found in the coastal mountains.

S. acaulis is often found at low levels along streams and rivers in company with, among others, *Saxifraga aizoides*. It also occurs as a pioneer plant on fresh moraines.

Silene wahlbergella Chowd.

(Syn. *Melandrium apetalum* (L.) Fenzl., *Silene uralensis* (Rupr.) Bocq. subsp. *apetala* (L.) Bocq.)

Maps: Hultén (1971a, no. 749). Total: Hultén (1968, 1971b, no. 54), Hultén & Fries (1986, no. 785). Local: Ouren (1952, 1966), Benum (1958), Lid (1959), Toftaker (1969), Skogen (1974), Hagen (1976), Elven (1979), Nordsteien (1982).

First Norwegian record: Gunnerus (1772 no. 680) reported the species from Nordland and Finnmark, but gave no localities and there are no specimens in the Gunnerus herbarium. There is a drawing of it in Flora Danica 14 (1780): "Hist og her, men sparsom, paa de høieste Fielde i Norge". The oldest herbarium collection dates back to C.

Smith (O), but lacks locality information. He reported it from Tron, Alvdal 1807 (see O. Dahl 1895), as did Hornemann (1807).

Excluded or doubtful stations: (1) Sør-Trøndelag: Tydal mountains. This locality was reported by Hornemann (1821), but no collection exists. A potential station in the vicinity might be the rich Sylane mountains, but it has not been reported there following very thorough investigations by Nordhagen (1928) or work by other botanists.

(2) Nordland: Hattfjelldal. In the first edition of Lid's Norsk Flora (Lid 1944), Hattfjelldal was given as a station, and this has been repeated in subsequent editions and other floras. However, no herbarium specimens exist and there is no written information about the species from Hattfjelldal. Nordhagen (1940) gave the northern area as Salten to West Finnmark, but in 1952 he mentioned Hattfjelldal (Nordhagen 1950-58), perhaps based on Lid (1944).

(3) Nordland: Saltdal, Balvatnet. Læstadius (1826) reported *Lychnis apetala* from the south side of Balvatn, arising from his journey in 1825. There are no herbarium specimens. Although there have been no subsequent reports of the species from this very well investigated area (see also Engegård 1970), it is difficult to question the report of Læstadius.

(4) East Finnmark. In herb. O there is a collection labelled "Finnmarkia orientali", herb. LD has one from "Varanger", and herb. UPS has one labelled "Norriges Fællesdistrict allt bort åt Kola allmän (Wg. scr.)", all are leg. P. V. Deinboll. These reports have been questioned. Blytt (1874) supposed that the species referred to might be *Vahlbergella affinis* (*Silene furcata* subsp. *angustifolia*), but this is unlikely as that species is not known from this area. O. Dahl (1934) considered the report to be dubious. It should, however, be borne in mind that *S. wahlbergella* is known from the Rybachiy Peninsula and the mouth of the River Kola. In 1820, Deinboll made a journey from East Finnmark to the Kola Peninsula travelling by sea via "Pejzen and Bomeni til Kola". According to the map of Russian Lapland (J. A. Friis 1871), Peisen Fjord is the old name for Petsamo Fjord (now changed again, to Pechenga Fjord). "Bomeni-ejd", mentioned in the travel itinerary, is the narrow isthmus connecting the Rybachiy Peninsula to the mainland. Deinboll did not travel round this peninsula, but followed the shorter route to 'Peisen Fjord' and then crossed the isthmus.

In a letter to J. W. Hornemann, dated 1822, Deinboll wrote: " De sjeldneste af de Planter, jeg paa denne Rejse forefandt, ere: Cineraria campestris (Retzii) [= Senecio integrifolius] der, saavidt jeg ved, ikke før er fundet saa højt i Norden. Den voxer paa de sydlige Side af Varanger-fjorden, især ved Bomeniejd og siden paa flere Steder henimot Kola. Primula integrifolia (Gun.) fandtes paa 3 Steder. Gentiana serrata ß detonsa især ved Pejzens Fjæld. Gentiana involucrata, men hyppig med blaae Blomster, var almindelig paa alle Strandbredder, Pulmonaria maritima ligeledes, og Lychnis apetala, samt Pisum maritimum". It is quite evident that Varanger, in the sense of Deinboll, included areas of Russian Lapland and that all the stations mentioned above most likely refer to this area. From an ecological point of view it may sound strange that *Lychnis apetala* is mentioned together with seashore plants, but the distance between seashore species and alpine ones may, in this area, be a question of a few steps.

Norwegian distribution: Bicentric. In South Norway, from Sesnuten in Bykle/Vinje to Rindal and Singsås. In North Norway, from Balvatn (?) and Langfjelltind in Skjerstad to Talvik in Alta.

Altitude limits: Low-alpine to middle-alpine. The Jotunheimen 1970 m (Torfinnstind and Øystre Bukkeholstindane (NAS)); down to 400 m in Vågå. In inner Troms up to 1080 m (Engelskjøn 1984). Down to almost sea level in Talvik.

Habitat: In damp heaths and meadows, and shallow rich fens, on solifluction soil. Calciphilous.

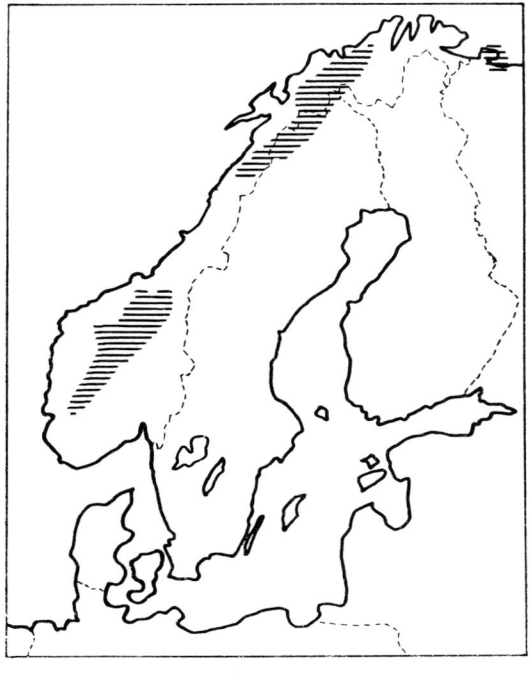

Fig. 90 The Fennoscandian distribution of *Silene wahlbergella*

Bocquet (1969) took the Fennoscandian plant to be a subspecies of a circumpolar taxon, *S. uralensis*. According to him, subsp. *apetala* should occur in Fennoscandia and southwest Alaska. I prefer to follow Hultén (1971b) and Hultén & Fries (1986), who viewed *S. wahlbergella* as a circumpolar species and the Fennoscandian plant as subsp. *wahlbergella*, endemic to this area. In Sweden, it is known from Pite Lappmark to Torne Lappmark, and in Finland, from Enontekis Lappmark. Then there is an isolated occurrence in Russian Lapland.

Compared with most bicentric species, it extends further east in the southern area. In the northern area, it shows almost the same distribution as *Cassiope tetragona*. In both

areas, the distribution displays a clear continental tendency. In South Norway, there is a single coastal locality at Storfjellet in Norddal.

Usually, *S. wahlbergella* grows dispersed, without any special sociological preference. It occurs in various kinds of damp habitats, in particular meadows, and also in *Festuca ovina*, *Poa arctica*, *Dryas octopetala* and *Rhododendron lapponicum* heaths, preferably on north-facing slopes. It may be regarded as an important species of the *Caricion bicolori-atrofuscae* alliance (cf. Nordhagen 1936a, 1943).

In the southern area, it is frequently met with in the low-alpine and middle-alpine belts, but in the northern area it is mainly low-alpine. It is also quite often found at lower levels, e.g. in pine forests, and in many places it follows stream beds down to fairly low altitudes. It may even occur as an apophyte in cultivated meadows. The species is calciphilous. Lunde (1962) made 114 tests in the northern area, obtaining a mean of pH = 6.21. It prefers mica schists and has also been reported from dolomite.

Stellaria crassipes Hultén

Maps: Nordhagen (1939), Hultén (1971a, no. 677). Total: Hultén (1958, no. 7), Hultén & Fries (1986, no. 732). Local: Gjærevoll (1963b).

First Norwegian record: Finnmark: Alta, Talvik, Vassbotnfjellet 1868. J. E. Zetterstedt (O). In the southern area: Oppdal, Knutshø 1909, E. Ekman (O).

Norwegian distribution: Bicentric. In South Norway, it is known from the mountains of Søndre, Midtre and Nordre Knutshø, Finshø and Sissihø. In North Norway, it is restricted to Jav'reoai'vit in Nordreisa and Vassbotnfjellet in Talvik.

Altitude limits: Middle-alpine. In South Norway, it has been reported from 1380 m (Finshø)) to 1690 m (Søndre Knutshø).

Habitat: The ecological range of the species is very narrow. It grows on solifluction soil, preferably on moss-covered margins of small solifluction lobes or patches of patterned ground.

S. crassipes belongs to the very critical, high-arctic, *S. longipes* group that has been studied by several workers (Hultén 1943c, Böcher 1951a, Porsild 1963). In agreement with Hultén & Fries (1986), the distribution of *S. crassipes* is looked upon as being circumpolar, but much uncertainty still remains. In Sweden, the species is known from the mountains of Nissontjårro and Pältsan. The Scandinavian occurrence is very isolated. The nearest localities are found in Svalbard and on Novaya Zemlya. The south Norwegian plant is taken as a variety, var. *dovrensis* (Hultén 1943c). It is dark green and somewhat hairy, whereas the northern plant is light blue-green and glabrous.

The bicentricity and occurrence of two races in Scandinavia were regarded by Nordhagen (1939) as the result of geographical separation and isolation during the last glaciation. Var. *dovrensis* rarely flowers, and propagation by seed has not been observed. Dispersal takes place by means of runners and the breaking-off of shoots. Taking into consideration its poor dispersal ability and mainly middle-alpine occurrence, Gjærevoll & Sørensen (1954) thought that var. *dovrensis* was an impoverished relic that might have survived on inland nunataks.

Ecologically, there is no difference between the southern and northern plants. Nordhagen (1943) regarded the species as a typical snow-bed plant growing on moist, schistose material affected by solifluction. This corresponds with my own experience. In the Knutshø area, where the species is fairly common above 1500 m, it is confined to gently sloping ground affected by solifluction and frost heaving. It may also be found on the relatively earthy strip between stone stripes and in patches on stabilised frost polygons. It avoids irrigated areas and the wet, active centres of polygons and solifluction lobes.

Sociologically, it is a characteristic species of the *Luzulion arcticae* alliance, in close relationship with communities containing *Poa stricta*, *Sagina caespitosa* and *Luzula arctica*.

Taraxacum dovrense (Dt.) Dt.

Maps: Dahlstedt (1928), Hultén (1971a, no. 1791), Gjærevoll (1963b, 1973). Total: Hultén & Fries (1986, no. 1922). Local: Schumacher & Løkken (1981), Nordsteien (1982).

First Norwegian record: Oppdal, Knutshø 1870, F. Ahlberg (Dahlstedt (1908).

Norwegian distribution: Endemic to the area mapped here.

Altitude limits: Middle-alpine. Ryggehø in the Jotunheimen 2045 m (NAS); down to 1200 m.

Habitat: Exposed heaths, calciphilous.

In 1870 and 1875, the Swedish botanist, F. Ahlberg, made the first collections of this *Taraxacum*, regarding it as a variety of *T. officinale* (Web.) Wigg. Dahlstedt (1908) described it as a subspecies of *T. reichenbachii* Huter (Dt.), subsp. *dovrense*, but subsequently gave it species rank (Dahlstedt 1928). *T. dovrense* belongs to the *Arctica* group, but is most closely related to *T. reichenbachii* from the Brenner Alps in Austria.

The distribution of *T. dovrense* is very restricted, and is divided between three areas. The species has a very concentrated occurrence in the Jotunheimen. Then there is a considerable gap to the next area which stretches from the mountains of the Grimsdalen district to Sissihø in Oppdal and westwards to Drugshø. The third area is a small one in the Trollheimen (Blåhø and Gjevilvasskammene). In

some places, such as Nordre Knutshø, it may be quite abundant.

Dahlstedt (1928) looked upon *T. dovrense* and *T. reichenbachii* as different taxa with a common origin, and thought that their differentiation must have taken place early. He also maintained that *T. dovrense* was more extensively distributed in the Norwegian mountains during the last interglacial and that it survived the last glaciation on nunataks. If *T. dovrense* had had close relatives in Norway it might have been a postglacial mutant. This solution seems very unlikely.

In contrast to all other *Taraxacum* species, which must be considered as mesophilous plants, *T. dovrense* grows in exposed lichen heaths. It always occurs on calcareous schists, together with species like *Campanula uniflora*, *Artemisia norvegica*, *Kobresia myosuroides*, *Carex rupestris*, *Draba fladnizensis*, *D. cacuminum* and *Papaver radicatum*.

It is a middle-alpine plant, rarely found below 1350 m. Ecologically, it may be expected to occur as a nunatak plant. Sørensen's reports of finding it on Ryggehø (2045 m) and Store Skagastølstind (2020 m) support this assumption.

Vahlodea atropurpurea
(Wahlenb.) Hartm.

Maps: Hultén (1971a, no. 178). Total: Hultén (1958, no. 186, 1968), Hultén & Fries (1986, no. 341). Local: Benum (1958), Ouren (1966), Toftaker (1969), Kristensen (1981), Holten (1984).

First Norwegian record: A drawing of the species is found in Flora Danica 17 (1790, no. 961) where it was named by Vahl, *Aira alpina*: "Paa de høieste Fielde i Wangs Præstegield i Walders, saaog i Finmarken i moradsige Egne især". Wahlenberg (1812) named it *Aira atropurpurea*, describing it on the basis of material collected by Vahl in Finnmark.

Norwegian distribution: Slightly bicentric. In the southern area, from Bygland in Aust-Agder to Fongen, a mountain between Tydal and Meråker. In the northern area, from Lierne to the Nordkinn Peninsula and Grense Jakobselv.

Altitude limits: Low-alpine to middle-alpine. Up to 1600-1700 m in the Jotunheimen (Søre Tverrhyttindane, Søre Uradalstindane, NAS). Haltdalen 470 m; frequently down to 800 m. Troms: Målselv, Vuomajav'ri 960 m. Down to sea level in Finnmark.

Habitat: Moist snow-beds, willow thickets and fen margins. Acidophilous.

V. atropurpurea s. str. is an amphi-Atlantic species. In addition to the Norwegian distribution, it is known from Dalarna to Torne Lappmark in Sweden (with a gap corresponding to that in central Norway). In Finland, it grows

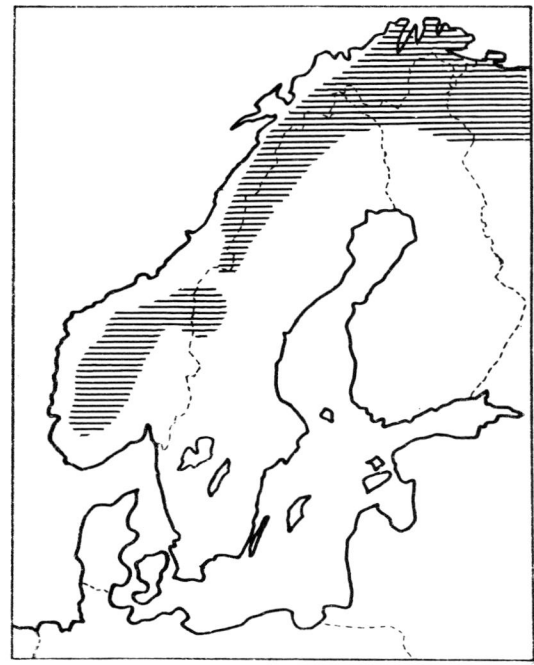

Fig. 91 The Fennoscandian distribution of *Vahlodea atropurpurea*

Veronica fruticans Jacq.

Maps: Hultén (1971a, no. 1546). Total: Hultén (1958, no. 70), Meusel et al. (1978), Hultén & Fries (1986, no. 1643). Local: Benum (1958), Ryvarden (1969), Toftaker (1969), Bråthen (1973), Kristensen (1981), Nordsteien (1982), Holten (1983).

First Norwegian record: Troms: Karlsøy, 18.7.1767. J. E. Gunnerus (TRH).

Norwegian distribution: From Hjelmeland in Rogaland to Magerøya and Sør-Varanger.

Altitude limits: Mainly low-alpine. Between 1800 and 1900 m in the Jotunheimen (Lauvhø, Søre Tverrbotntindane, Øystre Bukkeholstindane (NAS)). Down to 250 m in Nord-Fron, 150 m in Jølster and sea level in North Norway. Troms: Målselv, Njunis 1040 m (Engelskjøn 1986a).

Habitat: Rocks and ridges, cliff ledges, sunny dry slopes, and screes. Calciphilous.

south to Kuusamo. It is found scattered east to the Urals, and also occurs in southern Greenland and eastern Canada. From western North America to Kamchatka there is another population, subsp. *paramushirensis* (Kudo) Hult. The species is bipolar as it also grows in southernmost South America (subsp. *magellanica* (Hook. fil) Hult.).

The distribution in Norway is clearly continental, which is interesting since *V. atropurpurea* is an acidophilous species in contrast to most species with a similar distribution. The geological conditions should be favourable in the western mountains. It is very rare in the Trollheimen.

V. atropurpurea is most abundant in late snow-beds. Analyses of *V. atropurpurea* communities (Nordhagen 1943, Gjærevoll 1956) show that they belong to the *Nardo-Caricion bigelowii* alliance. The species has apparently found an ecological niche in snow-beds where the moisture content is too high for *Deschampsia flexuosa* and where the ground becomes exposed from snow a little later than usual for communities dominated by *Nardus stricta* and *Deschampsia flexuosa*. *V. atropurpurea* stands are often seen in the transition zone between *Nardus stricta* and *Salix herbacea* communities. The species is also frequently and abundantly found in open, low-growing willow thickets.

Nordhagen (1943) regarded *V. atropurpurea* as a calciphobous species. It is striking that it avoids the calcareous areas of central Dovre. Eight pH tests gave 4.3-5.0 (Gjærevoll 1956). This agrees with observations made by Selander (1950a) on the Sarek mountains:"*Deschampsia atropurpurea*: fairly common, at any rate locally, on the amphibolites and quartzites of the eastern part of the region, in all the Virihaure-Vastenjaure basin observed by me only on amphibolite, and in a few places, on leached glacifluvial material".

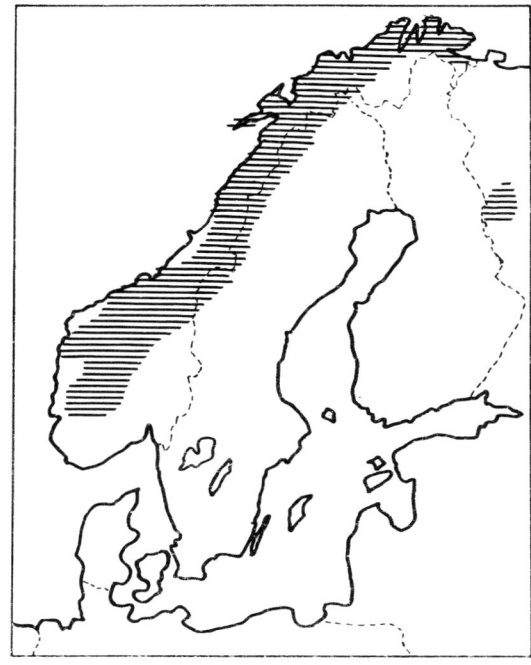

Fig. 92 The Fennoscandian distribution of *Veronica fruticans*

V. fruticans is an amphi-Atlantic, arctic-montane plant. It is widely distributed in central Europe from Spain to the Carpathians, and is known from the Faroes, Iceland, and southeast and southwest Greenland. In Sweden, it occurs from Härjedalen to Torne Lappmark. In Finland, it has only been found in Enontekis Lappmark. There is also an isolated occurrence from northern Karelia to the central part of the Kola Peninsula.

Apart from the southernmost mountains, *V. fruticans* has a wide distribution in Norway, occurring on coastal mountains as well as continental ones. Although generally a low-alpine species, it is often found at subalpine localities and even lower where exposed cliffs are available. It is sometimes found on shore cliffs in North Norway. It grows well on mica schists and all kinds of carbonate rocks.

Nordhagen (1943) regarded it as a characteristic species of his *Veronico-Poion glaucae* alliance, which inhabits low-alpine and subalpine screes, gravel and cliff ledges on calcareous substratum. Such localities usually have a mixture of alpine, subalpine and lowland species. Important companions are *Cardaminopsis petraea*, *Poa glauca*, *Roegneria borealis*, *Draba daurica* and *Potentilla nivea*.

Woodsia glabella R. Br.

Maps: Hultén (1955, 1971a, no. 51). Total: Hultén (1962, no. 38, 1968), Hultén & Fries (1986, no. 58). Local: Benum (1958), Ryvarden (1969), Aune (1980), Engelskjøn (1984).

First Norwegian record: Nordland: Fauske, "wid Langswands östra ända, 11. Juli 1807", G. Wahlenberg (UPS).

Norwegian distribution: Northern unicentric. From Søndre Bjellådalen in Rana to Båtsfjord.

Altitude limits: Chiefly low-alpine. Troms: Målselv, Kirkestinden 1200 m. Pältsan 1400 m (Hedberg et al. 1952). Frequently at low levels, down to sea level in Kåfjord, Alta.

Habitat: Exposed heaths, crevices and cliff ledges. Calciphilous, growing on all kinds of carbonate rocks.

W. glabella is a circumpolar, arctic-alpine species. In Sweden, it is known from Pite Lappmark to Torne Lappmark with some localities in the woodland east of the mountains. In Finland, there are some scattered localities south to Kuusamo. There are also some localities on the Kola Peninsula, in Russian Karelia and in northern Russia eastwards to the Urals. Then there is a considerable gap to

the nearest localities in northern and central Siberia. The isolated occurrence of *W. glabella* from northern Scandinavia to the Urals is congruent with that of several other

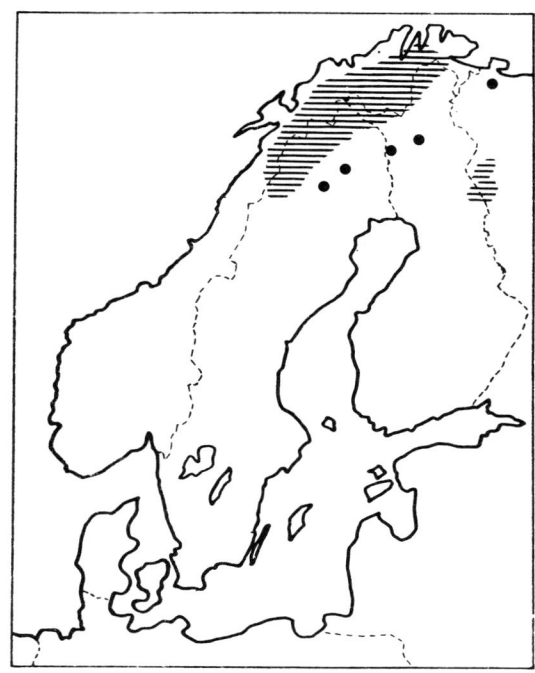

Fig. 93 The Fennoscandian distribution of *Woodsia glabella*

species, e.g. *Potentilla nivea*; these two species often grow together.

W. glabella keeps to the inland mountains. In contrast to most centric species, it has many localities on the Finnmarksvidda plateau, where it usually grows on south-facing cliffs.

W. glabella mainly grows dispersed and plays a minor role in the vegetation. It may be regarded as a characteristic species of the most exposed facies of *Kobresio-Dryadion*.

REFERENCES

Afzelius, K. 1932. Zur Kenntnis der Fortpflanzungsver-hältnisse und Chromosomenzahlen bei *Nigritella nigra*. – *Sv. bot. tidskr. 26:* 366- 369.

Afzelius, K. 1943. Zytologische Beobachtungen an einigen Orchidaceen. – *Sv. bot. tidskr. 37:* 365-369.

Alm, C. G. 1921. Om *Braya glabella* Richards. och dess utbredning i Skandinavien. – *Acta Florae Suecicae. Bd. 1:* 247-261.

Alm, C.G. 1929. Vår sällsyntaste orkidè och dess öden. – *Sveriges natur 20:* 134-137.

Alm, C. G. 1944. Om *Carex macloviana* d'Urv. och dess utbredning i Nord-Europa. – *The Svedberg 1884 30.8. 1944.* Uppsala och Stockholm. 578-600.

Alm, T., R. Elven & K. Fredriksen. 1987. Bidrag til karplantefloraen på Nordlandskysten. 2. – *Polarflokken 11:* 45-86.

Andersen, B. G. 1981. Late Weichselian ice sheets in Eurasia and Greenland. In Denton, G. H. & T. J. Hughes, (eds.): *The Last Great Ice Sheets,* 20- 27. Wiley-Inter-science, New York.

Arwidsson, Th. 1928. Bizentrische Arten in Skandinavien – eine terminologische Erörterung. – *Bot. not.* 49-53.

Arwidsson, Th. 1943. Studien über die Gefässpflanzen in den Hochgebirgen der Pite Lappmark. – *Acta Phyto-geographica Suecica XVII.* 1-274.

Aune, E. I. & O. Kjærem. 1978. Botaniske registreringar og vurderingar. Saltfjellet/Svartisen-prosjektet. Botanisk sluttrapport. – *K. norske vidensk. selsk. museet. Rapport, bot. ser. 6.* 1-78.

Aune, E. et.al. 1980. Botaniske undersøkingar i Kobbelv-og Hellemo-området, Nordland. – *K. norske vidensk. selsk. museet. Rapport, bot. ser. 1.* 1-122.

Baadsvik, K. 1974. Phytosociological and ecological in-vestigations in an alpine area at Lake Kamtjern, Troll-heimen Mts, Central Norway. Vegetation, snow and soil conditions. – *K. norske vidensk. selsk. skr. 5.* 1-61.

Benum, P. 1958. The flora of Troms fylke. – *Tromsø mus. skr. 6.* 1-402, 546 maps.

Berg, R. Y. 1963. Disjunksjoner i Norges fjellflora og de teorier som er framsatt til forklaring av dem. – *Blyttia 21:* 133-177.

Berg, R. Y. 1983. Bekkekløftfloraen i Gudbrandsdal. – *Blyttia 41:* 5-14.

Björkbäck, F., L. Imby, R. Lidberg, I. Sjöström & L. Österdahl. 1976. Något om brunkullans *(Nigritella nigra)* utbredning och ekologi i Sverige. Exempel på ADB-anpassad katalogisering och bearbetning. – *Fauna och flora 71:* 49-60.

Björkbäck, F. & J. Lundqvist. 1982. Aktion brunkulla – ett botaniskt WWF-projekt. – *Sv. bot. tidskr. 75:* 215-228.

Björkbäck, F., J. Lundqvist & C. O. Wetterhall. 1986. Brunkullan – en hotad ängs- og hagmarksväxt. – *Fauna och flora 81:* 192-194.

Björkman, G. 1939. Kärlväxtfloran inom Store Sjöfallets Nationalpark jämte angränsade delar av Norra Lule Lappmark. – *K. sv. vetensk.-akad. avh. i naturskydds-ärenden. No. 2:* 1-224.

Bjørlykke, B. 1938. Vegetasjonen på olivinstein på Sunn-møre. – *Nytt mag. naturv. 79:* 51-125.

Blytt, A. 1874. *Norges Flora eller Beskrivelse af de i Norge vildtvoxende Karplanter tilligemed angivelse af deres Udbredelse. 2.* – Christiania. 386 pp.

Blytt, A. 1876. *Essay on the immigration of the Norwegian flora during alternating dry and rainy periods.* – Christiania. 89 pp.

Blytt, A. 1893. Zur Geschichte der Nordeuropäischen, besonders der Norwegischen Flora. – *Englers Bot. Jahrb. 17, Beiblatt 41.* 30 pp.

Blytt, A. 1906. *Haandbok i Norges Flora. Udgivet ved Ove Dahl.* – Kristiania.

Blytt, M. N. 1838. Botanisk Reise i Sommeren 1836. – *Nyt mag. naturv. 1:* 257- 356.

114

Blytt, M. N. 1861. *Norges Flora.* – Christiania.

Bocquet, G. 1969. Revisio Physolychnidum (Silene sect. Psysolychnis). – *Phanerogamaruna monographiæ 1.*

Borgen, L. 1987. Postglasial evolusjon i Nordens flora – en oppsummering. – *Blyttia 45:* 147-187.

Borissova, A. G. 1959. The genus *Antennaria* Gaertn. – *In Flora of the USSR vol. 25.* Moskva/Leningrad.

Bowen, D. Q. & G. A. Sykes. 1988. Correlation of marine events and glaciations on the northeast Atlantic margin. – *Phil. Trans. R. Soc. London B. 318:* 619-635.

Braun-Blanquet, J. 1923. *L'origine et development des flores dans le Massif Central de France.* – Paris et Zürich. 282 pp.

Bretten, S. 1973. *Slekta Draba i Knutshø – Finshø-området på Dovre. Sider ved dens systematikk og autøkologi.* – Unpubl. thesis. University of Trondheim. 115 pp.

Bruun, H. 1938. Studies on heterostyled plants. 2. *Primula scandinavica* nov. sp., endemic in Scandinavia. – *Sv. bot. tidskr. 32:* 246-255.

Bråthen, G. 1973. *Karplantefloraen i kystområdene av Sør-Varanger.* – Unpubl. thesis. University of Oslo. 101 pp.

Böcher, T. W. 1938. Biological distributional types in the flora of Greenland. – *Med. om Grønland 106:2.* 1-339.

Böcher, T. W. 1951a. Studies on the distribution of the units within the collective species of *Stellaria longipes.* – *Bot. tidsskr. 48:* 401- 420.

Böcher, T. W. 1951b. Distribution of plants in the circumpolar area in relation to ecological and historical factors. – *Journal of Ecology 39:* 376- 395.

Böcher, T. 1954. Oceanic and continental vegetational complexes in Southwest Greenland. – *Med. om Grønland. Bd. 148.* 1-336.

Böcher, T. W. 1977. *Cerastium alpinum* and *C. arcticum,* a mature polyploid complex. – *Bot. not. 50:* 303-309.

Böcher, T., K. Holmen & K. Jakobsen. 1966. *Grønlands Flora.* – København.

Christophersen, E. 1925. Soil reaction and plant distribution in the Sylene National Park, Norway. – *Trans. of the Conn. Acad. of Arts and Sci. 27:* 471-577.

Chrtek, J. and Z. Pouzar. 1960. Observations on some Scandinavian species of the *Antennaria* Gaertn. genus. – *Nov. bot. Del. sem. Horti bot. Univ. Carol. Pragensis.* 11-15.

Clapham, A. R., T. G. Tutin & E. F. Warburg. 1962. *Flora of the British Isles.* – Cambridge.

Dahl, E. 1951. On the relation between summer temperature and the distribution of alpine vascular plants in the lowlands of Fennoscandia. – *Oikos 3:* 22-52.

Dahlstedt, H. 1908. *Taraxacum Reichenbachii* (Huter) Subsp. *dovrense.* – *Ark. bot. 7. Nr. 1:* 1-11.

Dahlstedt, H. 1928. De svenska arterna av släktet *Taraxacum.* – *K. sv. vetensk.-akad. 6:3.* 1-66.

Danielsen, A. 1971. Skandinavias fjellflora i lys av senkvartær vegetasjonshistorie. – *Blyttia 29:* 183-209.

Dierssen, K. 1982. *Die wichtigsten Pflanzengesellschaften der Moore NW-Europas.* – Geneve. 414 pp. 27 pls.

Dahl, E. 1955. Biogeographic and geologic indications of unglaciated areas in Scandinavia during the Glacial Ages. – *Bull. Geol. Soc. Am. 66:* 1499-1519.

Dahl, E. 1956. Rondane. Mountain vegetation in South Norway and its relation to the environment. – *Skr. Norske vidensk. akad. i Oslo I, Mat.-nat. kl. 1956, 3.* 1-373.

Dahl, E. 1958. Amfiatlantiske planter. – *Blyttia 16:* 93-121.

Dahl, E. 1961. Refugieproblemet og de kvartærgeologiske metodene. – *Svensk Naturvetenskap 14:* 81-96.

Dahl, E. 1963a. On the heat exchange of a wet vegetation surface and the ecology of *Koenigia islandica.* – *Oikos 14:* 190-211.

Dahl, E. 1963b. Plant migrations across the North Atlantic Ocean and their importance to the palaeogeography of the region. 137-188. In Löve & Löve: *North Atlantic Biota and their history.* Pergamon Press. Oxford.

Dahl, E. 1987a. Alpine-subalpine plant communities of South Scandinavia. – *Phytocoenologia 15 (4):* 455-484.

Dahl, E. 1987b. The nunatac theory reconsidered. – *Ecol. Bull. 38:* 77-94.

Dahl, E., R. Elven, A. Moen & A. Skogen, 1986. *Vegetasjonskart over Norge 1:1500 000. Nasjonalatlas for Norge.* Statens Kartverk, Ringebu.

Dahl, O. 1888-1911. Biskop Gunnerus's virksomhet fornemmelig som botaniker tilligemed en oversigt over botanikens tilstand i Danmark og Norge indtil hans død. – *K. norske vidensk. selsk. skr.*

Dahl, O. 1895. Breve fra norske botanikere til prof. J. W. Hornemann. – *Arch. Math. Naturv. 17. Nr. 4.* 99 pp.

Dahl, O. 1906-07. Botaniske undersøgelser i indre Ryfylke I og II. – *Forh. Vidensk. Selsk. Chra 1906* (36 pp) *og 1907* (58 pp).

Dahl, O. 1934. Floraen i Finnmark fylke. – *Nyt mag. naturv. 69.* 1-430. 17 pl.

Du Rietz, G. E. 1940. Problems of bipolar plant distribution. – *Acta Phytogeographica Suecica XIII:* 215-282.

Dyring, J. 1911. Flora grenmarensis. Et bidrag til kundskaben om vegetationen ved Langesundsfjorden. – *Nyt mag. naturv. 49:* 99-276.

Edvardsen, H., Nilssen, E. & Skifte, O. 1983. Tidlig sommerbesøk på Jav'reoai'vit i Nordreisa med gjenfunn av "fjellvalmue". – *Polarflokken 7:* 112-116.

Ekman, E. 1917. Zur Kenntnis der nordischen Hochge-birgs-Drabae. – *K. sv. vetensk.-akad. handl. N.F. 57, Nr. 3.* 68 pp.

Ekman, E. 1926. Zur Kenntnis der nordischen Hochge-birgs-Drabae II. – *K. sv. vetensk.- akad. handl. Ser. 3.2. Nr. 7.* 56 pp.

Ekman, E. 1927a. Three new bicentric plants in the south of Norway. – *Nyt mag. naturvid. Bd. 66:* 93-95.

Ekman, E. 1927b. Några växtlokaler för *Antennaria alpina* (L.) Gaertn. – *Sv. bot. tidskr. 21:* 93-94.

Ekman, E. 1927c. Notes on some Greenland *Antennariae.* – *Sv. bot. tidskr. 21:* 49-57. 1 pl.

Ekman, E. 1932. Contribution to the *Draba* flora of Green-land. IV. – *Sv. bot. tidskr. 26:* 431-447.

Eldholm, O. et al. 1986. Above the Arctic Circle. Reflector identified, glacial onset seen. – *Geotimes. March 1986:* 12-15.

Elfstrand, M. 1927. Var hava fanerogama växter överlevat sista istiden i Skandinavien? – *Sv. bot. tidskr. 21:* 269-284.

Elven, R. 1975. Plant communities on recently deglaciated moraines at Finse, southern Norway. – *IBP in Norway. Methods and Results Sections PT-UM Grazing projekt, Hardangervidda, Botanical Investigations. Annual Report 1974. Appendix I:* 381-467.

Elven, R. 1978. Vegetasjonen ved Flatisen og Østerdalsisen, Rana, Nordland, med vegetasjonskart over Vesterdalen i 1:15000. Saltfjellet/Svartisen-prosjektet. Botanisk del-rapport nr. 3. – *K. norske vidensk. selsk. museet. Rapport, bot. ser. 1.* 1-83.

Elven, R. 1979. Botaniske verneverdier i Røros, Sør-Trøn-delag. – *K. norske vidensk. selsk. museet. Rapport, bot. ser. 6.* 1-159.

Elven, R. 1984. *Floraen i Røros-området.* – Univ. Tromsø, Inst. biol. geol. Unpubl. 531 pp.

Elven, R. & V. Johansen. 1983. Havstrand i Finnmark. Flora, vegetasjon og botaniske verneverdier. – *Rapport T-541. Inst. biol. geol. Universitetet i Tromsø.* 1-357.

Elven, R. & A. Aarhus. 1984. A study of *Draba cacuminum* (Brassicaceae). – *Nord. J. Bot. 4 (4):* 425-441.

Engegård, G. 1970. Blindurt *(Melandrium apetalum)* i Nordland fylke og om sentriske arter i Salten-området. – *Blyttia 28:* 183-186.

Engelskjøn, T. 1965. Nye funn av *Arenaria humifusa* Wg. i Nordland og Lule Lappmark. – *Blyttia 23:* 105-124.

Engelskjøn, T. 1967. Contribution to the cytotaxonomy of *Erigeron humilis* Grah., *E. uniflorus* L. and their hybrid. – *Nytt mag. naturv. 14:* 77-85.

Engelskjøn, T. 1970. Flora of Nord-Fugløy, Troms. – *Astarte, Vol. 3:2:* 63-82.

Engelskjøn, T. 1979. Chromosome numbers in vascular plants from Norway, including Svalbard. – *Opera Bot. 52.* 38 pp.

Engelskjøn, T. 1984. Barduvassdraget. Flora og vegeta-sjon i Barduvassdraget ovenfor Altevatn. – *Tromura. Naturvitenskap. Nr. 36. Universitetet i Tromsø.* 199 pp.

Engelskjøn, T. 1985. "Isdalstinden" i Troms – en forveks-ling av lokaliteter. – *Blyttia 43:* 2-6.

Engelskjøn, T. 1986 a. *Patterns of distribution in the mountain flora of north Scandinavia.* – Oslo (unpubl.). 1-81 + maps.

Engelskjøn, T. 1986 b. Zonality of climate and plant distributions in some Arctic and Antarctic regions. – *Norsk Polarinstitutt Rapportserie 30.* 31 pp.

Engelskjøn, T., S. Sivertsen & O. Skifte. 1968. Nytt om *Sagina caespitosa* (J. Vahl) Lge. i Nord-Norge. – *Blyttia 26:* 146-156.

Engelskjøn, T. & G. Knaben. 1971. Chromosome numbers of Scandinavian arctic-alpine plant species. II. – *Acta Borealia Scientia A, 21.* 1-50.

Engelskjøn, T. & O. Skifte. 1984. Forekomsten av *Nigri-tella nigra* i Nordreisa, Troms. – *Blyttia 42:* 138-142.

Engelskjøn, T. & O. Skifte. 1987. Distribution and ecology of *Trichophorum pumilum* (Vahl) Sch. & Th. *(Cyperaceae)* in Norway. – *Rep. Kevo subarctic. res. st. 20:* 9-19.

Engler, A. 1924. *Pflanzenreich IV, H. 86.*

Erlandsson, S. 1942. Fjällväxter i södra Sverige. – *Bygd och natur. Årsbok 1942:* 127-133.

Fernald, M. L. 1925. Persistence of plants in unglaciated areas of Boreal America. – *Mem. Acad. Arts and Sci. Vol. 15:* 237-342.

Fernald, M. L. 1926. Some relationships of the floras of the northern hemisphere. – *Proc. Intern. Cong. of Plant Sci:* 1487-1507.

Fernald, M. L. 1934. *Draba* in temperate Northeastern America. – *Rhodora 36:* 241-404.

Fernald, M. L. 1937. Local plants of the inner coastal plain of southeastern Virginia.- *Rhodora 39:* 321-491.

Fernald, M. L. 1943. Studies in North American species of *Scirpus.* – *Rhodora 45:* 279-296.

Flatberg, K. I. 1976. Plantesamfunn fra Lofoten. – *Blyttia 34:* 23-45.

Forman, S. L. & Miller, G. H. 1984. Time-dependent soil morphologies and pedogenic processes on raised be-aches, Brøgger-halvøya, Spitsbergen, Svalbard Archi-pelago. – *Arctic. Alp. Res. 16:* 381-394.

Frenzl, B. 1987. Grundprobleme der Vegetationsgeschichte Mitteleuropas während des Eiszeitalters. – *Mitt. Natur-forsch. Ges. Luzern 29:* 99-122.

Fries, E. M. 1817. *Novitiarum Florae Suecicae. ed 1.* – Upsaliae.

116

Fries, E. M. 1839. *Novitiarum Florae Suecicae.* – Upsaliae.

Fries, E. M. 1843. Plantæ Suecanae ex Illustr. Kochii Floræ Germanicae Synopsi, ed. 2, illustratæ. – *Bot. not.* 113-121.

Fries, Th. C. E. 1913. *Botanische Untersuchungen in nördlichsten Schweden. Vetensk. och prakt. unders. i Lappland.* – Stockholm.

Fries, Th. C. E. 1919. Antennaria alpina (L.) Gaertn. och dess skandinaviska elementararter. – *Sv. bot. tidskr. 13:* 179-193.

Fries, Th. C.E. 1921. Die skandinavischen Formen der *Euphrasia salisburgensis.* – *Ark. bot. 17:6.* 1-17.

Friis, J. A. 1871. *En Sommer i Finmarken, Russisk Lapland og Nordkarelen.* – Christiania.

Funder, S. 1982. Planterefugierne i Grønland. – *Naturens Verden:* 242-255.

Fægri, K. 1958. *Norges planter.* – Oslo.

Fægri. K. 1960. *Maps of distribution of Norwegian vascular plants. Volume 1. Coast plants.* – *Univ. Bergen skr. 26:* 1-134.

Gaare, E. 1963. *Sølendet i Brekken. En plantegeografisk beskrivelse av ei godgrasmyr.* – Unpubl. thesis. University of Oslo. 76 pp.

Gabler, H.-M. & Elvestad, M. 1988. Undersøkelser i Jav'reoai'vit naturreservat i Nordreisa 1988. – *Polarflokken 12:* 261-266.

Gartner, C. 1694. *Horticultura.* – København.

Gauslaa, Y. 1985. Fjellplantenes avhengighet av klimaet. – *Blyttia 43:* 75-86.

Gjærevoll, O. 1948. Et nytt funn av *Draba crassifolia* Grah. – *Sv. bot. tidskr:* 182-184.

Gjærevoll, O. 1950. Contribution to the ecology of *Carex bicolor* All. in Scandinavia. – *K. norske vidensk. selsk. forh, bd. XXIII, nr. 4:* 11-15.

Gjærevoll, O. 1956. The plant communities of the Scandinavian alpine snow-beds. *K. norske vidensk. selsk. skr. 1956 nr. 1.* 1-405.

Gjærevoll, O. 1959. Overvintringsteoriens stilling i dag. – *K. norske vidensk. selsk. forh. 32:* 36-71.

Gjærevoll, O. 1962. Reinrose, *Dryas octopetala* L. – *Trondhj. Turistfor. årb. 1962:* 1-8.

Gjærevoll, O. 1963a. Botanical investigations in Central Alaska, especially in White Mts. II. – *K. norske vidensk. selsk. skr. 4.* 1-63.

Gjærevoll, O. 1963b. Survival of plants on nunataks in Norway during the Pleistocene Glaciation 261-283. – In Löve & Löve: *North Atlantica Biota and their history.* Pergamon Press. Oxford.

Gjærevoll, O. 1973. *Plantegeografi.* – Oslo.

Gjærevoll, O. 1979. Oversikt over flora og vegetasjon i Oppdal kommune, Sør-Trøndelag. – *K. norske vidensk. museet. Rapport. bot. ser. 2.* 1-44.

Gjærevoll, O. 1980. Oversikt over flora og vegetasjon i Trollheimen. – *K. norske vidensk. selsk. museet. Rapport, bot. ser. 2.* 1-42.

Gjærevoll, O. & R. Jørgensen. 1952-1987 (several editions). *Fjellflora.* (Editions also in Swedish, English and German). – Trondheim.

Gjærevoll, O. & N. A. Sørensen. 1954. Plantegeografiske problemer i Oppdalsfjellene. – *Blyttia 12:* 117-152.

Gjærevoll, O. & K.-G. Bringer. 1965. Plant cover of the alpine regions. – *Acta Phytogeographica Suecica 50:* 255-268.

Gjærevoll, O. & L. Ryvarden. 1977. Botanical investigations on J. A. D. Jensens Nunatakker in Greenland. – *K. norske vidensk. selsk. skr. No. 4.* 1-40.

Grønlie, A. 1950. Litt om Trollheimen under den siste istid. – *Norsk Geol. Tidsskr. 32:* 168-190.

Grønlie, O. T. 1927. The Folden Fiord. Quaternary Geology. – *Tromsø mus. skr. Vol. 1, Part II.* 1-73.

Gunnerus, J. E. 1766, 1772. *Flora Norvegica.* – København.

Hadac, E. 1946. The plant-communities of Sassen Quarter, West-Spitsbergen. – *Studia bot. Cechoslov. 7:* 127-164.

Hafsten, U. 1958. Finds of subfossil pollen of *Koenigia islandica* from Scandinavia. – *Bot. not.* 333-335.

Hagen, M. 1976. Botaniske undersøkelser i Grøvu-området i Sunndal kommune, Møre og Romsdal. – *K. norske vidensk. selsk. museet. Rapport bot. ser. 5.* 1-57.

Hagen, M. 1976. *Flora og vegetasjon i Grøvuområdet på Nordmøre.* – Unpubl. thesis. University of Trondheim. 188 pp. 1 map.

Halvorsen, R. & P. H. Salvesen. 1983. Bidrag til Vest-Hardangerviddas karplanteflora. – *Blyttia 41:* 93-106.

Hämet-Ahti, L. & al. 1984. *Retkeilykasvio.* Helsinki.

Hartman, C. J. 1879. *Handbok i Skandinaviens flora, 11.uppl.* – Stockholm.

Hatlelid, S. Aa. 1980. *Mellomalpin vegetasjon på Knutshø i Oppdal herred.* – Unpubl. thesis. University of Trondheim. 142 pp.

Hedberg, O. 1947. Bidrag till Torne lappmarks flora. – *Bot. not.* 178-181.

Hedberg, O., O. Mårtensson & S. Rudberg. 1952. Botanical investigations in the Pältsa region of northernmost Sweden. – *Bot. not. suppl. vol. 3:2.* 1-209.

Hegi, G. 1962. *Illustrierte Flora von Mitteleuropa, Band III.* – Berlin-Hamburg.

Heltzen, A. M. 1949. Lauparenområdet i den siste istiden. – *Norsk geogr. tidsskr. 12:* 32-41.

Heltzen, A. M. & R. Nordhagen. 1944. En vestlig utpost av *Saxifraga hieraciifolia*. – *Naturen 1968:* 125-128.

Hermann, F. 1939. Eine Fahrt durch Finnland, Nord-Norwegen und Lappland. – *Jahresbericht Preuss. Bot. Verein 1937/38.* 101-118.

Heywood, V. H. 1973. Flora Europaea, Notulae systematicae ad Floram Europaeam spectantes, No. 14. – *Bot. Journ. Linn. Soc. 67:* 275-283.

Hiitonen, I. 1947. Über die ostfennoskandischen Formen und Bastarde der kollektivart *Potentilla nivea* L. nebst Erörterung einiger anderen Arten der Niveae-Gruppe. – *Arch. Soc. zool. bot. Fenn. "Vanamo" 2:* 23-33.

Holmboe, J. 1905. Studier over norske planters historie. I. *Gentiana purpurea* L. – *Nyt mag. naturv. 43:* 33-60.

Holmboe, J. 1912. *Vegetationen ovenfor barskogsgrænsen i Karesuando og Jukkasjärvi nord for Torneträsk, særlig med sigte paa reinbeitet.* – Bergen. 139 pp.

Holmboe, J. 1923. Søteroten *(Gentiana purpurea)* gjenfundet nordenfjelds. – *Naturen 47:* 95-96.

Holmboe, J. 1932. Spredte bidrag til Norges flora. II. – *Nyt mag. naturv. 71:* 147-184.

Holmboe, J. 1934. Spredte bidrag til Norges flora. III. – *Nyt mag. naturv. 74:* 71-116.

Holmboe, J. 1936. Über *Nigritella nigra* (L.) Rchb., ihre Verbreitung und Geschichte in Skandinavia. – *Ber. Schweiz. Bot. Gesellsch. 46:* 102-116.

Holmboe, J. 1937. The Trondheim district as a centre of Late Glacial and Postglacial plant migrations. – *Avh. Norske vidensk.-akad. i Oslo. Mat.-nat. kl. 1936. I:9.* 59 pp.

Holmen, K. & H. Mathiesen. 1953. *Luzula Wahlenbergii* in Greenland. – *Bot. tidsskr. 49:* 233-238.

Holmsen, G. 1913. Oversikt over Hatfjelddalens geologi. – *Norges geol. unders. Nr. 61:* 2-34. 4 pl. 1 map.

Holten, J. 1983. Flora- og vegetasjonsundersøkelser i nedbørfeltene for Sanddøla og Luru i Nord-Trøndelag. – *K. norske. vidensk. selsk. museet. Rapport, bot. ser. 2.* 1-148.

Holten, J. 1984. Flora- og vegetasjonsundersøkelser i Raumavassdraget, med vegetasjonskart i M 1:50 000 og 1:150 000. – *K. norske vidensk. selsk. museet. Rapport, bot. ser. 4.* 1-141.

Horn, K. 1938. Cromosome numbers in Scandinavian Papaver species. – *Avh. Vidensk.-akad. Oslo. I. Mat.-nat. kl. 1938, No. 5.* 13 pp.

Hornemann, J. W. 1806, 1821. *Forsøg til en dansk oeconomisk Plantelære.* – København.

Hultén, E. 1937. *Outline of the history of arctic and boreal biota during the Quaternary Period.* – Stockholm. 1-168. 42 pl.

Hultén, E. 1943a. Vår sällsyntaste orkidè och dess amerikanska släktning. – *Flora och Fauna 30:* 166-174.

Hultén, E. 1943b. On the races in the Scandinavian Flora. – *Sv. bot. tidskr. 43:* 383-406.

Hultén, E. 1943c. *Stellaria longipes* Goldie and its allies. – *Bot. not.* 251-270.

Hultén, E. 1945. Studies in the *Potentilla nivea* group. – *Bot. not.* 127-148.

Hultén, E. 1950. *Atlas of distribution of vascular plants in NW Europe.* – Stockholm.

Hultén, E. 1954. *Artemisia norvegica* Fr. and its Allies. – *Nytt mag. bot. 3:* 63-82.

Hultén, E. 1955. The isolation of the Scandinavian mountain flora. – *Acta Soc. fauna et flora Fennica 72 N:o 8.* 1-22.

Hultén, E. 1956. The *Cerastium arcticum* complex. – *Sv. bot. tidskr. 50:* 411- 495.

Hultén, E. 1958. The amphi-Atlantic plants and their phytogeographical connections. – *K. Sv. vetensk.-akad. handl. Ser. 4, 7 (1).* 1-340.

Hultén, E. 1959. Studies in the genus *Dryas.* – *Sv. bot. tidskr. 53:* 507-542.

Hultén, E. 1962. The circumpolar plants I. – *K. sv. vetensk.-akad. handl. Ser. 4,8 (5).* 1-275.

Hultén, E. 1968. *Flora of Alaska and neighbouring territories.* – Stanford.

Hultén, E. 1971a. *Atlas of distribution of vascular plants in NW Europe.* – Stockholm.

Hultén, E. 1971b. The circumpolar plants II. – *K. sv. vetensk.-akad. handl. Ser. 4, 13 (1).* 1-463.

Hultén, E. & M. Fries. 1986. *Atlas of North European vascular plants.* – Koeltz Scientific Books. Königstein.

Hylander, N. 1947. *Cardaminopsis suecica* (Fr.) Hiit., a northern amphiploid species. – *Bull. Jard. Bot. Bruxelles 27:* 591-604.

Hylander, N. 1953. *Nordisk kärlväxtflora I.* – Stockholm.

Høiland, K. 1986a. Utsatte planter i Nord-Norge. Generell del. – *Økoforsk rapport 1986:1.* 1-33.

Høiland, K. 1986b. Utsatte planter i Nord-Norge. Spesiell del. – *Økoforsk rapport 1986:2.* 3-163.

Høiland, K. 1988. Purpurkarse *(Braya purpurascens)* – vår mest arktiske plante. – *Blyttia 46:* 113-117.

Johansen, B. E. 1979. Svalbardvalmue *(Papaver dahlianum* Nordh.), femte funn i Europa og dens relasjoner til de andre valmuene i Finnmark. – *Polarflokken 3:* 11-17.

Jungner, H., Landvik, J. Y. & Mangerud, J. 1989. Thermoluminescence dates of Weichselian sediments in western Norway. – *Boreas 18:* 21-29.

118

Jørgensen, R. 1933. Karplantenes høidegrenser i Jotunheimen. – *Nytt mag. naturv. 72:* 1-130.

Jørgensen, R. 1937. Die Höhengrenzen der Gefässpflanzen in Troms fylke. – *K. norske vidensk. selsk. skr. 1936,8.* 106 pp.

Kalela, A. 1944. Systematische und pflanzengeographische Studien an der *Carex*-Subsection *Alpinae* Kalela. Ein Beitrag zur Kenntnis der pleistozänen und holozänen Pflanzenwanderungen im holarktischen Raum. – *Ann. Bot. Soc. zool. bot. Fenn. "Vanamo" 19:* 1-218. 7 pl.

Kalliola, R. 1939. Pflanzensoziologische Untersuchungen in der alpinen Stufe Finnisch-Lapplands. – *Ann. Bot. Soc. zool. bot. Fenn. "Vanamo" 13 (2):* 1-328.

Knaben, G. 1954. *Saxifraga osloensis* n. sp., a tetraploid species of the Tridactylites Section. – *Nytt mag. bot. 3:* 117-138. 1 pl.

Knaben, G. 1959a. On the evolution of the Radicatum-group of the Scapiflora Papavers as studied in 70 and 56 chromosome species. A. Cytotaxonomical aspects. – *Opera Bot. 2(3):* 1-74.

Knaben, G. 1959b. On the evolution of the Radicatum-group of the Scapiflora Papavers as studied in 70 and 56 chromosome species. B. Experimental studies. – *Opera Bot. 3(3):* 1-96.

Knaben, G. 1966a. Studies on the life form of some *Sedum* species. – *Blyttia 24:* 232-243.

Knaben, G. 1966b. Om kromosomvariasjon og rasedann-else i den norske flora. – *Blyttia 24:* 65-79.

Knaben, G. 1966c. Cytotaxonomical studies in some *Draba* species. – *Bot. not. 119:* 427-444.

Knaben, G. 1970. Om artsbegrepet hos fjellvalmuer. – *Blyttia 28:* 187-193.

Knaben, G. 1979. Additional experimental studies in the *Papaver radicatum* group. – *Bot. not. 132:* 483-490.

Knaben, G. 1985. Neo-polyploids in the North Atlantic region. – *Botanica Helvetica 95:* 177-191.

Knaben, G. & T. Engelskjøn. 1967. Chromosome numbers of Scandinavian arctic-alpine plant species II. – *Acta Borealia. A. Scientia. 21.* 1-50. 3 pl.

Knaben, G. & Hylander, N. 1970. On the typification of *Papaver radicatum* Rottb. and its nomenclatural consequences. – *Bot. not. 123:* 338-345.

Kristensen, T. S. 1981. *En plantegeografisk undersøkelse av Nordkynhalvøya i Finnmark.* – Unpubl. thesis. University of Trondheim. 315 pp.

Kytövuori, I. 1969. *Epilobium davuricum* Fisch. *(Onagraceae)* in Eastern Fennoscandia compared with *E. palustre* L. A morphological, ecological and distributional study. – *Ann. Bot. Fenn. 6:* 35-58.

Lagerberg, T. & J. Holmboe. 1937-40. *Våre ville planter.* – Oslo.

Landvik, J. Y. & J. Mangerud, 1985. A Pleistocene sandur in western Norway; facies relationship and sedimentological characteristics. – *Boreas 14:* 161-174.

Lange, J. 1880. Bemærkninger ved det 50de Hæfte af Flora Danica. – *Overs. K. danske vidensk. selsk. forh.* 111-131.

Lid, J. & A. R. Zachau. 1928. Utbredningen av *Viscaria alpina* (L.) G. Don, *Alchemilla alpina* L. og *Rhodiola rosea* L. i Skandinavien. – *Med. Göt. Bot. Trädg. IV:* 69-144.

Lid, J. 1944. *Norsk flora.* – Oslo.

Lid, J. 1954. *Carex bicolor* in Southern Norway. – *Nytt mag. bot. 3:* 147-158.

Lid, J. 1958. Two Glacial relics of *Dryas octopetala* and *Carex rupestris* in the forests of Southeastern Norway. – *Nytt mag. bot. 6:* 5-9.

Lid, J. 1959. The vascular plants of Hardangervidda, a mountain plateau of Southern Norway. – *Nytt mag. bot. 7:* 61-128.

Lid, J. 1963. *Norsk og svensk flora.* – Oslo.

Lindblom, A. E. 1837-38. Strödda botaniska anteckningar til upplysande af Norges Flora. – *Physiogr. Sällsk. Tidskr. 1:* 315-360.

Lindblom, A. E. 1838. Om O. Sperling och G. Fuiren samt deras bidrag til Skandinaviens flora. – *Physiogr. Sällsk. Tidskr. I. Lund 1837-38:* 360-389.

Lindblom, A. E. 1842. Strödda anteckningar öfver Norges vegetationsförhållanden. – *Bot. not.* 193-208.

Lindeberg, C. J. 1855. Fortsatta excursioner i Norge 1855. – *Nya bot. not. 1855:* 161-167.

Lindroth, C. 1970. Survival of animals and plants on ice-free refugia during the Pleistocene glaciations. – *Endeavour, Vol. XXIX, no. 108:* 129-134.

Longva, O. et al. 1983. Beskrivelse til kvartærgeologisk kart 1019 II-M 1:50000. – *Norges Geol. Unders. Nr. 393.* 66 pp.

Lund, N. 1846a. Foreløbig Beretning om en botanisk Reise i Ost-Finmarken i Sommeren 1842. – *Bot. not.* 33-48.

Lund, N. 1846b. Første Anhang til Beretningen i Botaniska Notiser 1846. – *Bot. not.* 65-95.

Lunde, T. 1962. An investigation into the pH-amplitude of some mountain plants in the county of Troms. – *Acta Borealia Ser. A. 20.* 103 pp.

Læstadius, L. L. 1822. Botaniska Anmärkningar gjorda i Lappmarken och tillgränsande landsorter. – *K. sv. vetensk.-akad. handl.* 327-342.

Læstadius. L. L. 1826. Beskrifning öfver några sällsyntare växter från norra delarna af Sverige, jemte anmärk-ningar i Växt-geographien. – *K. sv. vetensk.-akad. handl.* 169-174.

Løkken, S. 1969. Noen nye funn av *Braya linearis* Rouy, spesielt fra Sør-Norge, og noen bemerkninger til den bisentriske utbredelse av denne art i Skandinavia. − *Blyttia 27:* 107-117.

Löve, A. 1950. Some innovations and nomenclature suggestions in the Icelandic flora. − *Bot. not.* 24-60.

Löve, A. 1961. *Hylandra* − a new genus of Cruciferae. − *Svensk. Bot. Tidskr. 55:* 211-217.

Löve, A. 1962. Typification of *Papaver radicatum* − a nomenclatural detective story. − *Bot. not. 115:* 113-136.

Löve, A. & Löve, D. 1947. Thrjár nyfundar jurtategundir. − *Náttúrufr. 17:* 166-174.

Löve, A. & D. Löve. 1951. Studies on the origin of the Icelandic flora. II. *Saxifragaceae.* − *Sv. bot. tidskr. 45:* 368-399.

Maguire, B. 1943. Monograph of the Genus *Arnica.* − *Brittonia 4:* 386-510.

Malme, L. 1971. Oseaniske skog- og heiplantesamfunn på fjellet Talstadhesten i Fræna, Nordvest-Norge. − *K. norske vidensk. selsk. museet. Miscellanea 2.* 39 pp + tab.

Manabe, S. & D. G. Hahn. 1977. Simulation of the Tropical Climate of an Ice Age. − *Jour. Geoph. Res. 82:* 3889-3911.

Mangerud, J. 1973. Isfrie refugier i Norge under istidene. − *Norges Geol. Unders. Nr. 297.* 23 pp.

Mangerud, J. et al. 1981. A Middle Weichselian ice-free period in Western Norway: the Ålesund Interstadial. − *Boreas 10:* 447-462.

Marret, L. 1911-24. *Icones Florae Alpinae Plantarum. Sr. 1-3.* − Paris.

Mejland, Y. 1939. Om floraen på Javreoaive i Nordreisa. − *Nytt mag. naturv. 79:* 165-191.

Mellor, R. & Wilson, M. J. 1987. Processes and products of mineral weathering in soils of the SWAP catchments in Scotland and Norway. 99-105. − In: *Surface Water Acidification Programme Mid-Term Review Conference.* Bergen, Norway 22-26 June 1987. Royal Society.

Meusel, H. et al. 1965, 1978. *Vergleichende Chorologie der zentraleuropäischen Flora.* − Jena.

Miller, G. H. 1982. Quaternary depositional episodes, Western Spitsbergen, Norway: Aminostratigraphy and glacial history. − *Arct. Alp. Res. 14:* 321-340.

Moe, B. 1985. *Fjellflora og -vegetasjon i Midtre Hordaland. Analyse av en botanisk øst-vest gradient.* − Unpubl. thesis. University of Bergen. 1-194.

Moen, A. 1970. *Myr- og kildevegetasjon på Nordmarka, Nordmøre.* − Unpubl. thesis. University of Trondheim. 245 pp. 35 pls.

Moen, A. 1976. Botaniske undersøkelser på Kvikne i Hedmark, med vegetasjonskart over Innerdalen. − *K. norske vidensk. selsk. museet. Rapport bot. ser. 2.* 1-100.

Moen, A. 1990. The plant cover of the boreal uplands of Central Norway. I. Vegetation ecology of Sølendet nature reserve; haymaking fens and birch woodlands. *Gunneria 63:* 1-451, 1 map.

Moen, A. & M. Selnes. 1979. Botaniske undersøkelser på Nord-Fosen, med vegetasjonskart. − *K. norske. vidensk. selsk. museet. Rapport, bot. ser. 4.* 1-96.

Moen, A. & L. Kjelvik. 1981. Botaniske undersøkelser i Garbergelva/Rotla-området i Selbu, Sør-Trøndelag, med vegetasjonskart. − *K. norske vidensk. selsk. museet. Rapport, bot. ser. 3.* 1-106.

Moen, J. & A. Moen. 1977. Flora og vegetasjon i Tromsdal i Verdal og Levanger, Nord-Trøndelag, med vegetasjonskart. − *K. norske vidensk. selsk. museet. Rapport, bot. ser. 6.* 1-85. 1 map.

Møller, O. 1966. Utbredelse, ytre morfologi og cytologi hos *Gentiana purpurea* L., *Campanula barbata* L. og *Phyteuma spicatum* L. − Dissert. paper, unpubl. University of Oslo. 140 pp.

Mølster, L. 1981. Flora og vegetasjon i Syltefjordsvassdraget (Vesterelva), Varangerhalvøya, Finnmark, Nord-Norge. − *Tromura, nat. vit. 19:* 1-87. 1 map.

Nannfeldt, J. A. 1935. Taxonomical and plant-geographical studies in the *Poa laxa* group. − *Symb. bot. Ups. 5.* 1-113. 4 pl.

Nannfeldt, J. A. 1940. On the polymorphy of *Poa arctica* R. Br. with special reference to its Scandinavian forms.- *Symb. bot. Ups. IV:4.* 1-86.

Nannfeldt, J. A. 1941. *Sagina caespitosa* (J. Vahl) Lge. funnen i Lule Lappmark. − *Bot. not.* 279-284.

Nannfeldt, J.A. 1947. Några synpunkter på den skandinaviska fjällflorans ålder. − *K. vet. soc. årsb. 1947:* 51-58.

Nannfeldt, J. A. 1958. Den skandinaviska fjällfloran och nedisningarna. − *Statens naturvet. forskn. råd. årsb. Svensk naturvetensk. II. 1957-58:* 119-133.

Nannfeldt, J. A. 1963. Taxonomic differentiation as an indication of the migratory history of the North Atlantic flora with special regard to the Scandes. 87-97. In Löve & Löve. *North Atlantic biota and their history.* Pergamon Press, Oxford.

Nathorst, A. G. 1892. Über den gegenwärtigen Stand unserer Kenntnis der Verbreitung fossiler Glazialpflanzen. − *Bihang till Kungl. sv. vet.-akad. handl. 17. III:* 1-35.

Nazarov, M. J. 1924. Der Adams'sche Nachlass. (Nachtrag zu den Resultaten der botanischen Forschungsreise Adams ins arktische Sibir). − *Bull. Soc. Nat. de Moscow 32.*

Nesje, A. et al. 1988. Block fields in southern Norway: Significance for the Late Weichselian ice sheet. − *Norsk Geol. Tidsskr. 68:* 149-169.

Neuman, L. M. 1901. *Sveriges Flora (Fanerogamer).* − Lund.

120

Neuman, L. M. 1905. Bidrag til kännedomen om floran vid Saltenfjord och på Sulitälma-området i Norge. – Bot. not. 251-282.

Nilsson, Ö. 1976. Bidrag till kärlväxtfloran i västra Härjedalen. – Sv. bot. tidskr. 70: 233-250.

Nilsson, Ö. 1986. Norges fjellflora. – J. W. Cappelens Forlag. Oslo.

Nilsson, Ö. & Gustafsson, L.-Å. 1979. Projekt Linnè rapporterar 93-105. – Sv. bot. tidskr. 73: 71-85.

Nordhagen, R. 1923. Botaniske notiser. 1. Om skuddbygningen hos Sedum villosum. – Vid. selsk. i Kristiania. Skr. 1922. 1. Mat. nat. kl. nr. 15. 16 pp.

Nordhagen, R. 1928. Die Vegetation und Flora des Sylenegebietes. I. Die Vegetation. – Skr. Norske vidensk. akad. i Oslo. I. Mat. nat. klasse 1927:1. 1-612.

Nordhagen, R. 1929. Bredemte sjøer i Sunndalsfjellene. – Norsk geogr. tidsskr. 2: 281-356.

Nordhagen, R. 1931a. Studien über die skandinavischen Rassen des Papaver radicatum Rottb. sowie einige mit denselben verwechselten neuen Arten. Vorläufige Mitteilung. – Berg. mus. årb. 1931. Naturv. rekke, nr. 2: 1-50.

Nordhagen, R. 1931b. En botanisk ekskursjon i Eikisdalen. – Bergens mus. årb, 1930. 35 pp.

Nordhagen, R. 1933. De senkvartære klimavekslinger i Nordeuropa og deres betydning for kulturforskningen. – Inst. for sammenl. kulturforsk. Ser. A., No. 12. 246 pp.

Nordhagen, R. 1935. Om Arenaria humifusa Wg. og dens betydning for utforskningen av Skandinavias eldste floraelement. – Bergens mus. årb. 1935:1. 1-185.

Nordhagen, R. 1936a. Versuch einer neuen Einteilung der subalpinen-alpinen Vegetation Norwegens. – Bergens mus. årb. 1936:7. 1-88.

Nordhagen, R. 1936b. Skandinavias fjellflora og dens relasjoner til den siste istid. – Nordiska (19.skand.) naturforskarmötet. Helsingfors: 93-124.

Nordhagen, R. 1939. Bidrag til fjellet Pältsas flora. Et nytt funn av Stellaria crassipes. – Bot. Not. 691-700.

Nordhagen, R. 1940. Norsk flora. – Oslo.

Nordhagen, R. 1943. Sikilsdalen og Norges fjellbeiter. – Bergens mus. skr. 22. 1-607.

Nordhagen, R. 1952. Bidrag til Norges flora. II. Om nyere funn av Euphrasia lapponica Th. Fr. fil. i Norge. – Blyttia 10: 29-50.

Nordhagen, R. 1954a. Apologi for Poa stricta Lindeb. – Sv. bot. tidskr. 48: 1-18.

Nordhagen, R. 1954b. Some new observations concerning the geographic distribution and the ecology of Arenaria humifusa Wg. in Norway as compared with Arenaria norvegica Gunn. – Bot. tidsskr. 51: 248-262.

Nordhagen, R. 1955. Kobresieto-Dryadion in Northern Scandinavia. – Sv. bot. tidskr. 49: 248-262.

Nordhagen, R. 1950-58. Våre ville planter 1-8. – Johan Grundt Tanum. Oslo.

Nordhagen, R. 1963a. Recent discoveries in the south Norwegian flora and their significance for the understanding of the history of the Scandinavian mountain flora during and after the last glaciation. 241-260. – In Löve & Löve: North Atlantic biota and their history. Pergamon Press. Oxford.

Nordhagen, R. 1963b. Om Crepis multicaulis (Led.) og dens utbredelse i Norge, arktisk Russland og Asia. – Blyttia 21: 1-42.

Nordhagen, R. 1964. Om Oxytropis lapponica (WG.) Gaud. og O. deflexa (Pall.) DC. subsp. norvegica Nordh. – Sv. bot. tidskr. 58: 129-166.

Nordhagen, R. 1965a. Taxonomiske og økologiske studier over Saxifraga aizoon Jacq. i Norge. – Blyttia 23: 145-162.

Nordhagen, R. 1965b. Om vestgrensen for Rhododendron lapponicum (L.) Wg. i Syd-Norge. – Norske vidensk.-akad. i Oslo. avh.l. Mat. nat. kl. Ny ser. 6. 35 pp.

Nordhagen, R. 1973. Über ein spontanes Vorkommen von Rheum rhaponticum L. in Aurland im inneren Sognefjord-Gebiet, Norwegen, sowie über das Vorkommen der Art in Bulgarien. – Norske vidensk.-akad. i Oslo. skr. 1. Mat.-nat. kl. Ny ser. 31. 90 pp.

Nordsteien, J. H. 1982. Flora og vegetasjon i den alpine delen av Blåhø – Gjevilvasskamman. – Unpubl. thesis. University of Trondheim. 130 pp.

Norman, J. M. 1851. Beretning om en i Gudbrandsdalen foretagen botanisk Reise. – Nyt mag. naturv. 6: 17-43.

Norman, J. M. 1894 -1901. Norges Arktiske Flora. – Kristiania.

Notø, A. 1905. Fjeldfloraen mellem Altevand og Kirkesdalen. – Tromsø Museums Aarsh. 1904. 1-19.

Nygren, A. 1936. Carex holostoma Drej., en för Sverige ny fanerogam, funnen i Torne Lappmark. – Sv. bot. tidskr. 30: 137-153.

Nygren, A. 1950. Cytological and embryological studies in Arctic Poae. – Symb. bot. Ups. X:4. 1-64. 15 pl.

Odland, A. 1986. Utbredelse av fjellburkne (Athyrium distentifolium) i relasjon til klimafaktorer. – K. norske vidensk. selsk. museet. Rapport, bot. ser. 2: 15-30.

Oeder, C.G. 1761-1770. Flora Danica. I-IX. København.

Ostenfeld, C. H. 1917. De nordiske former av kollektivarten Arenaria ciliata L. – Nyt mag. nat. 55: 215-225.

Ouren, T. 1950. Gentiana purpurea L. i Trøndelag. – K. norske vidensk. selsk. årb. 1949: 68-80.

Ouren, T. 1952. Floraen i Budal herred i Sør-Trøndelag. – K. norske vidensk. selsk. skr. 1952. nr. 1: 1-101.

Ouren, T. 1959. Floraen i Soknedal herred i Sør-Trøndelag. – *K. norske vidensk. selsk. museet. årb. 1959:* 71-121.

Ouren, T. 1961. Floraen i Singsås herred i Sør-Trøndelag. – *K. norske vidensk. selsk. museet. årb. 1961:* 5-73.

Ouren, T. 1964. Floraen i Støren herred i Sør-Trøndelag. – *K. norske vidensk. selsk. museet årb, 1964:* 7-78.

Ouren, T. 1966. Floraen i Haltdalen herred i Sør-Trøndelag. – *K. norske vidensk. selsk. museet årb, 1966:* 25-102.

Ouren, T. 1979. Søterot i Gauldalen. – *Gauldalsminne, band 4:* 18-24.

Pojarkova, A. I. 1956. *Flora Murmanskoi oblasti III.* – Akad. Nauk SSSR. Moscow – Leningrad.

Porsild, A. E. 1958. *Dryas Babingtoniana,* nom. nov. An overlooked species of the British Isles and western Norway. – *Dept. of north. aff. Bull. 160:* 133-145.

Porsild, A. E. 1963. *Stellaria longipes* and its allies in North America. – *Nat. Museum of Canada Bull. No. 186:* 1-35.

Printz, H. 1921. The vegetation of the Siberian-Mongolian frontiers (the Sayansk region). – Contribution ad floram Asiae interioris pertinentes. Ed. Henrik Printz. III. – *K. norske vid. selsk.*

Ramus, J. 1719. *Norriges beskrivelse etc.* – København.

Raymond, M. 1950. Esquisse Phytogeographic de Quebec. – *Mem. du jardin bot. de Montreal:* 6-147.

Resvoll-Holmsen, H. 1932. Om planteveksten i grensetrakter mellem Hallingdal og Valdres. – *Skr. utg. av Det norske vidensk.-akad. I. Mat. nat. kl. 1931, No. 9.* 1-50. 10 pl.

Reusch, H. 1910. Norges Geologi. – *Norges Geol. Unders. 50.* 1-196.

Roaldset, E. et al. 1982. Remnants of pregacial weathering in western Norway. -*Norsk Geol. Tidsskr. 62:* 169-178.

Rokoengen, K. & Rønningsland, T. M. 1983. Shallow bed geology and Quaternary thicknesses in the Norwegian sector of the North Sea between 60 30'N and 62 N. – *Norsk Geol. Tidsskr. 63:* 83-102.

Rottböll, C. F. 1770. Afhandling om en Deel enten gandske nye eller vel forhen bekiendte, men dog for os rare Planter, som i Island og Grønland ere fundne. – *Skr. Københavnske Selsk. Lærd. Vid. Elskere 10:* 393-462.

Rouy, G. 1898. *Illustrationes Plantarum Europæ rariorum, Fasc. XII.* – Paris.

Rune, O. 1945. Några anmärkningsvärda växtfynd i södra Lapplands fjäll. – *Sv. bot. tidskr. 39:* 299-303.

Rune, O. 1948. Nya växtfynd i Lycksele lappmarks fjällområde. – *Sv. bot. tidskr. 42:* 494-497.

Rune, O. 1950. *Draba cacuminum* i Sverige. – *Sv. bot. tidskr. 44:* 497-503.

Rune, O. 1953. Plant life on serpentine and related rocks in the north of Sweden. – *Acta Phytogeographica Suecica 31.* 139 pp.

Rune, O. 1954. *Arenaria humifusa* on serpentine in Scandinavia. – *Nytt mag. bot. 3:* 183-196.

Rune, O. 1955. *Arenaria humifusa* i Sverige. – *Sv. bot. tidskr. 49:* 197-216.

Rune, O. 1957. De serpentinicola elementen i Fennoskandiens flora. – *Sv. bot. tidskr. 51:* 43-105.

Rune, O. & O. I. Rønning. 1954. A new variety of *Euphrasia lapponica.* – *Bot. not.* 297-303.

Rune, O. & O. I. Rønning. 1956. *Antennaria nordhagiana* nova species. – *Sv. bot. tidskr. 50:* 115-128. 2 pl.

Ryvarden, L. 1966. *Saxifraga paniculata* Miller (syn. *S. aizoon* Jacq.) i Ryfylke. – *Blyttia 24:* 322-330.

Ryvarden, L. 1969. The vascular plants from the Rastigaissa area (Finnmark, Northern Norway). – *Acta Borealia A. Scientia 26.* 1-56.

Ryvarden,. I. 1974. *Pedicularis hirsuta* og *P. flammea,* frøkapasitet og utbredelse. – *Blyttia 32:* 139-142.

Ryvarden, L. & P. E. Kaland. 1968. *Artemisia norvegica* Fr. funnet i Rogaland (foreløpig meddelelse). – *Blyttia 26:* 75-84.

Ryvarden, L. & Sivertsen, S. 1969. Noen plantefunn fra Nord-Norge 1968. – *Blyttia 27:* 210-215.

Rønning, O. I. 1956a. Nye funn av *Carex holostoma.* – *Blyttia 14:* 100-102.

Rønning, O. I. 1956b. *Draba crassifolia* in Scandinavia. – *Acta Borealia A. Scientia, No. 11.* 1-20.

Rønning, O.I. 1959. The vascular flora of Bear Island. – *Acta Borealia. A. Scienta. 15.* 1-53.

Rønning, O. I. 1965. Studies in *Dryadion* of Svalbard. – *Norsk Polarinst. skr. 134.* 1-52.

Samuelsson, G. 1917. Studien über die Vegetation bei Finse im inneren Hardanger. – *Nyt mag. naturv. 55.* 1-108. 7 pl.

Samuelsson, G. 1921. *Carex dioeca*-gruppen i den nordiska floran. – *Actae Florae Suecicae I:* 219-244. 1 pl.

Samuelsson, G. 1943. Die Verbreitung der *Alchemilla*-arten aus der Vulgaris-Gruppe in Nordeuropa. – *Acta Phytogeographica Suecica XVI.* 253 pp. 6 pl.

Saxer, A. 1955. *Fagus-, Abies-* und *Picea-* gürtelarten in der Kontaktzone der Tannen- und Fichtenwälder der Schweiz. – *Beitr. z. Geobot. Landesaufnahme der Schweiz. H. 36.* 198 pp.

Schilling, A. D. & Pollard, D. W. F. 1964. Floraen i Sørdalen i Lyngen. – *Blyttia 22:* 53-65.

Schumacher, T. & S. Løkken. 1981. Vegetasjon og flora i Grimsavassdragets nedbørfelt. – *Kontaktutvalget for vassdragsreguleringer, Universitetet i Oslo. Rapport 31.* 114 pp. 1 map.

122

Selander, S. 1950a. Floristic phytogeography of South-Western Lule Lappmark. – *Acta Phytogeographica Suecica 27.* 200 pp.

Selander, S. 1950b. Kärlväxtfloran i sydvästra Lule Lappmark. – *Acta Phytogeographica Suecica 28.* 152 pp. 488 maps.

Sernander, R. 1896. Några ord med anledning av Gunnar Andersson: Svenska vaxtvarldens historia. – *Bot. not.* 114-128.

Skifte, O. 1985. Nye funn av grønlandsstarr – *Carex scirpoidea* Michx. – i Nordland fylke. – *Blyttia 43:* 16-21.

Skifte, O. 1988. Feltarbeid i vårt nordligste utbredelsesområde for grønlandsstarr *(Carex scirpoidea).* – *Blyttia 46:* 15-22.

Skogen, A. 1970. Plantegeografiske undersøkelser på Frøya, Sør-Trøndelag. III. Alpine og nordlige innslag i floraen. – *Blyttia 28:* 108-124.

Skogen, A. 1971. Bidrag til karplantefloraen i Grotli-Tafjordfjellene. – *K. norske vidensk. selsk. museet.* 1-46 + 16 maps.

Skogen, A. 1974. Fjellfloraen på Storfjellet i Tafjord og forbindelsen mellom Sunnmørsfjellenes og Jotunheimens plantesentra. – *Blyttia 32:* 199-210.

Skogen A. 1976. Nye plantefunn i devon-områdene i Hyen, i relasjon til fjellfloraen i Nordfjord. – *Blyttia 34:* 173-187.

Skogen, A. 1979. Vegetasjon og fjellplanteflora i Stavbrekkene på Geirangerfjellet, et rikt fjell i Vestfjellenes fattigområde. – *Blyttia 37:* 109-125.

Skogen, A. 1981. Lappmarksrublom, *Draba lactea,* i Indre Sogn. – *Blyttia 39:* 189-192.

Smith, H. 1940. *Carex arctogena* nova species. – *Acta Phytogeographica Suecica XIII.* 191-200.

Sommerfelt, S. C. 1826. *Supplementarum Florae Lapponicae.* – Chra.

Sommerfelt, S. C. 1833. Bidrag til Spitsbergens og Beeren-Eilands Flora, efter Herbarier medbragte av M. Keilhau. – *Mag. Naturv. 11:* 232-252.

Strauch, F. 1970. Die Thule Landbrücke als Wanderweg und Faunenscheide zwischen Atlantik und Skandik im Tertiär. – *Geol. Rundschau 60:* 381-417.

Strauch, F. 1983. Geological history of the Iceland-Faeroe-ridge and its influence on Pleistocene glaciations. pp 601-606. – In Bott et al. (eds.): *Structure and development of the Greenland – Scotland ridge.* Plenum Press, London.

Størmer, P. 1952. Dansk Botanisk Forenings ekskursjon til Norge 1951. – *Blyttia 10:* 17.

Sunding, P. 1962. Høydegrenser for høyere planter på Svalbard. – *Norsk Polarinstitutt Årbok 1962.* 32-59.

Sæther, B., T. Klokk & H. Taagvoll. 1980. Flora og vegetasjon i Gaulas nedbørfelt, Sør-Trøndelag og Hedmark. Botaniske undersøkelser i 10-års-verna vassdrag. Delrapport 2. – *K. norske vidensk. selsk. museet. Rapport bot. ser. 7.* 1-154.

Sørensen, N. A. 1949. Gjevilvasskammene – nunatakker i Trollheimens midte? – *Naturen 73:* 65-81.

Tengwall, T. Å. 1913. De sydliga skandinaviska fjällväxterna och deras invandringshistoria. – *Sv. bot. tidskr. 7:* 258-274.

Toftaker, H. 1969. *Floristiske undersøkelser i Oppdal herred, Sør-Trøndelag.* – Unpubl. thesis. University of Oslo. 149 pp.

Tolmatchev, A. I. 1923. Über die europäischen Rassen von *Papaver radicatum* Rottb. (in Russian, not seen). – *Notulae systematicae ex Herb. Horti Bot. Petropolit. 4 (11-12),* 87.

Tolmatchev, A. I. 1927. Über die Formen von *Papaver radicatum* Rottb. und ihre Verbreitung in Scandinavien. – *Sv. bot. tidskr. 21:* 73-83.

Tolmatchev, A. I. 1975. *Arktitcheskaya flora SSSR. VII. Papaveraceae-Cruciferae.* – Akad. Nauk SSSR. Moscow – Leningrad.

Tralau, H. 1961. De europeiska arktiskt-montana växternas arealutveckling under kvartärperioden. – *Bot. not, 114:* 213-238.

Tralau, H. 1963. The recent and fossil distribution of some boreal and arctic montane plants in Europe. – *Arkiv bot. ser. 2,5:* 533-582. 8 pl.

Turesson, G. 1927. Contributions to the genecology of glacial relics. – *Hereditas 9.*

Tutin, T. G. et al. 1964-80. *Flora Europaea I-V.* – Cambridge University Press.

Urbanska, K. M. 1986. Some differention pattern within the *Antennaria carpatica* group. – *Acta Univ. Symb. Bot. Ups. XXVII:2.* 207-221.

Urbanska-Worytkiewicz, K. 1967. Cytological investigations in *Antennaria* Gaertn. from North Scandinavia. – *Acta Borealia A. Scientia. No. 22:* 123-131.

Vogt, T. 1913. Landskabsformene i det ytterste av Lofoten. – *Norsk Geogr. Selsk. Årb. 23:* 1-50.

Vogt, T. 1944. *Arenaria norvegica* fra Røros og noen andre plantefunn. – *Blyttia 2:* 37-41.

Vold, L. E. 1982. *Autoøkologiske og synøkologiske studier over Artemisia norvegica* Fr. – Unpubl. thesis. University of Trondheim. 123 pp.

Vorren, T. O., K.-D. Vorren, T. B. Alm, S. Gulliksen & R. Løvlie. 1988. The last deglaciation (20 000 – 11 000 B.P.) on Andøya, northern Norway. – *Boreas 17:* 41-77.

Wahlenberg, G. 1812. *Flora lapponica.* – Berolini.

Wahlenberg, G. 1824, 1826. *Flora Svecica.* – Upsaliae.

Warming, E. 1888. Om Grønlands Vegetation. – *Medd. Grønl. 12.* 1-233.

Warming, E. 1890. Biologiske Optegnelser om grønlandske Planter. 3. *Schrophulariaceae. – Bot. tidsskr. 17.*

Weimarck, G. 1976. Karyotypes and population structure in aneuploid *Hierochloë alpina* ssp. *alpina (Gramineae)* in northern Scandinavia. – *Hereditas 82:* 149-156.

Weimarck, G. 1981. Numerical analysis of the floristic composition of localities including *Hierochloë (Poaceae)* species in northern Europe. – *Vegetatio 44:* 101-135.

Wesenberg, J. 1988. Primærlokaliteter for skjeggklokke, *Campanula barbata* L. i Norge? – *Blyttia 46:* 154-159.

Wikström, J. E. 1827. Öfversikt af Botaniska arbeten och Upptäckter uti Norrige för år 1826. – *Årsber. framstegen uti botanik för år 1826.* 250-251, 281-286.

Wille, N. 1905. Om Indvandringen af det arktiske Flora-element til Norge. – *Nyt mag. naturv. 43:* 315-338.

Wille, N. & J. Holmboe. 1903. *Dryas octopetala* bei Langesund. Eine glaciale Pseudorelikte. – *Nyt mag. naturv. 41:* 27-43.

Wilmann, B. 1983. *Økologiske studier av Pedicularis oederi Vahl og noen andre plantegeografisk interessante arter i Tifjellområdet på Nordmøre.* – Unpubl. thesis. University of Trondheim. 174 pp.

Yeo, P.F. 1978. A taxonomic revision of *Euphrasia* in Europe. – *Bot. Journ. Linn. Soc. 77:* 223-334.

Zetterstedt, J. E. 1822. *Resa genom Sweriges och Norriges Lappmarker, förrättad År 1821. I och II.* Lund. 227-231.

Zetterstedt, J. E. 1854. Om vegetationen ved Altefjord. – *Öfvers. K. sv. vetensk.akad. förh. 31. no. 10:* 33-51.

INDEX OF PLANT NAMES

Only the names of the species treated are included.
Italicized figures indicate the page where the species have been discussed in connection with their maps.

Legends to the Maps

Herbarium material Literary records

● ○ Ordinary stations

✳ ☼ Localization inexact

⊕ ⊕ Extinct

● Rejected stations

Antennaria alpina (L.) Gaertn.♂ *Antennaria nordhageniana*   *Antennaria porsildii Elis. Ekman*
                                    Rune & Rønning

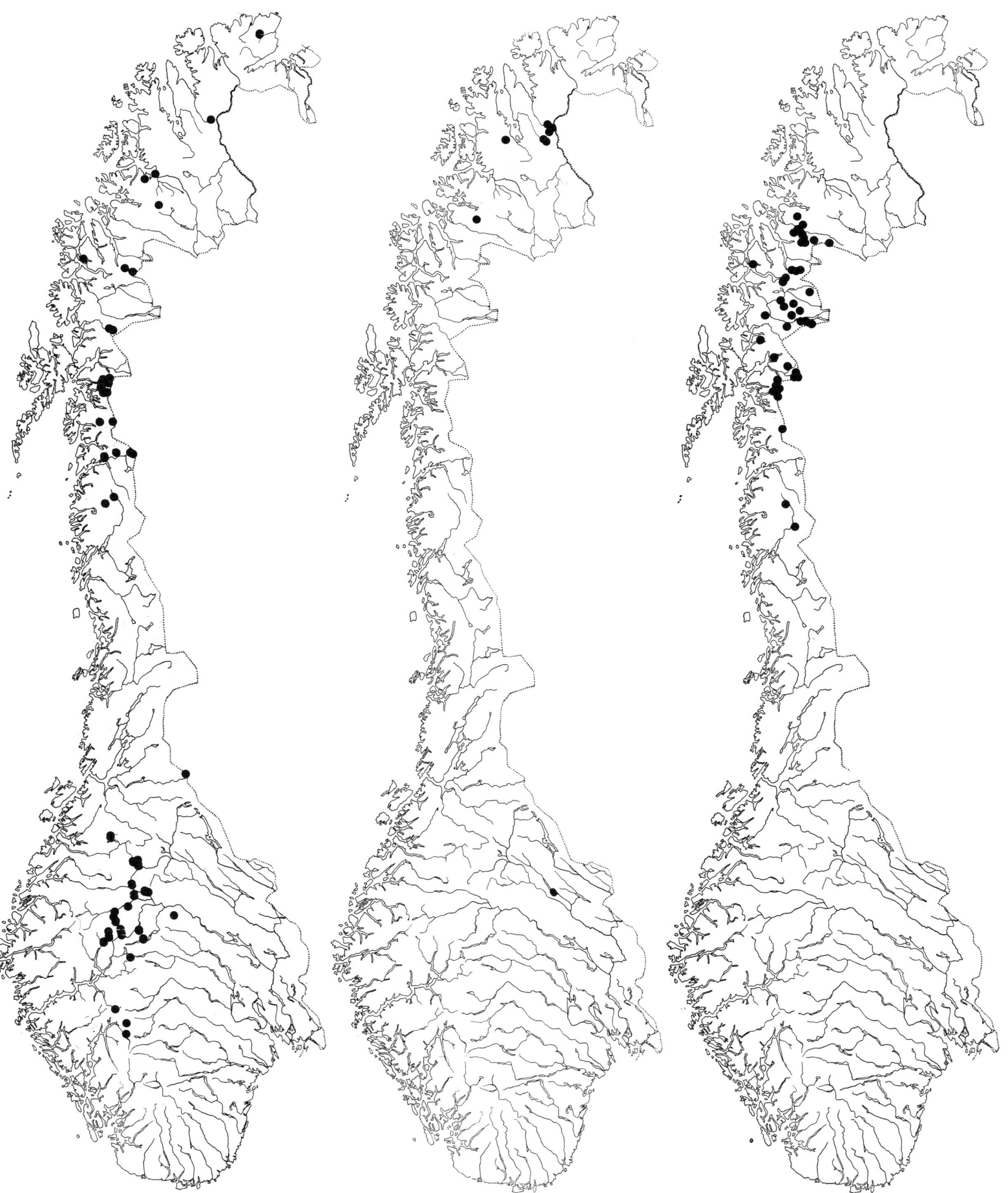

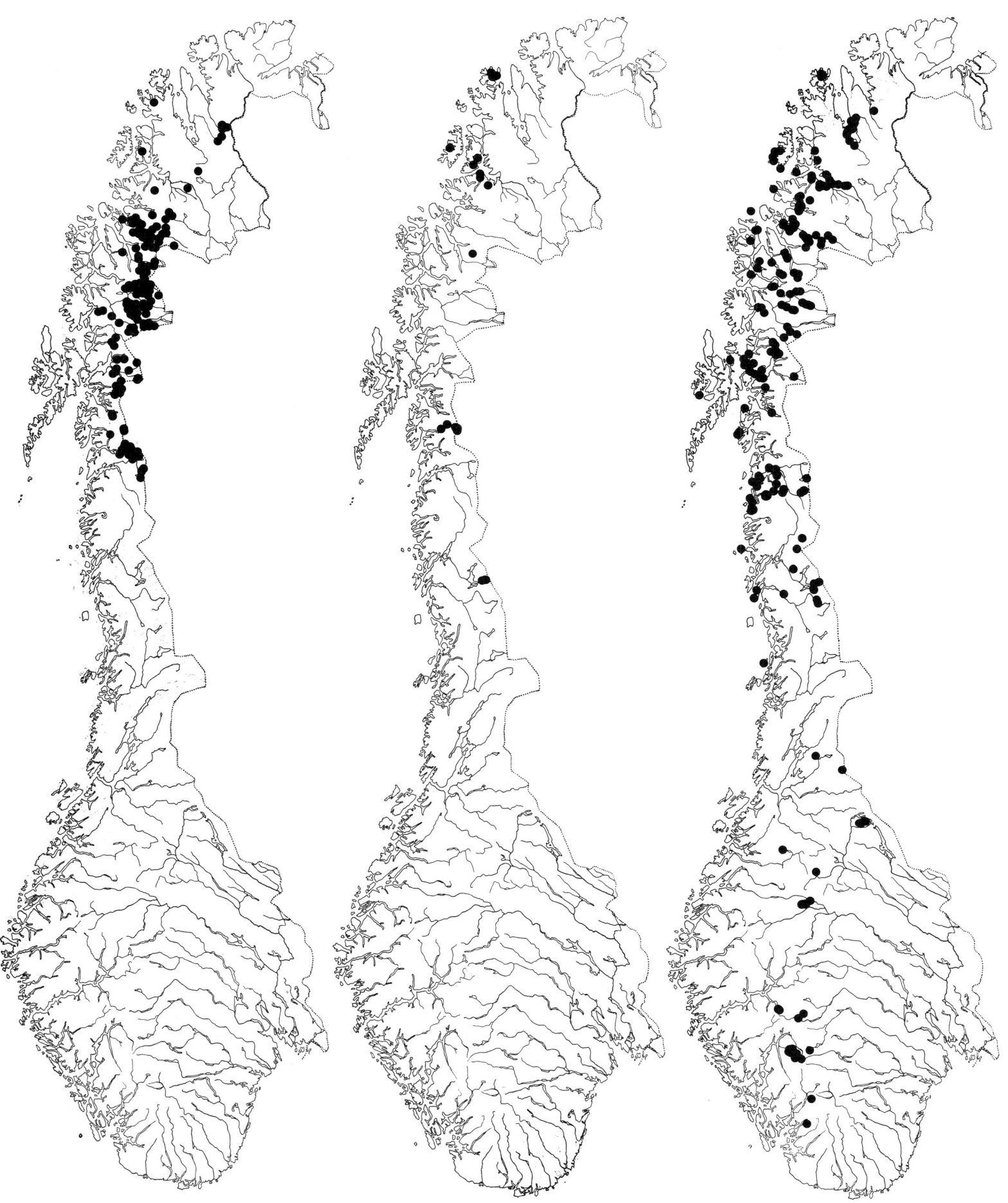

Antennaria villifera Boriss.   *Arenaria humifusa Wahlenb.*   *Arenaria norvegica Gunn.*

Arenaria pseudofrigida
(Ostenf. & Dahl) Juz.

Armeria scabra Pall.

Arnica angustifolia Vahl subsp.
alpina (L.) I.K. Ferguson

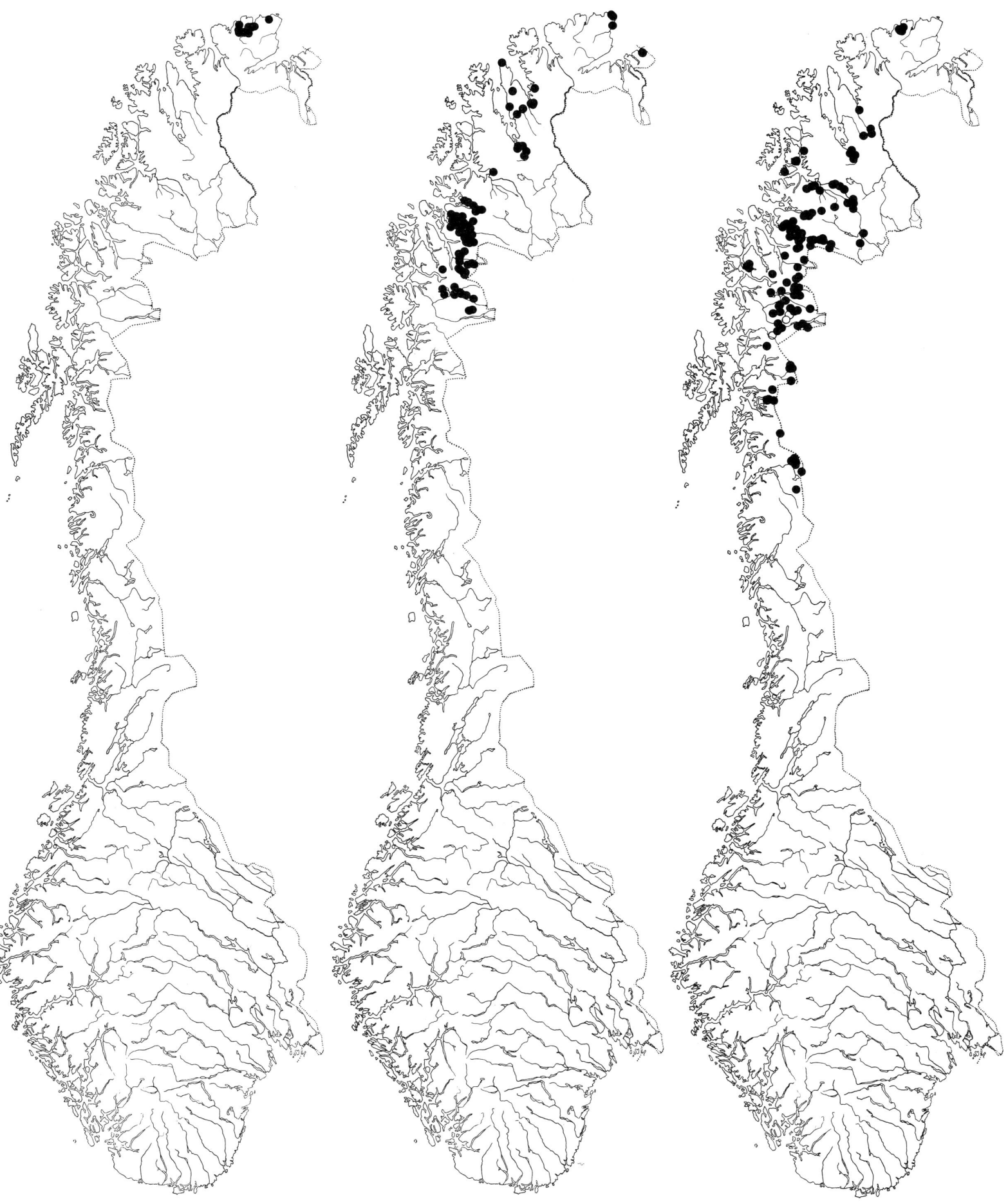

Artemisia norvegica Fr.   *Astragalus frigidus (L.) A. Gray*   *Astragalus norvegicus Weber*

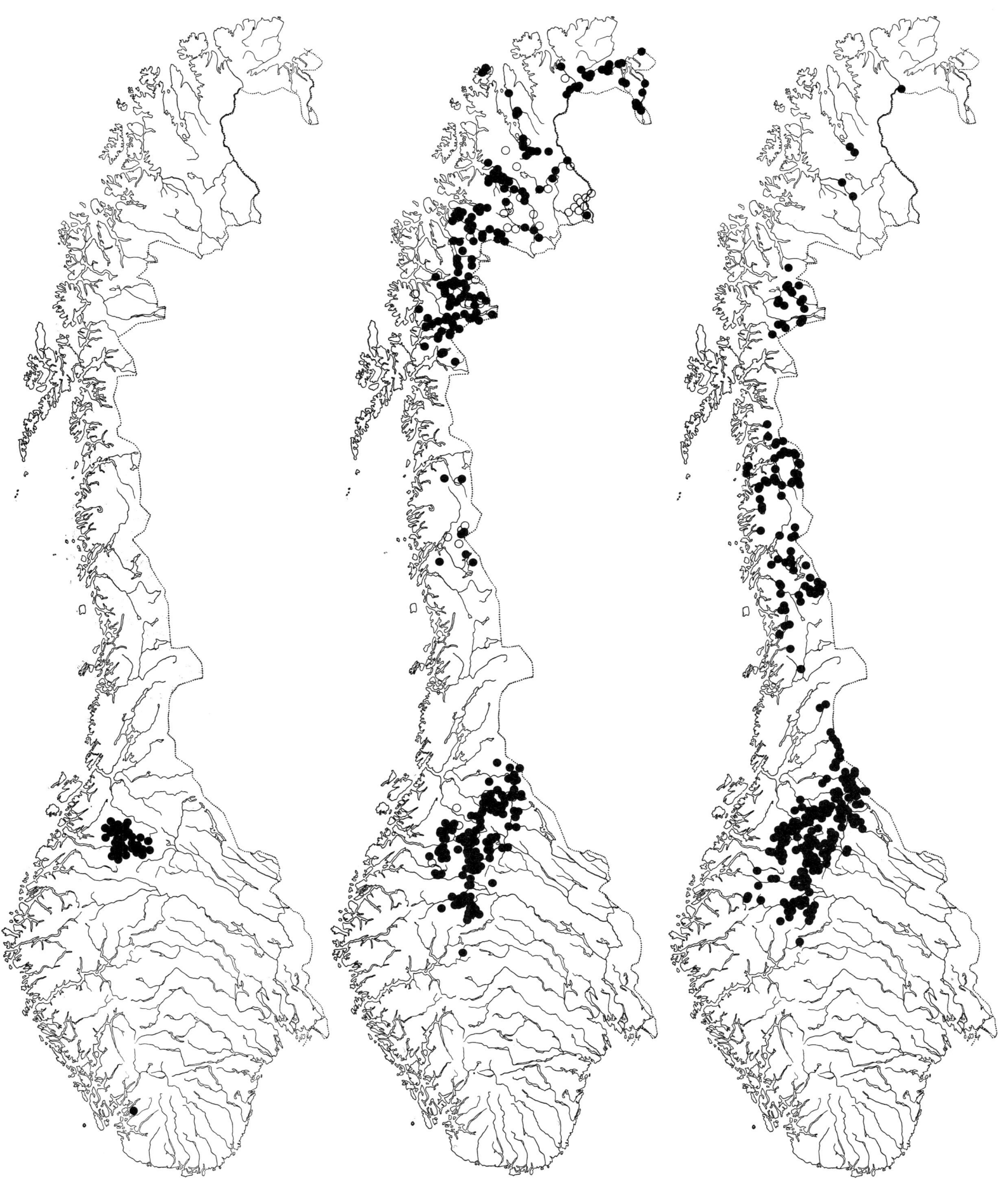

Athyrium distentifolium Opiz   *Botrychium boreale Milde*   *Braya linearis Rouy*

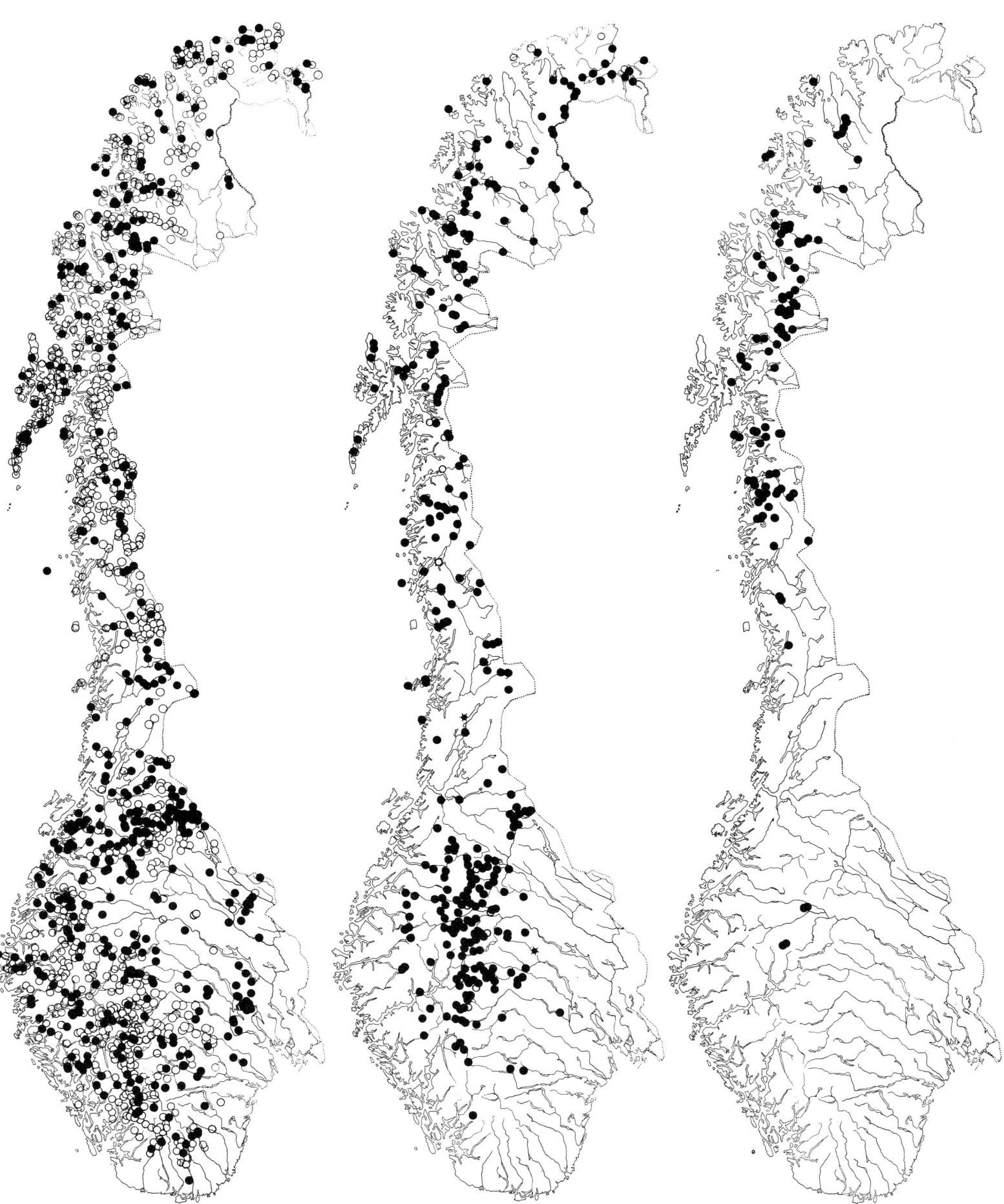

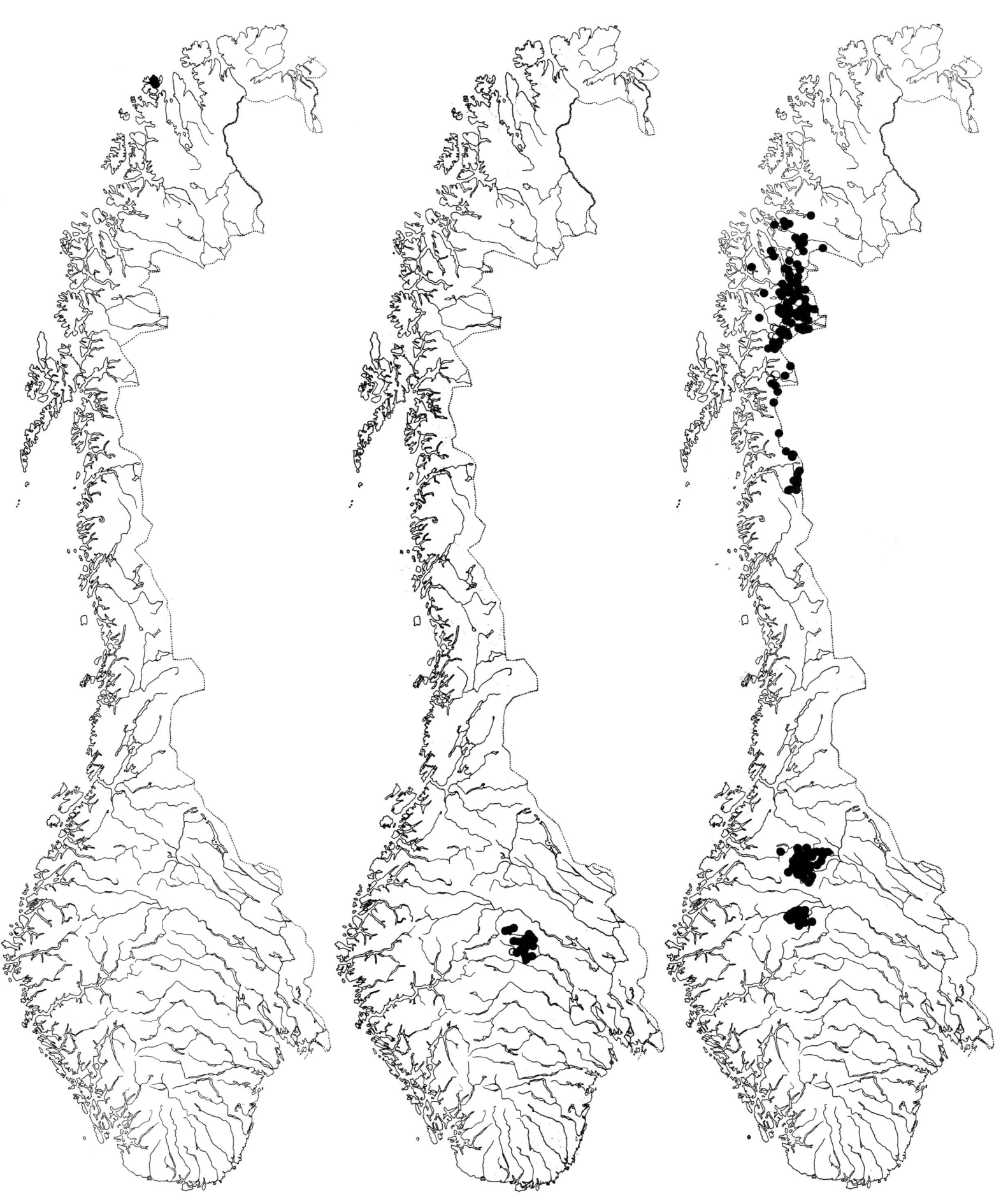

Braya purpurascens (R.Br.) Bge.   *Campanula barbata L.*   *Campanula uniflora L.*

Cardaminopsis petraea (L.) Hiit.   *Carex arctogena H. Sm.*   *Carex atrofusca Schkuhr*

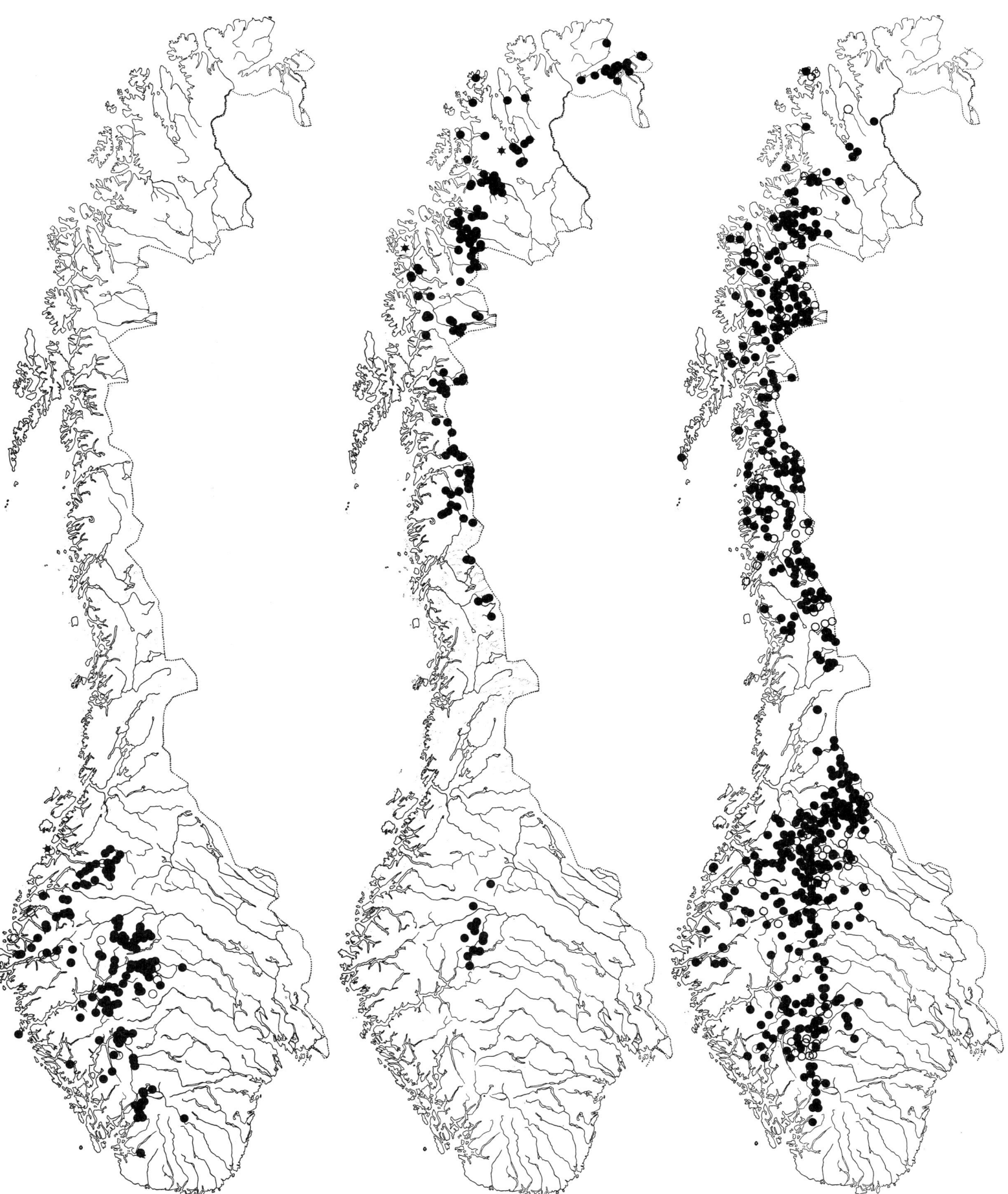

Carex bicolor All.   *Carex capitata L.*   *Carex glacialis Mack.*

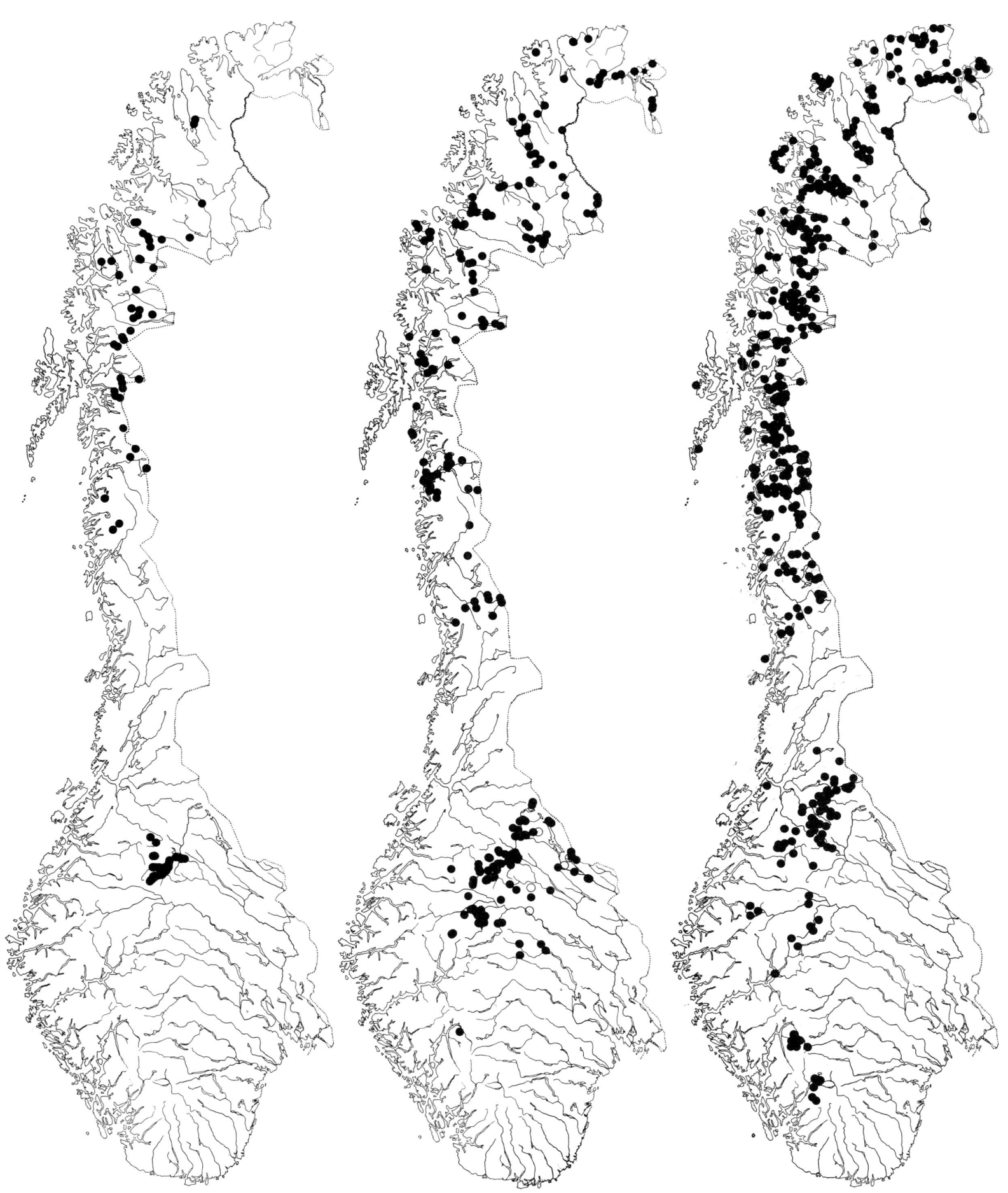

Carex holostoma Drej.   *Carex macloviana D'Urv.*   *Carex microglochin Wahlenb.*

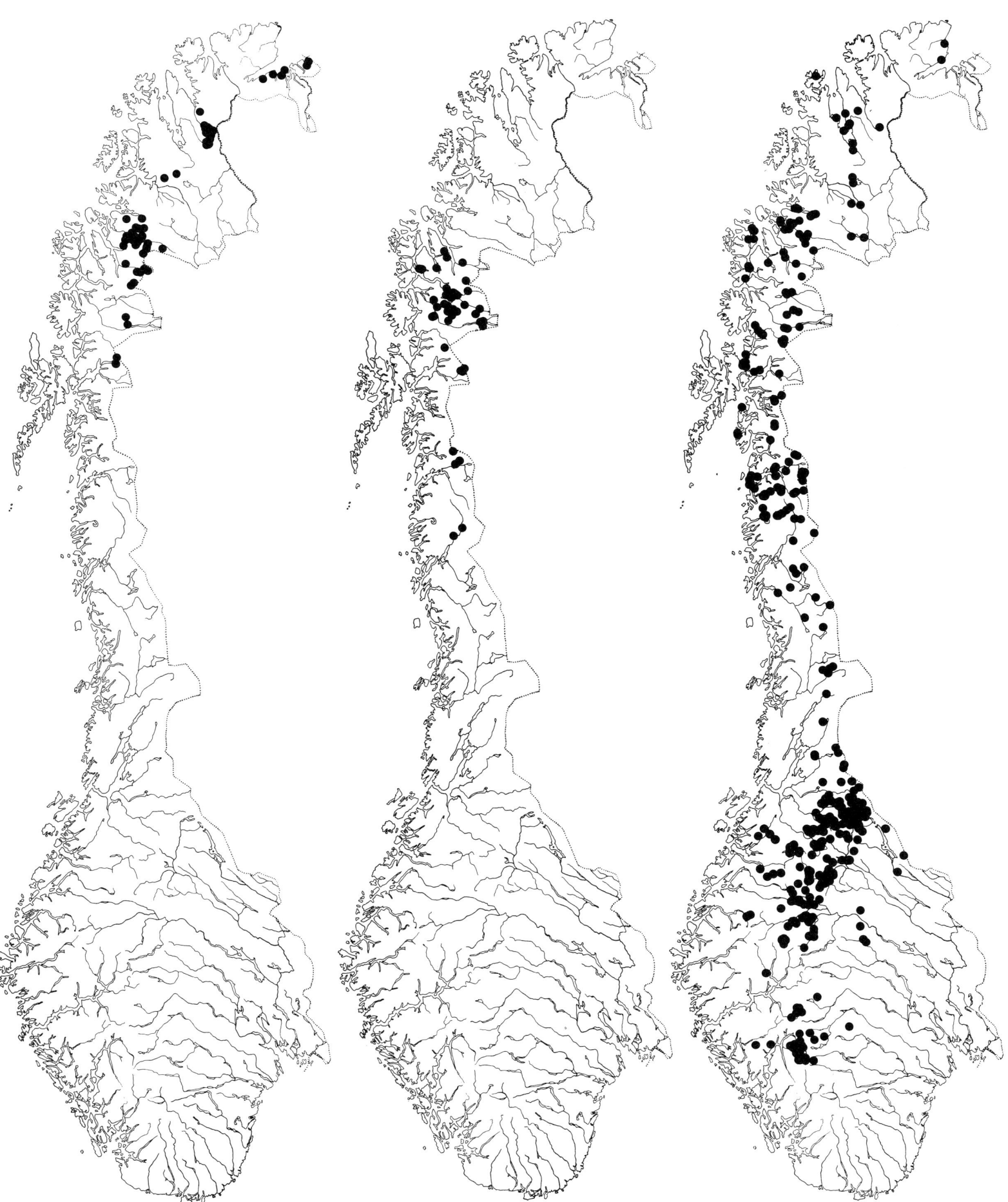

Carex misandra R.Br.　　　*Carex nardina Fr.*　　　*Carex parallela (Læst.) Sommerf.*

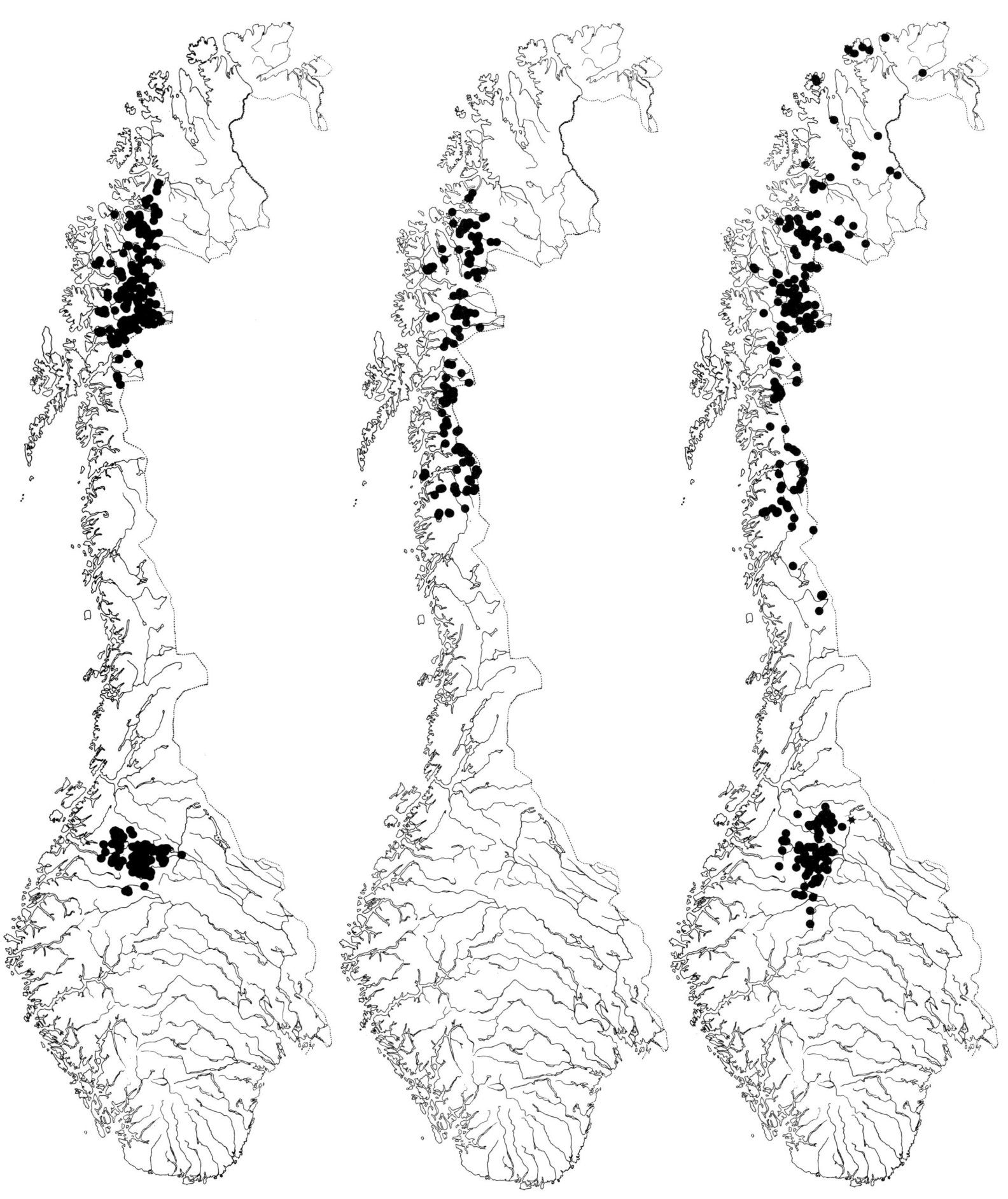

Carex rufina Drej.   *Carex rupestris All.*   *Carex scirpoidea Michx.*

Cassiope tetragona (L.) D.Don   *Cerastium arcticum Lge.*   *Chamorchis alpina (L.) Rich.*

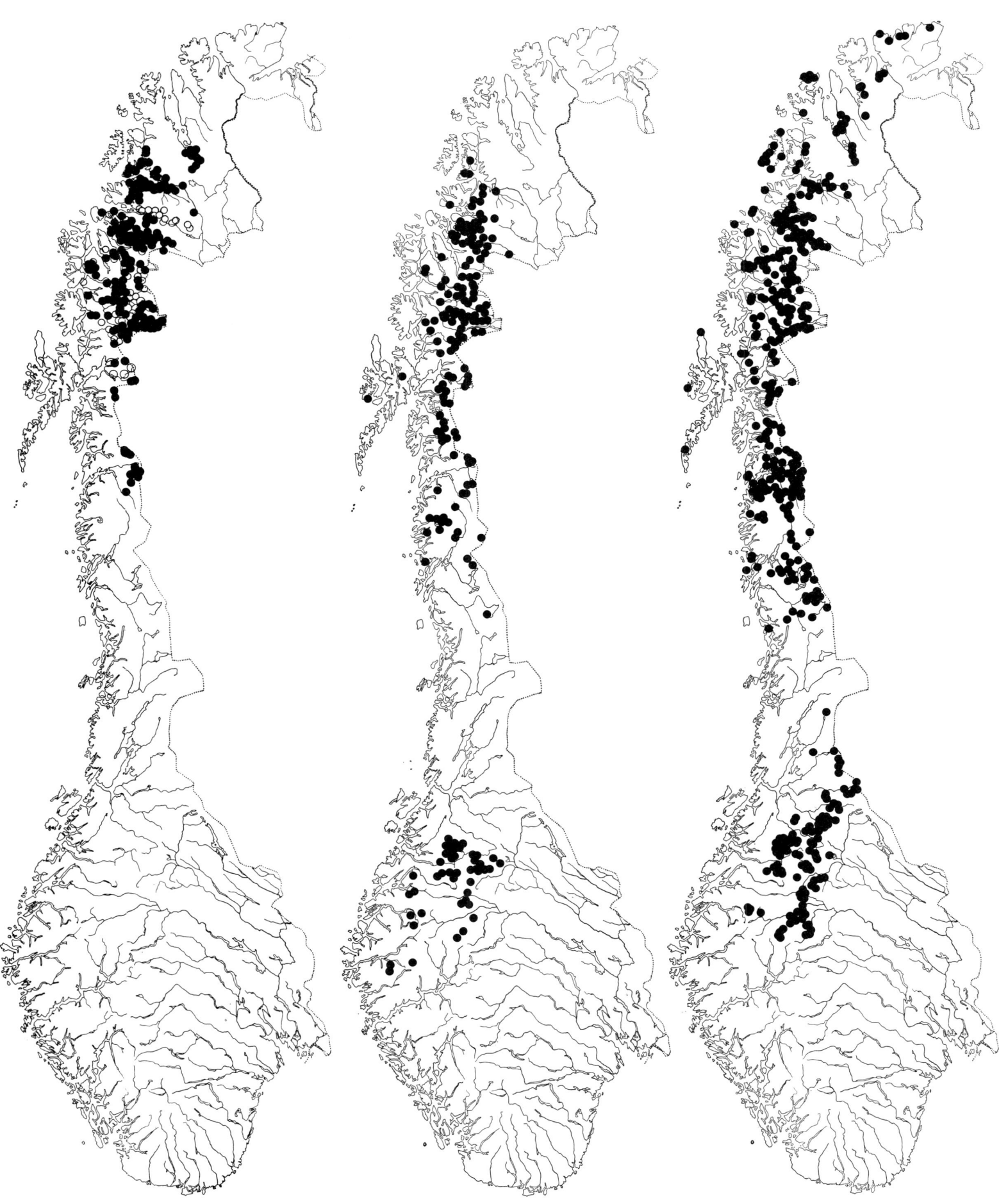

Cystopteris montana (Lam.) Desv.   *Diapensia lapponica L.*   *Draba alpina L.*

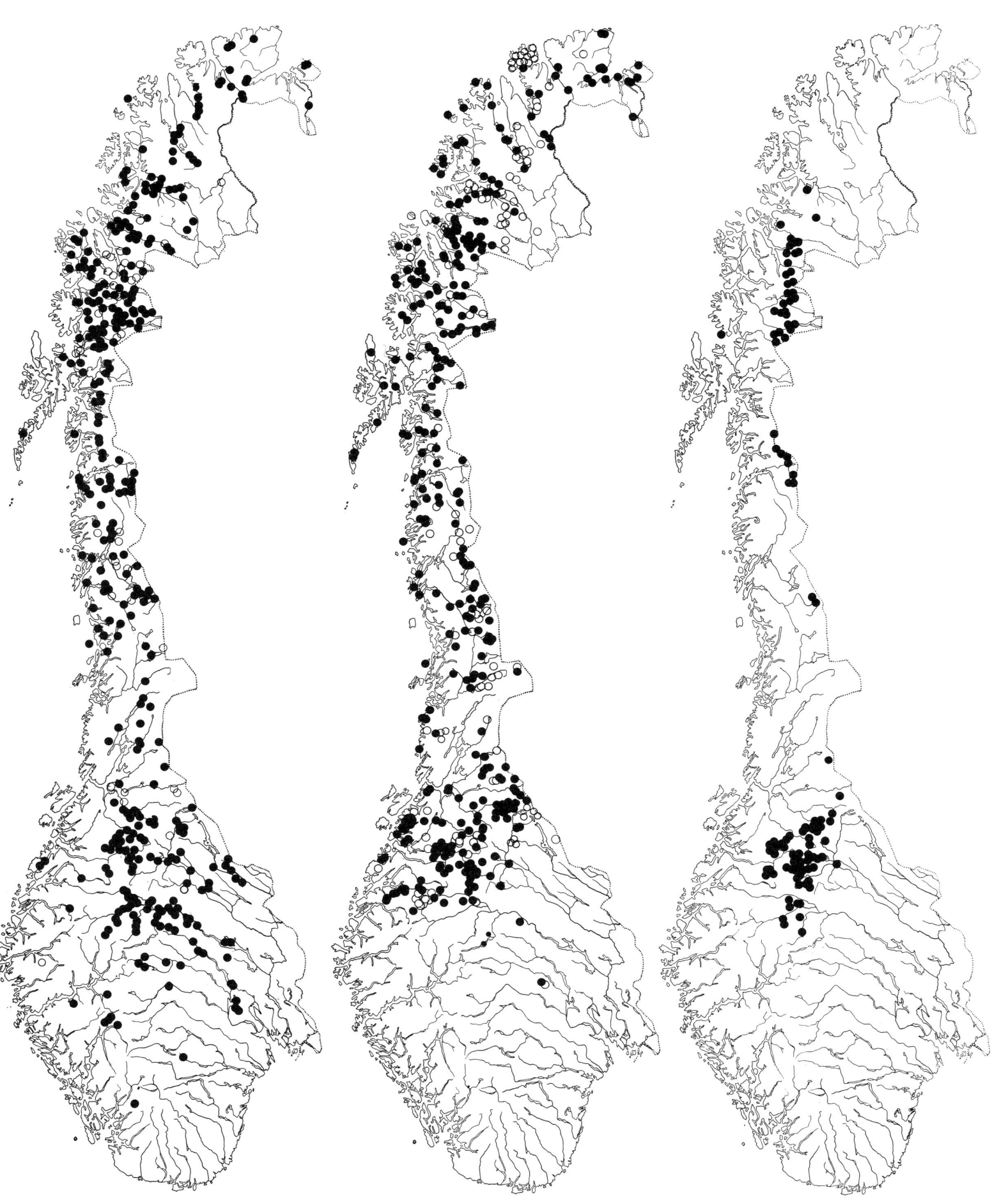

Draba cacuminum Elis. Ekman   *Draba crassifolia Grah.*   *Draba fladnizensis Wulf.*

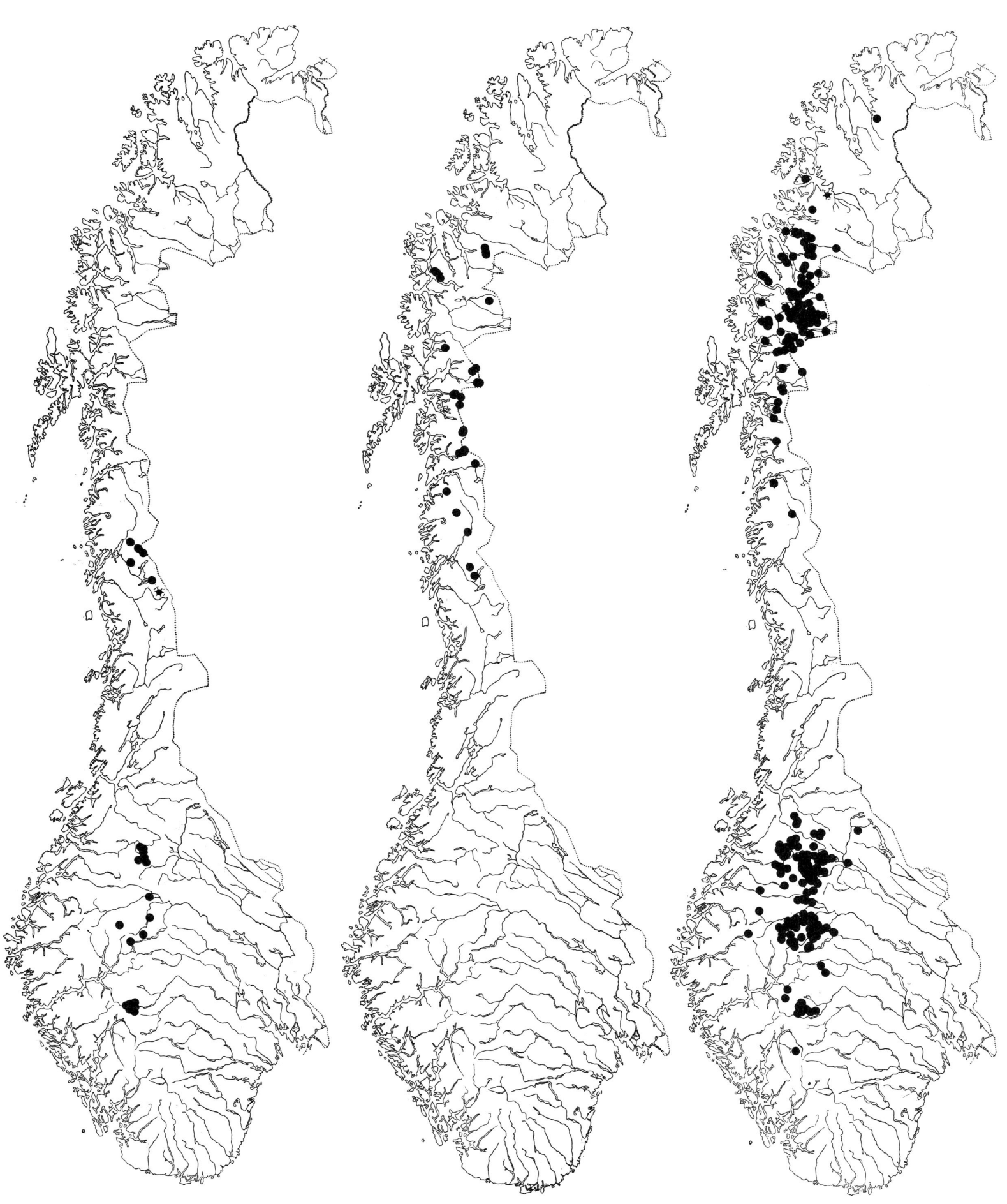

Draba lactea Adams   *Draba nivalis Liljebl.*   *Draba oxycarpa Sommerf.*

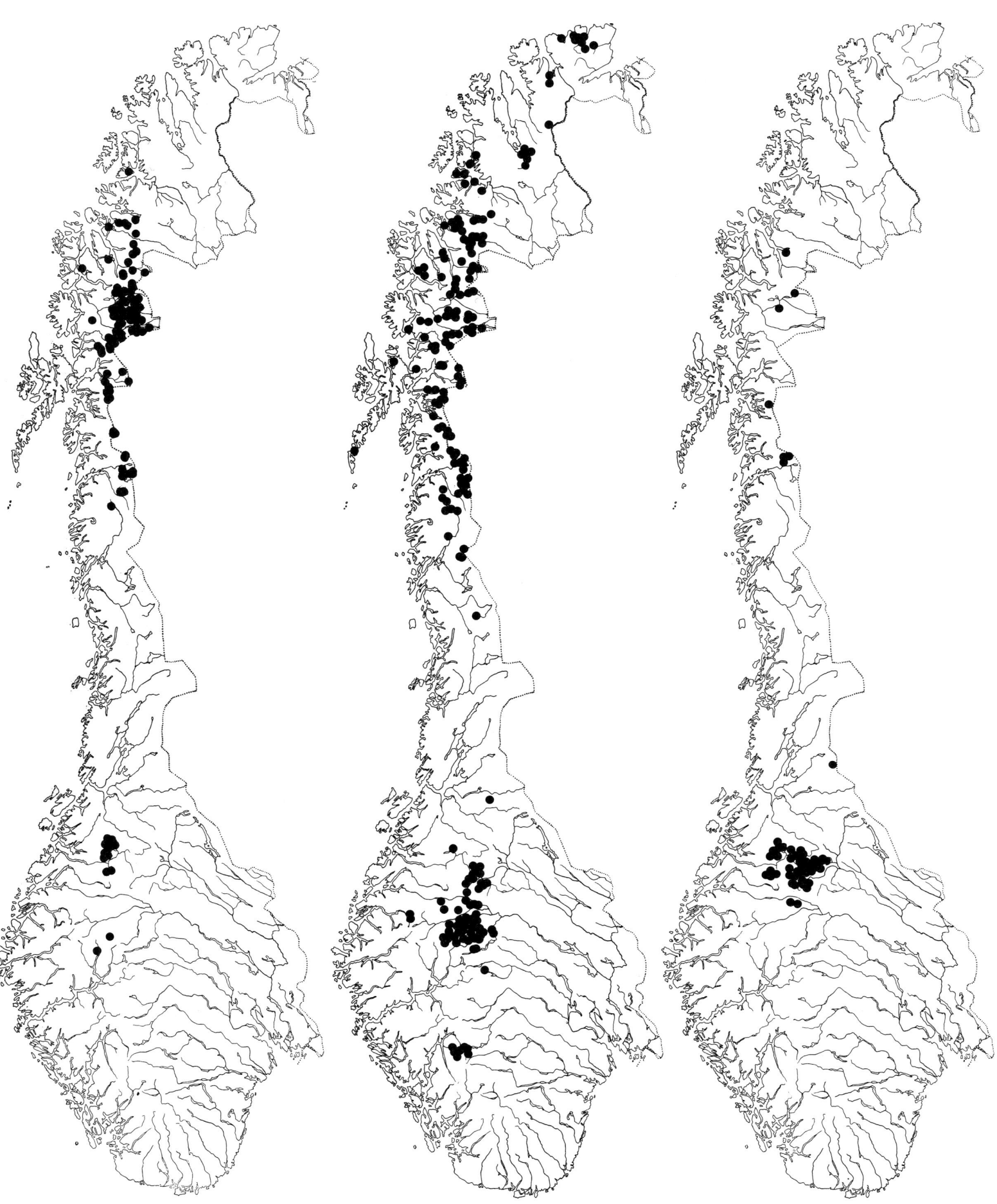

Dryas octopetala L.   *Epilobium davuricum Horn.*   *Equisetum scirpoides Michx.*

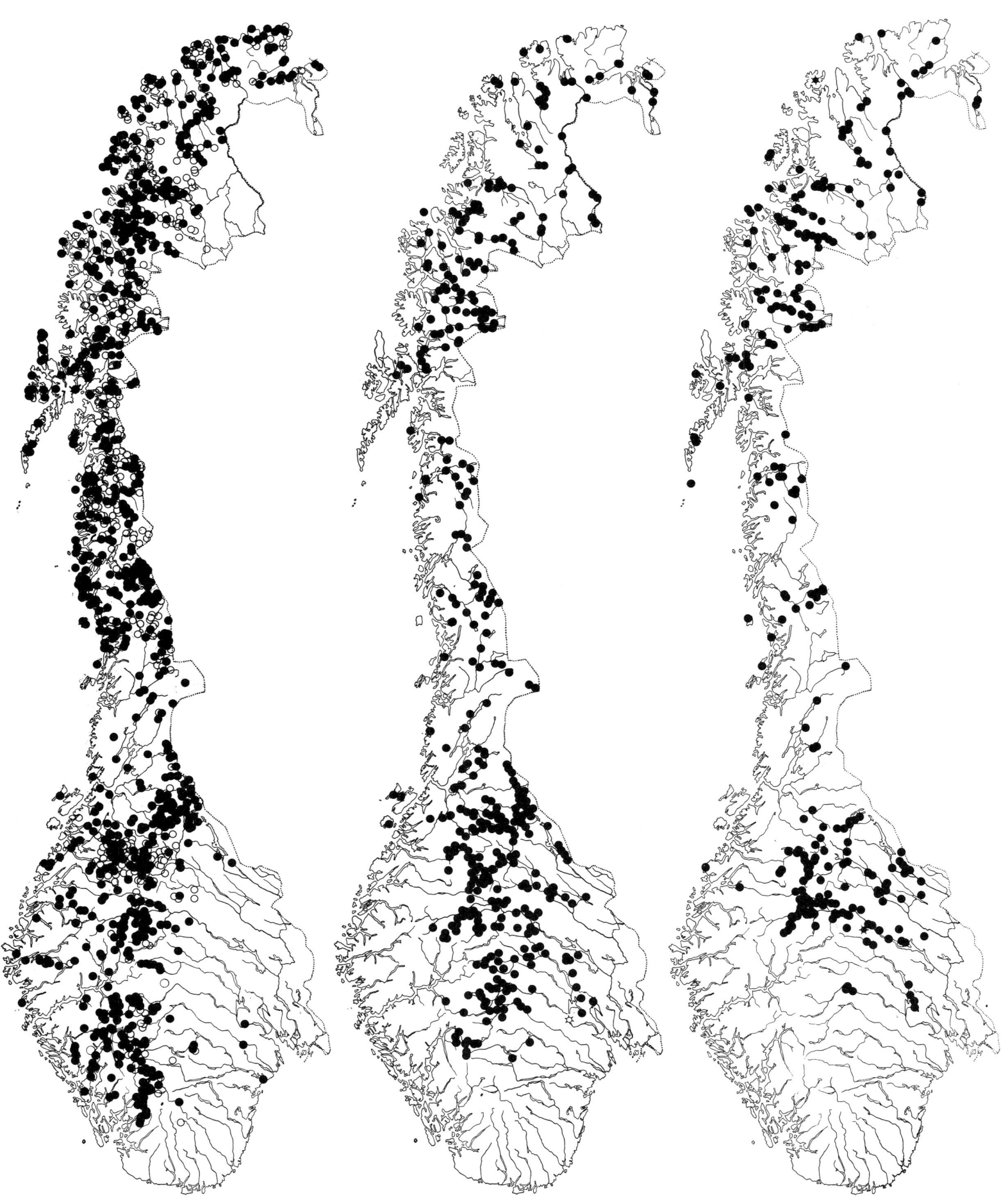

Equisetum variegatum Weber & Mohr   *Erigeron humilis Grah.*   *Erigeron politus Fr.*

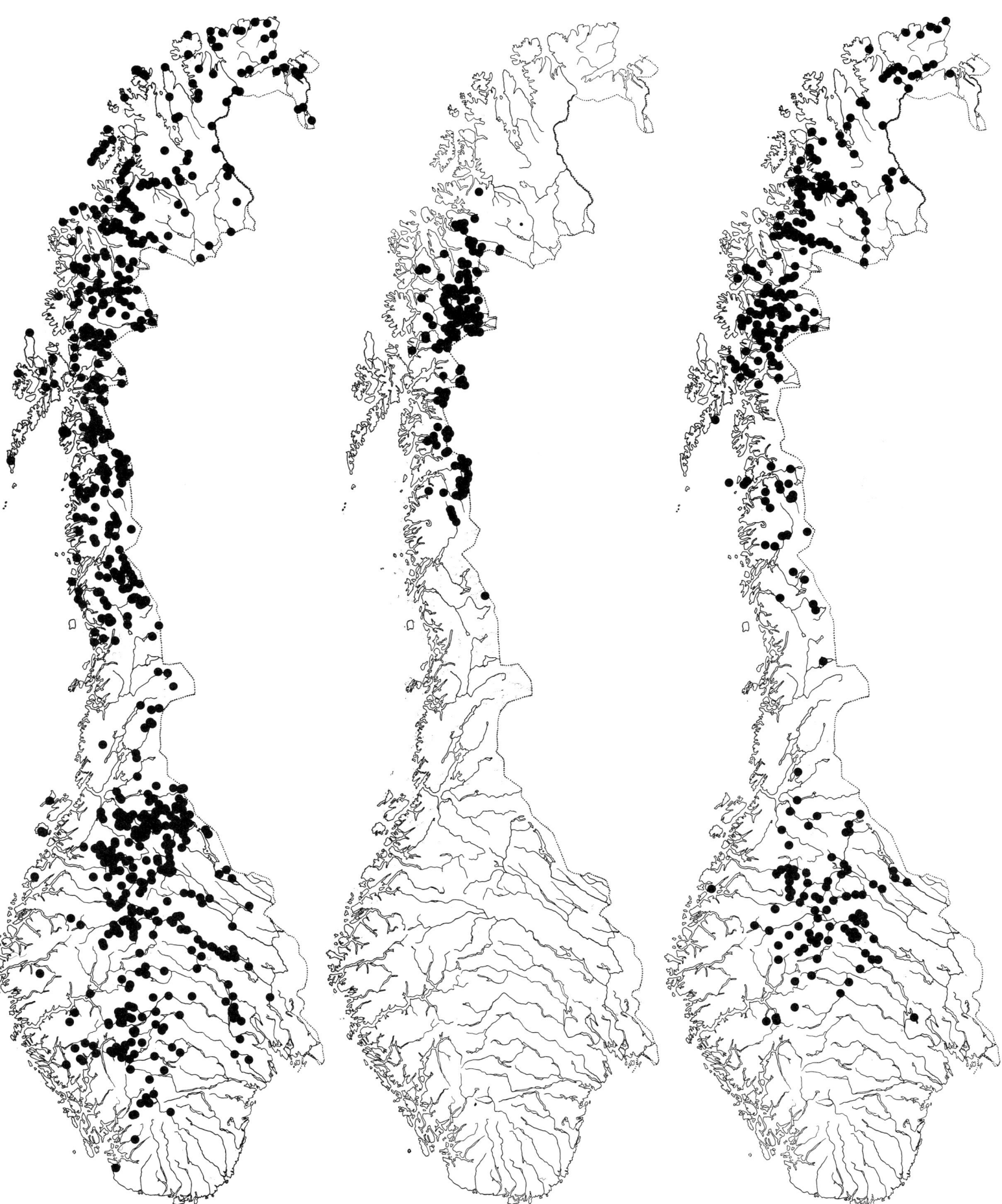

Euphrasia salisburgensis Funck   *Gentiana purpurea L.*   *Gentianella tenella (Rottb.) Börner*

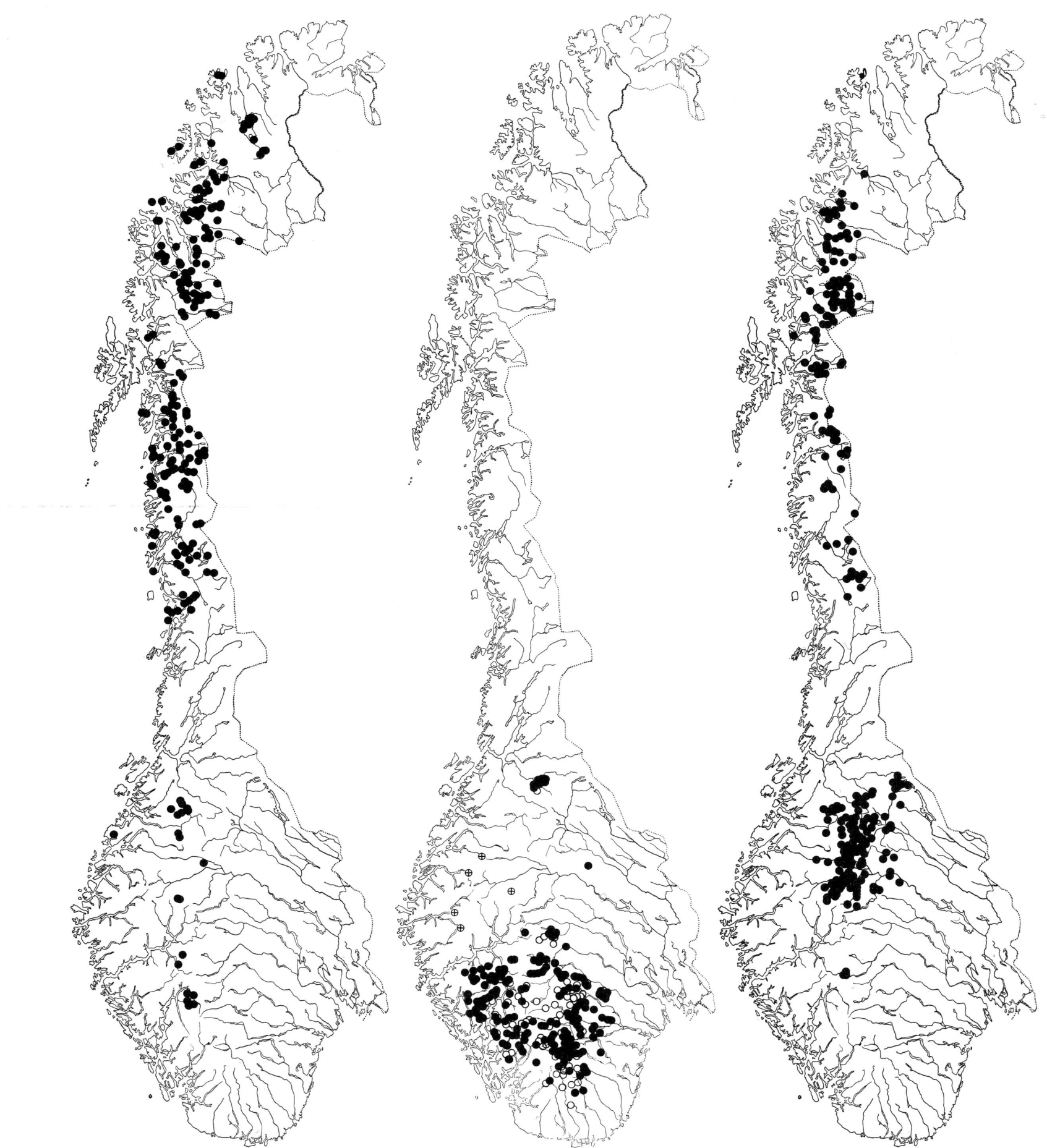

Hierochloë alpina (Willd.)
Roemer & Schultes

Juncus arcticus Willd.

Juncus castaneus Sm.

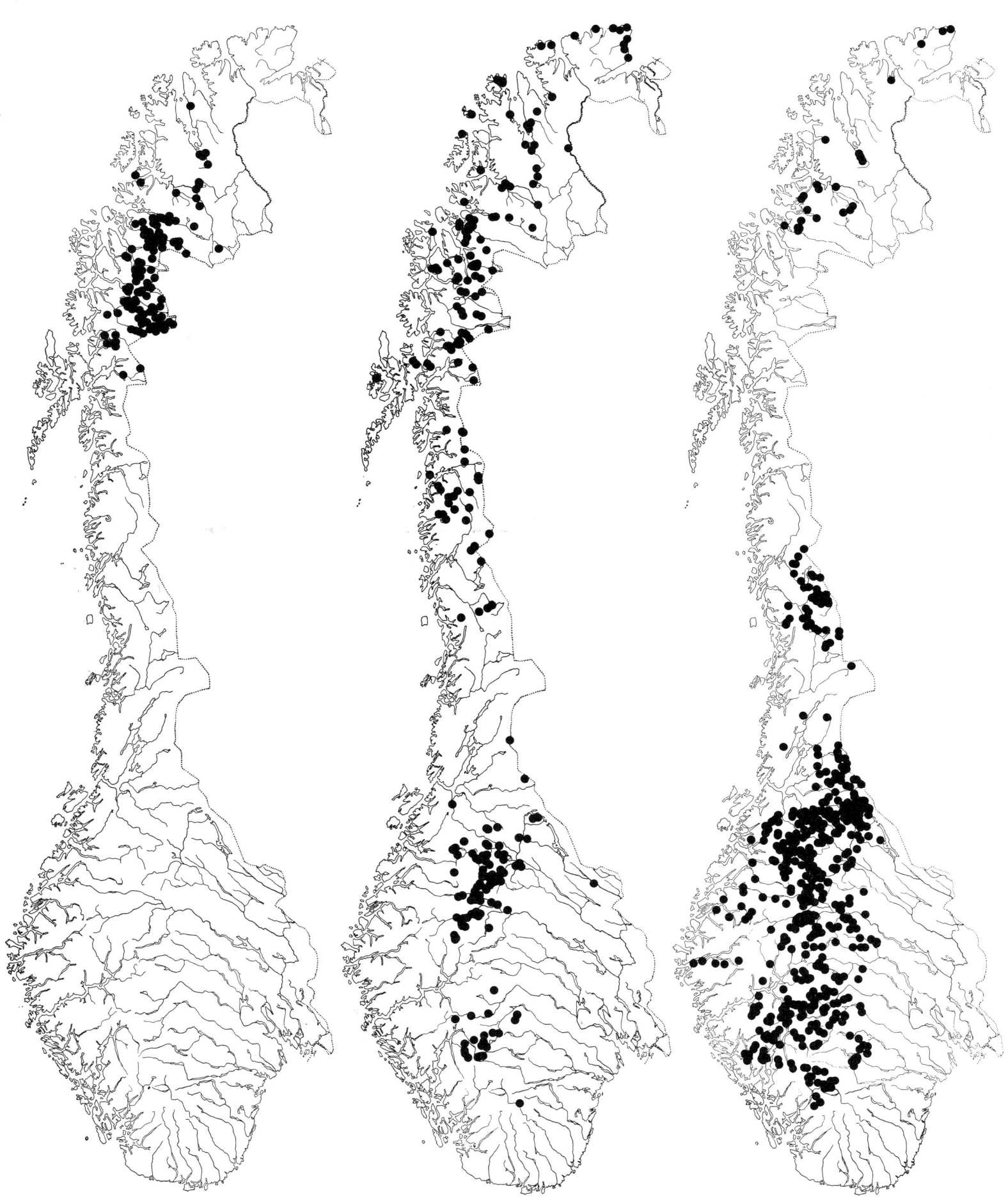

Kobresia myosuroides (Vill.) Fiori *Kobresia simpliciuscula* *Koenigia islandica L.*
(Wahlenb.) Mack.

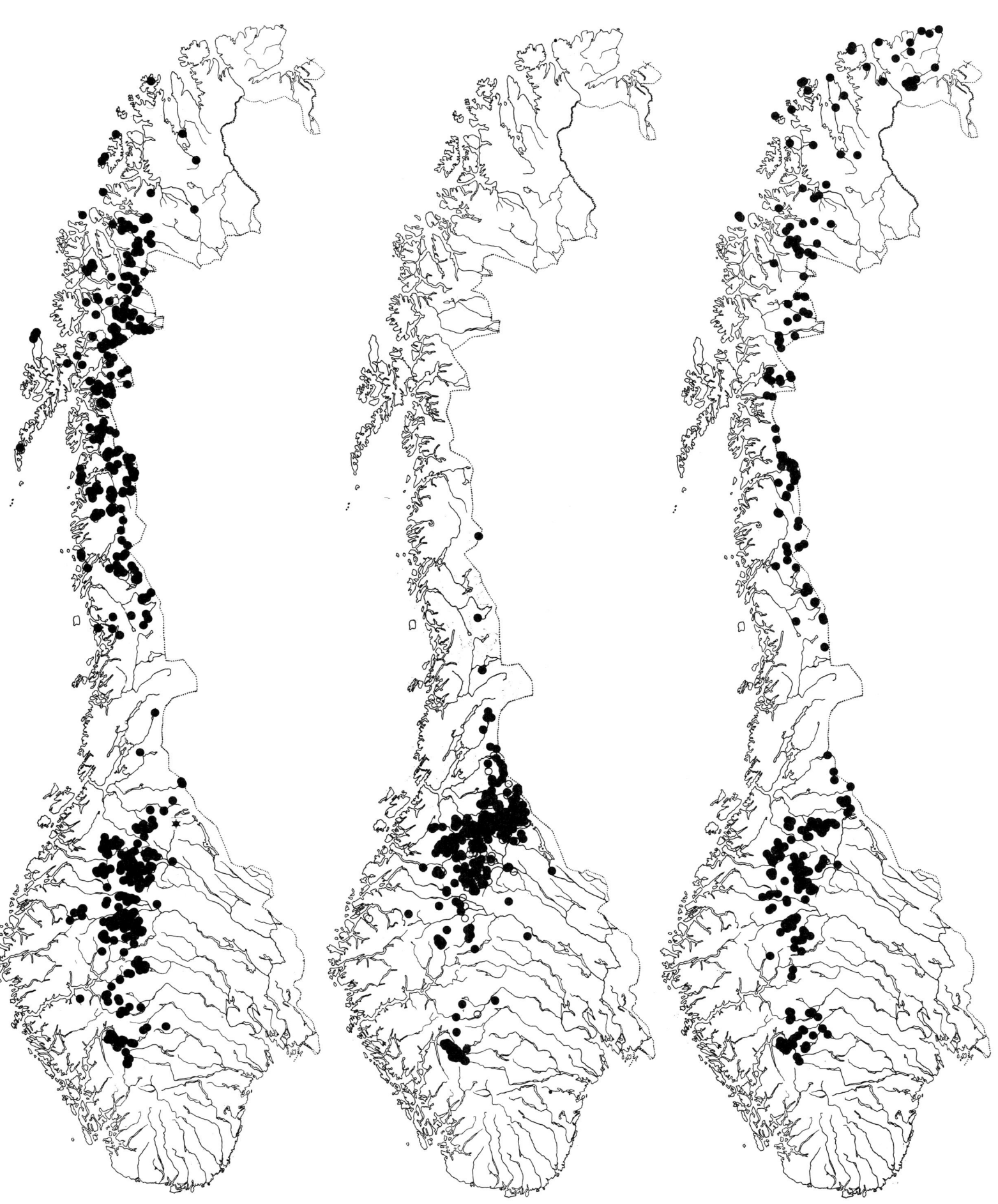

Leucorchis straminea (Fern.) A. Löve   *Luzula arctica Blytt*   *Luzula parviflora (Ehrh.) Desv.*

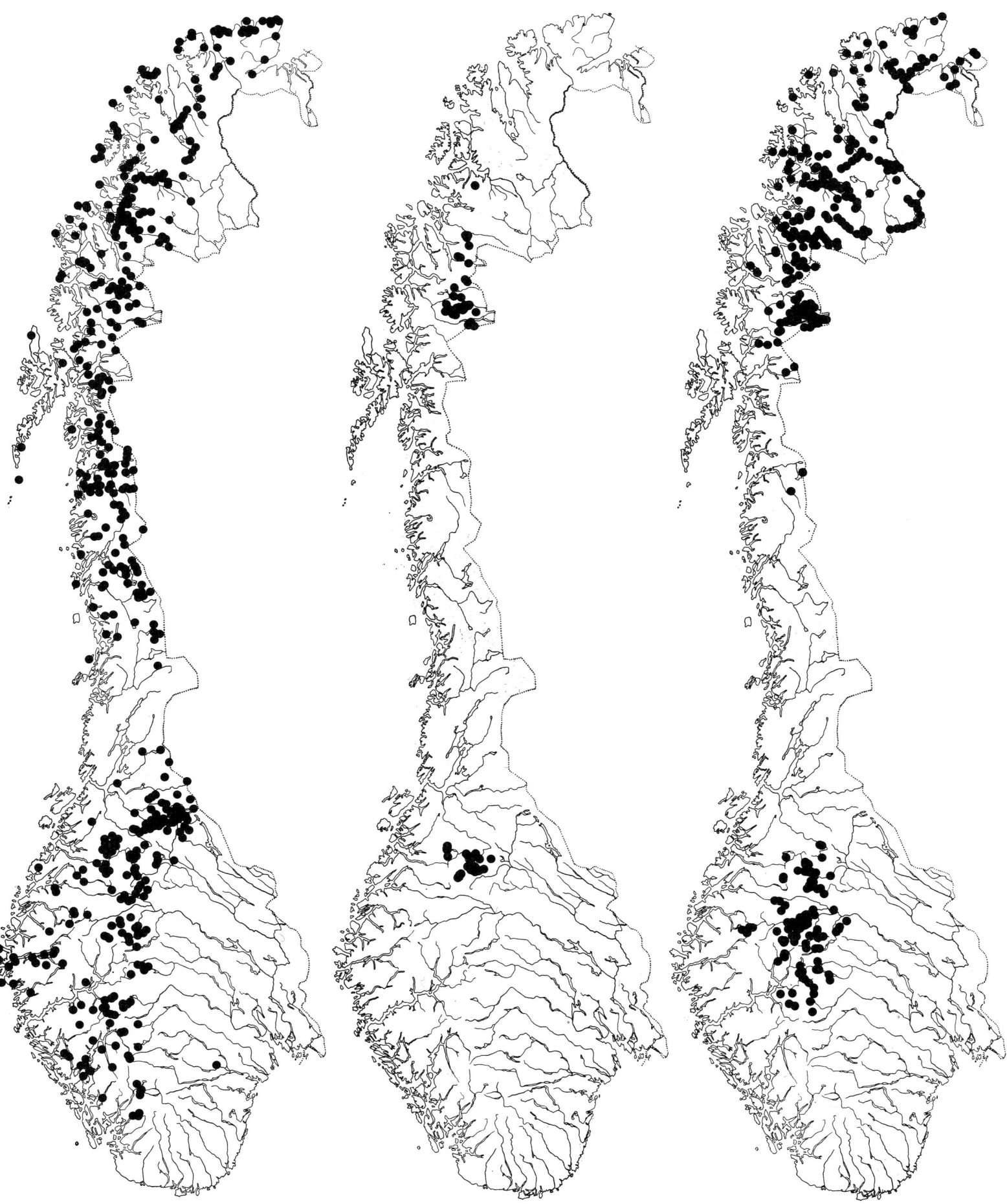

Luzula wahlenbergii Rupr. *Lychnis alpina L.* *Minuartia rubella (Wahlenb.) Hiern.*

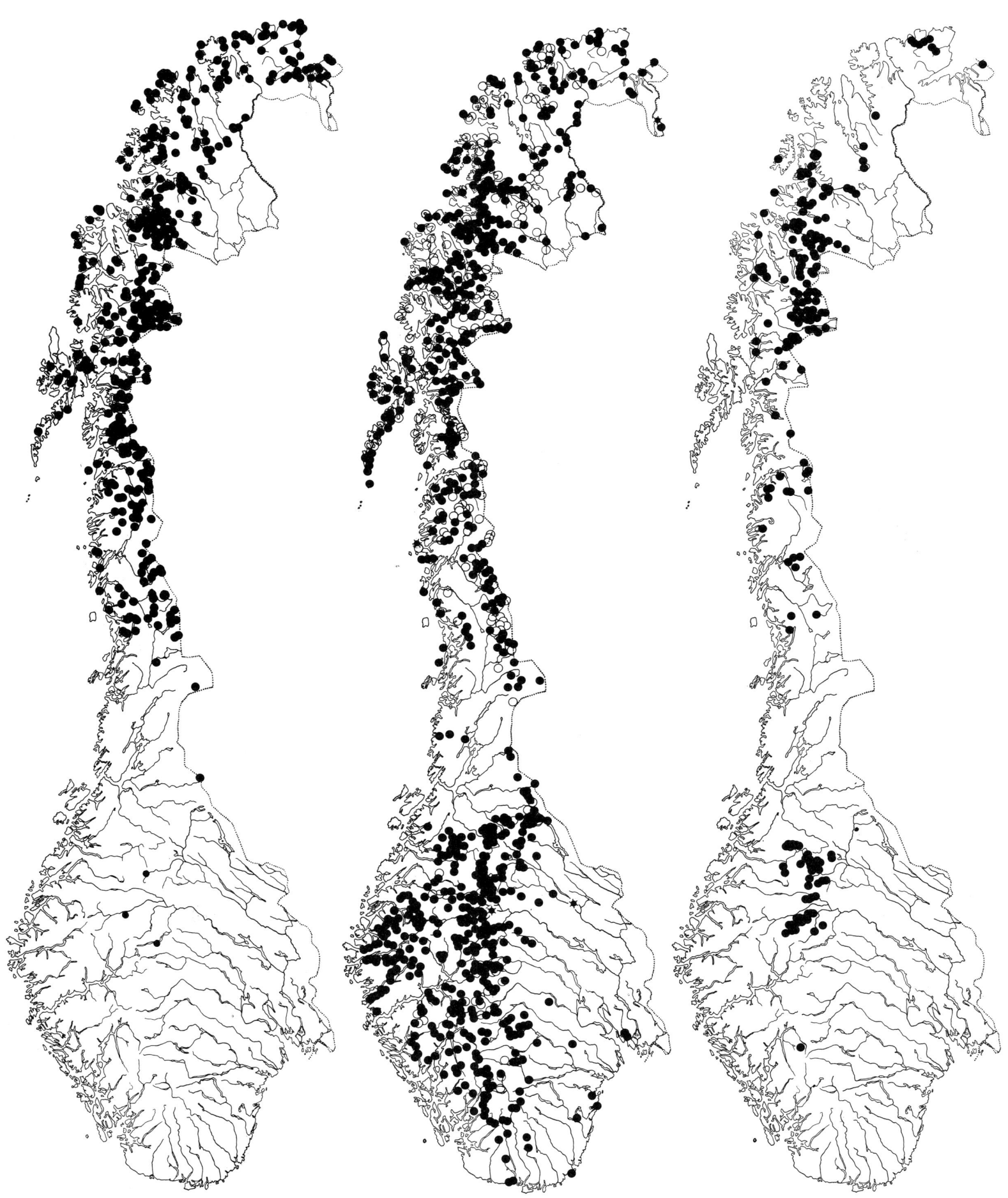

Minuartia stricta (Sw.) Hiern. *Nigritella nigra (L.) Rchb. fil.* *Oxyria digyna (L.) Hill*

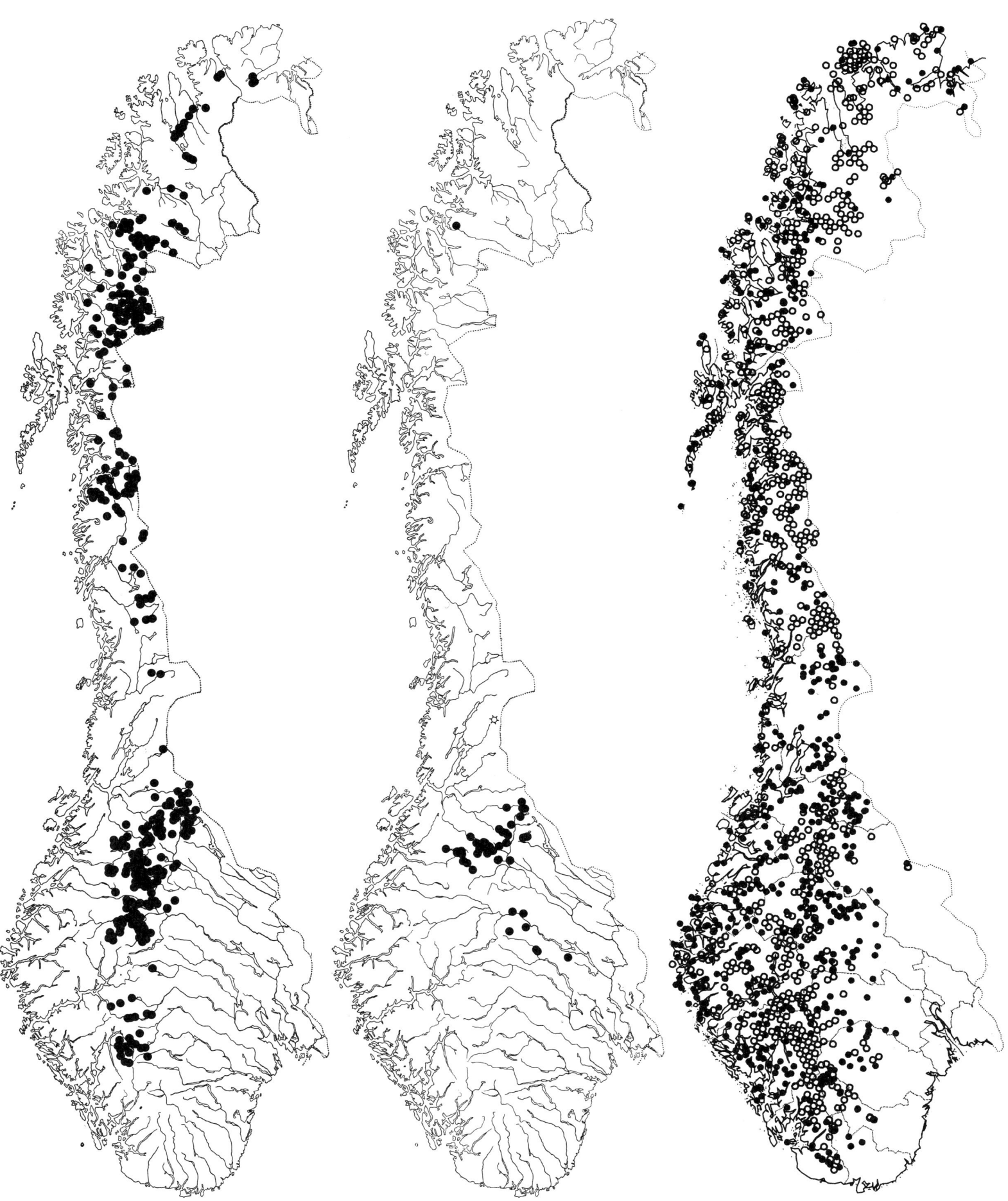

Oxytropis lapponica (Wahlenb.) Gay *Papaver dahlianum Nordh.* *Papaver læstadianum (Nordh.) Nordh.*

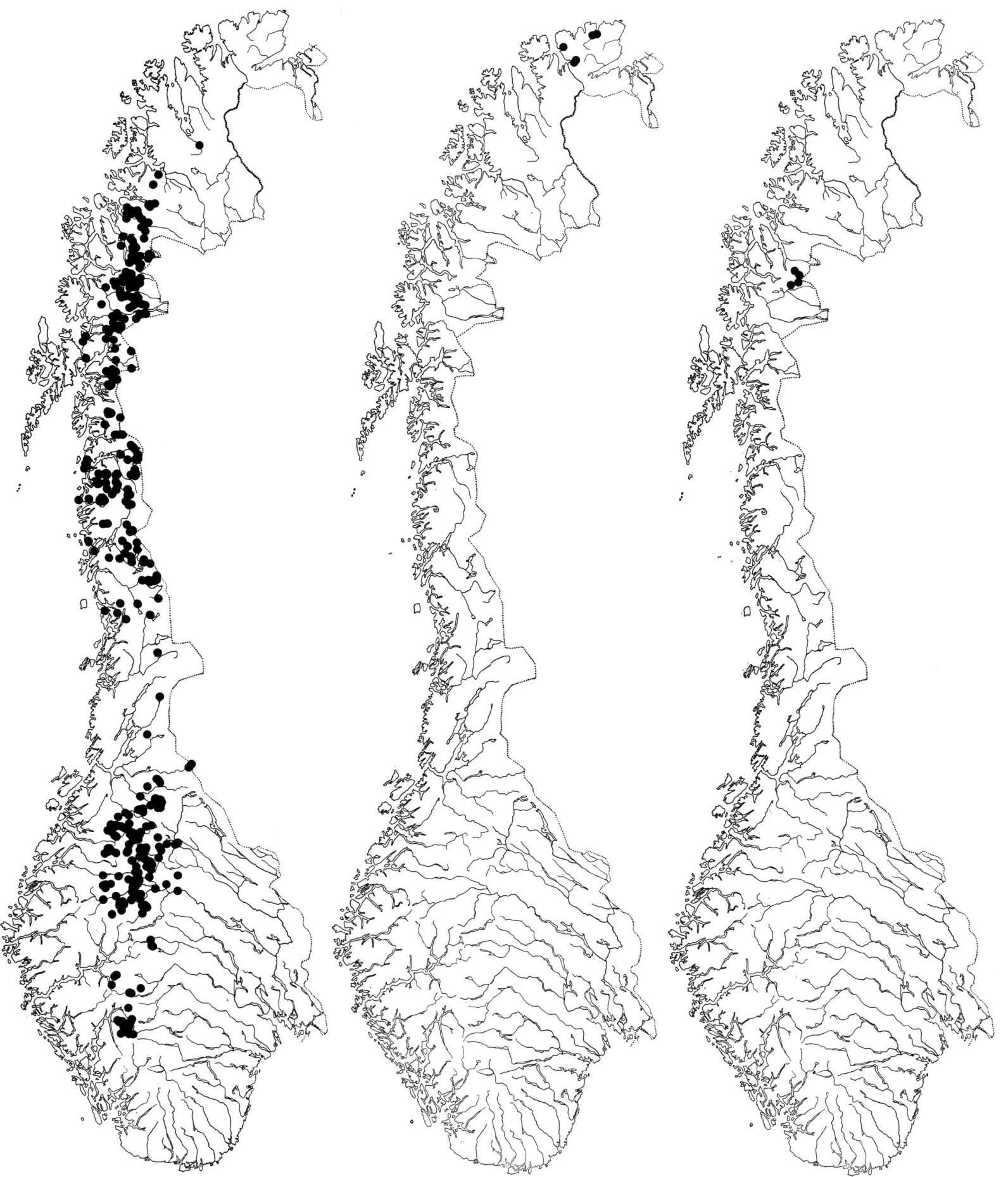

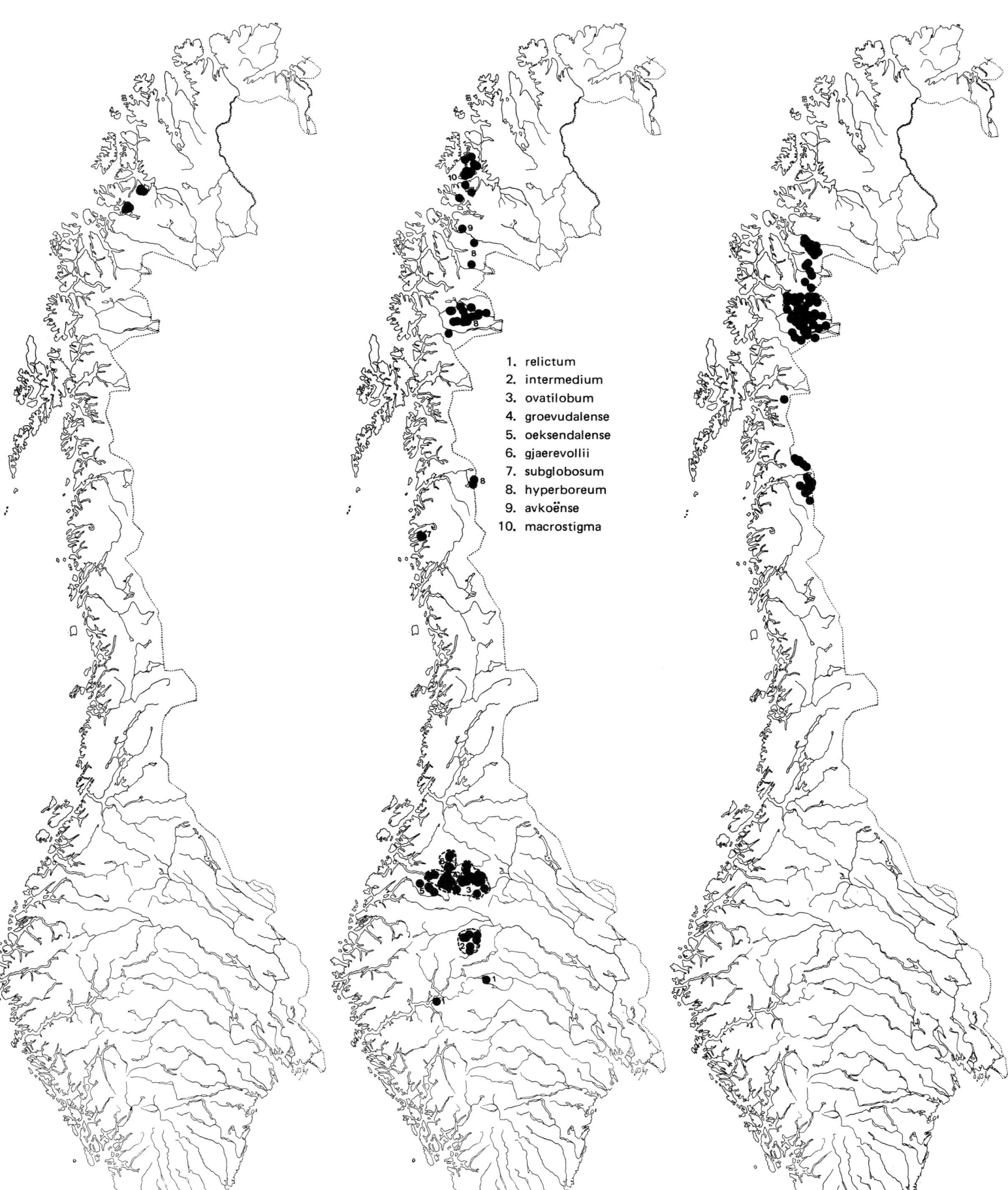

Papaver lapponicum (Tolm.) Nordh.

Papaver radicatum Rottb.

Pedicularis flammea L.

1. relictum
2. intermedium
3. ovatilobum
4. groevudalense
5. oeksendalense
6. gjaerevollii
7. subglobosum
8. hyperboreum
9. avkoënse
10. macrostigma

Pedicularis hirsuta L.   *Pedicularis oederi Vahl*   *Petasites frigidus (L.) Fr.*

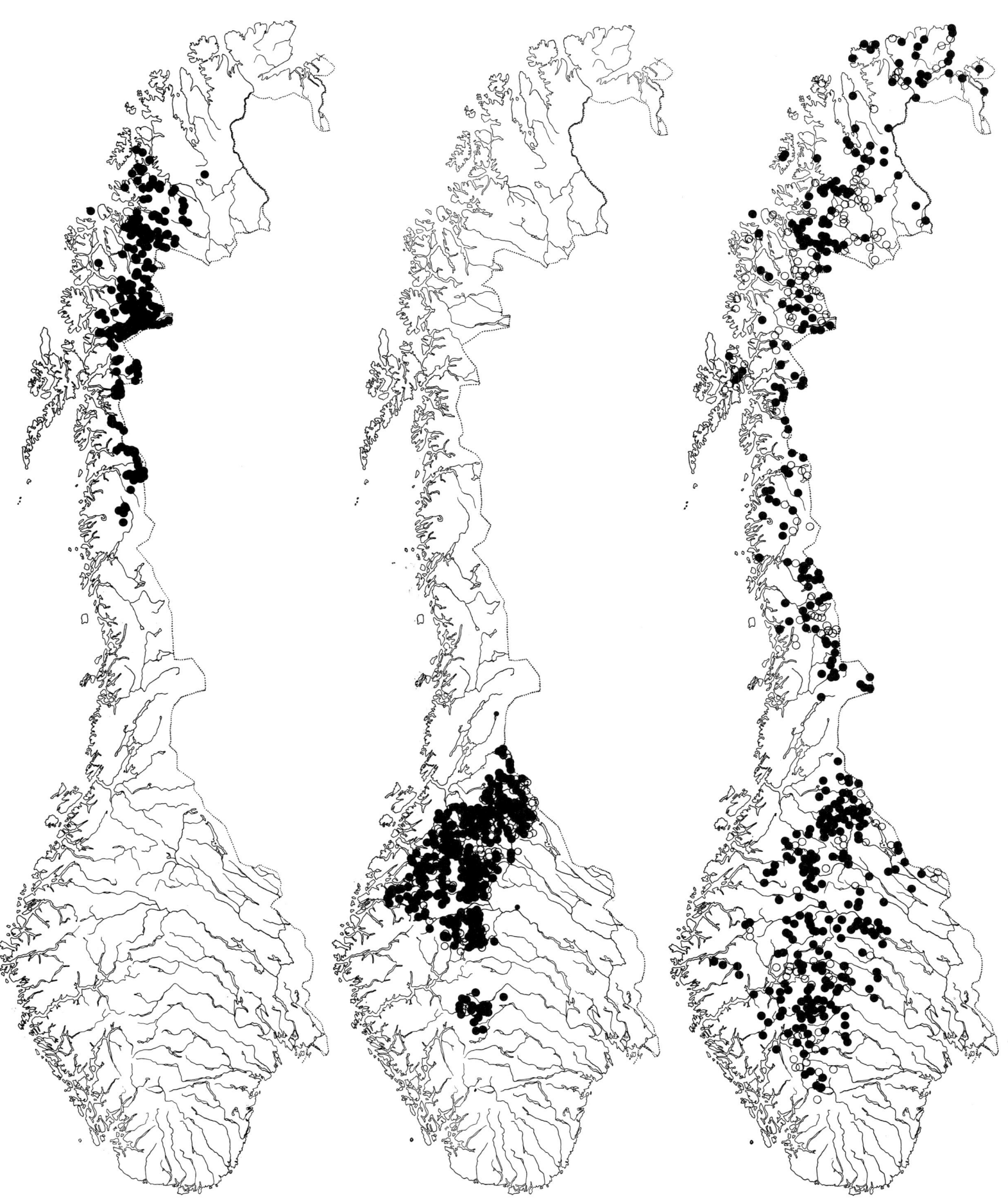

Pinguicula alpina L.   *Platanthera obtusata (Pursh.) Lindl.*   *Poa arctica R.Br.*
subsp. oligantha (Turcz.) Hultén

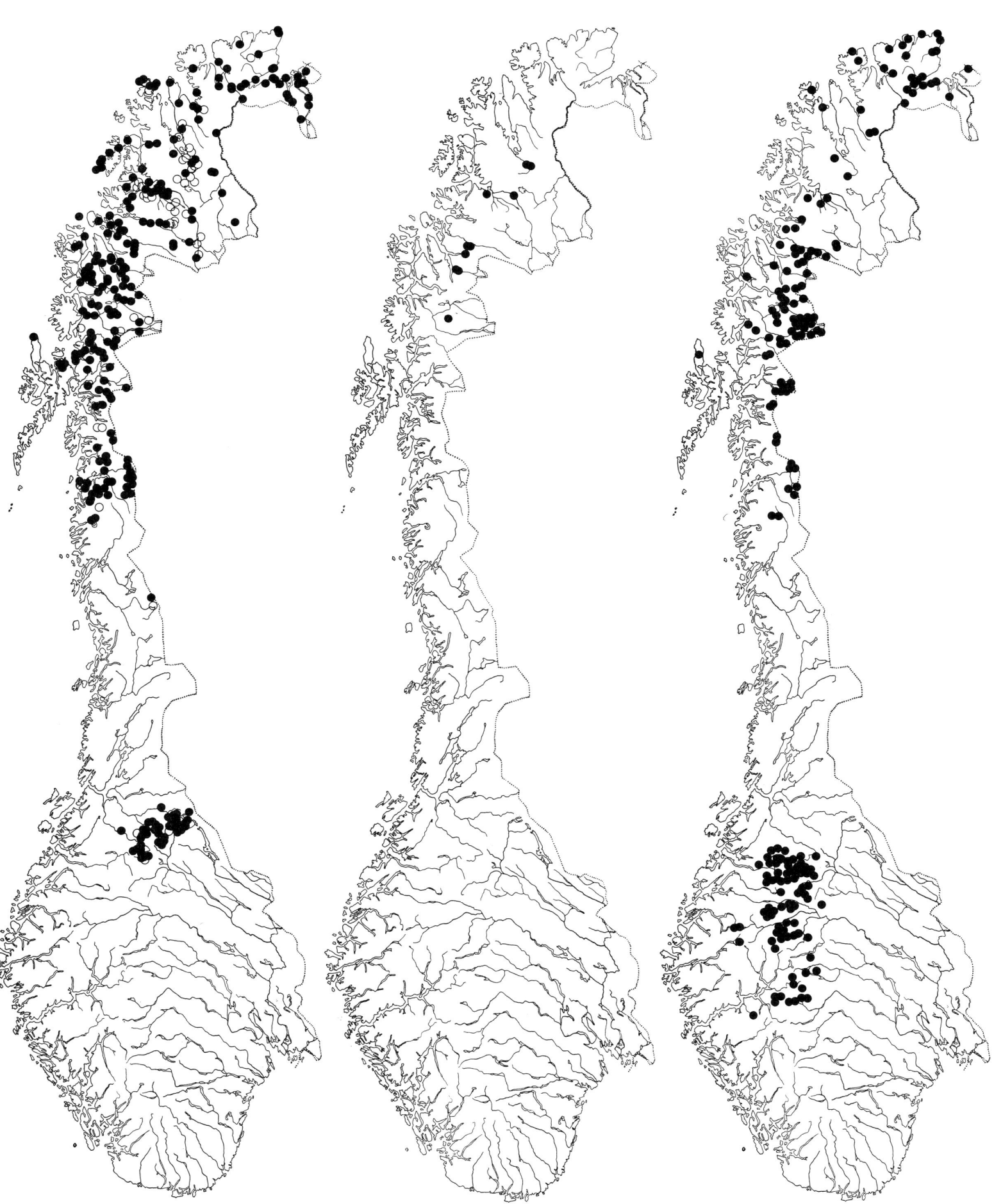

Poa flexuosa Sm.　　　　*Poa stricta Lindeb.*　　　　*Potentilla chamissonis Hultén*

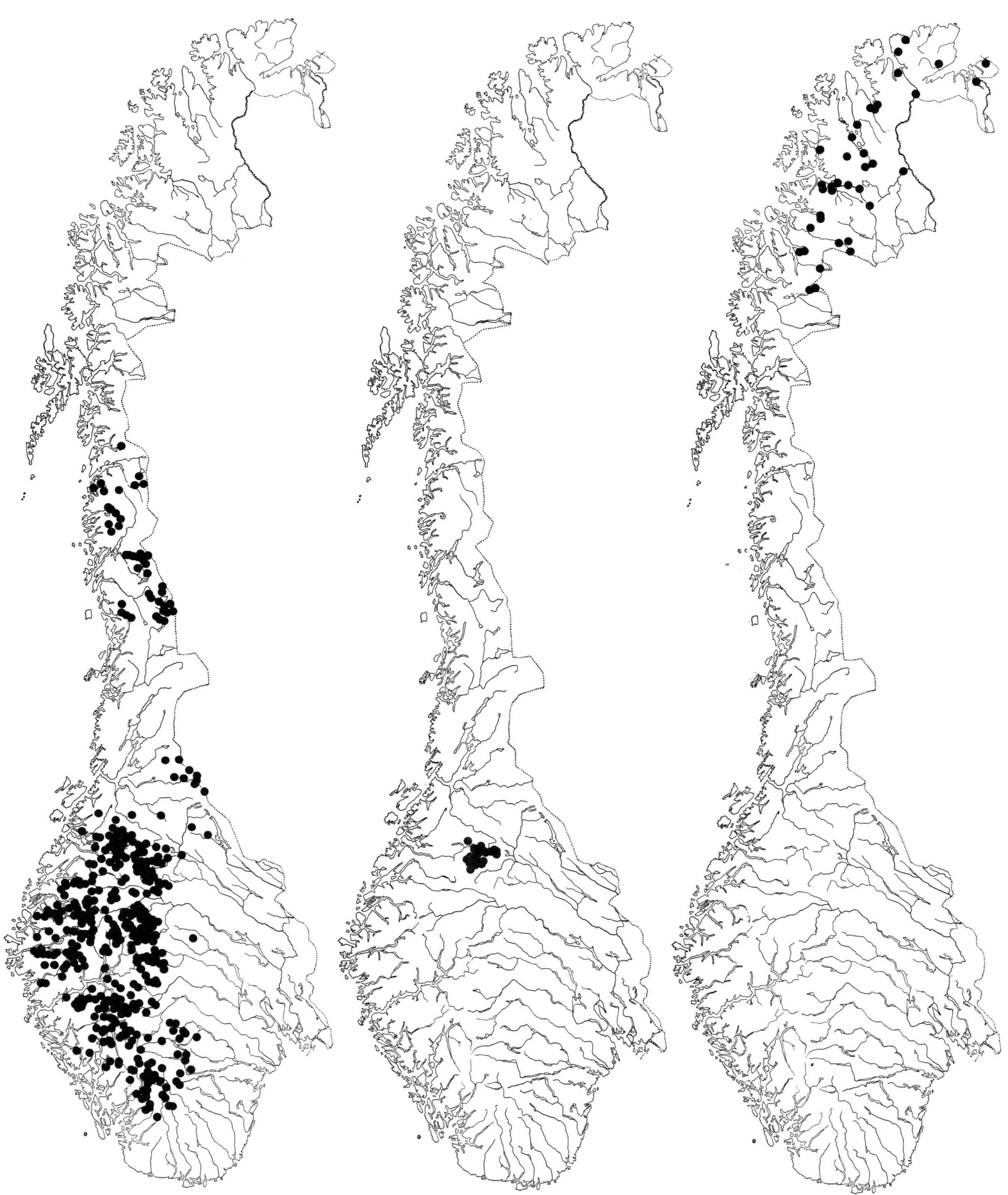

Potentilla nivea L.   *Primula scandinavica H. Bruun*   *Ranunculus glacialis L.*

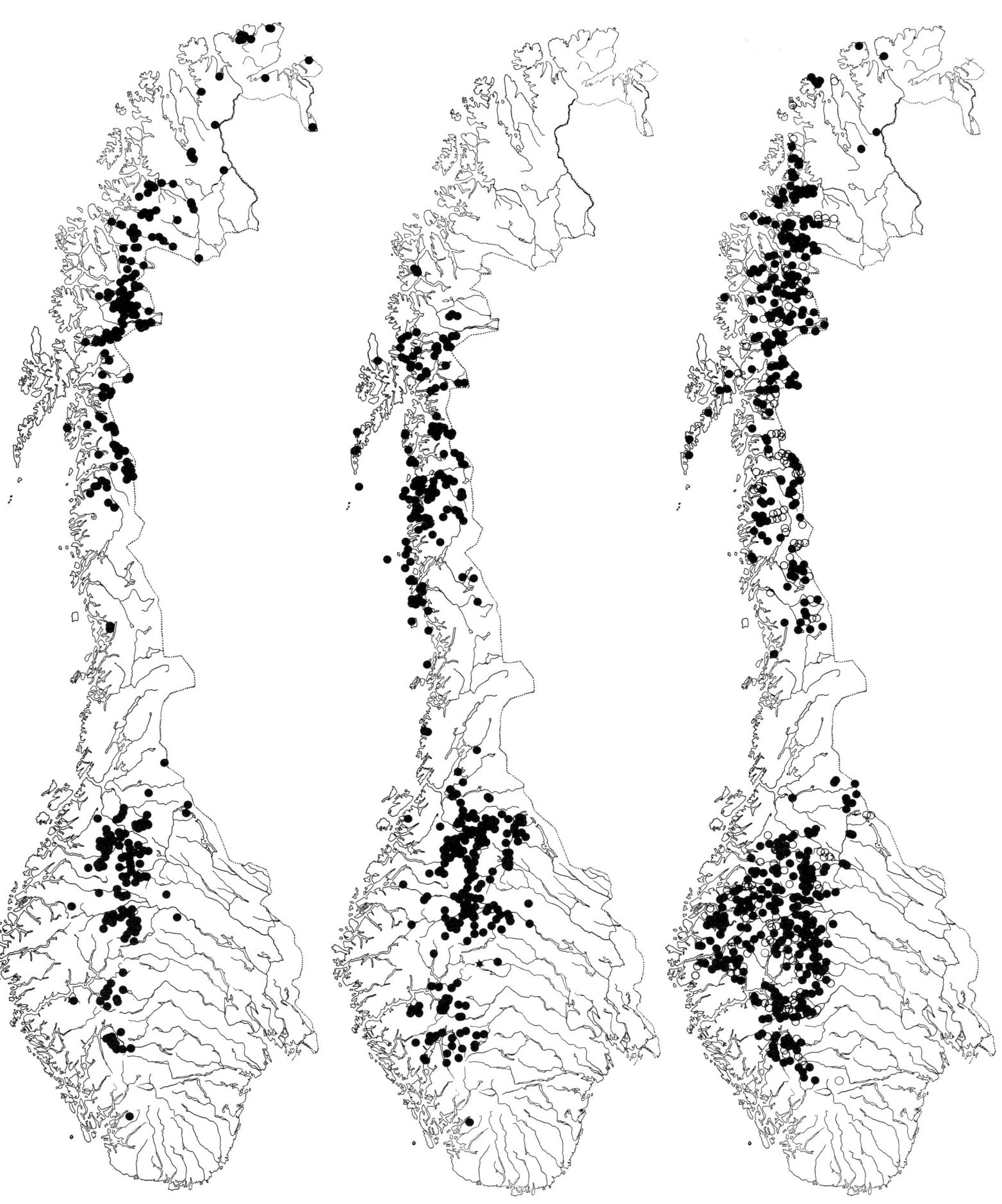

Ranunculus hyperboreus Rottb. *Ranunculus nivalis L.* *Ranunculus sulphureus C.J.Phipps*

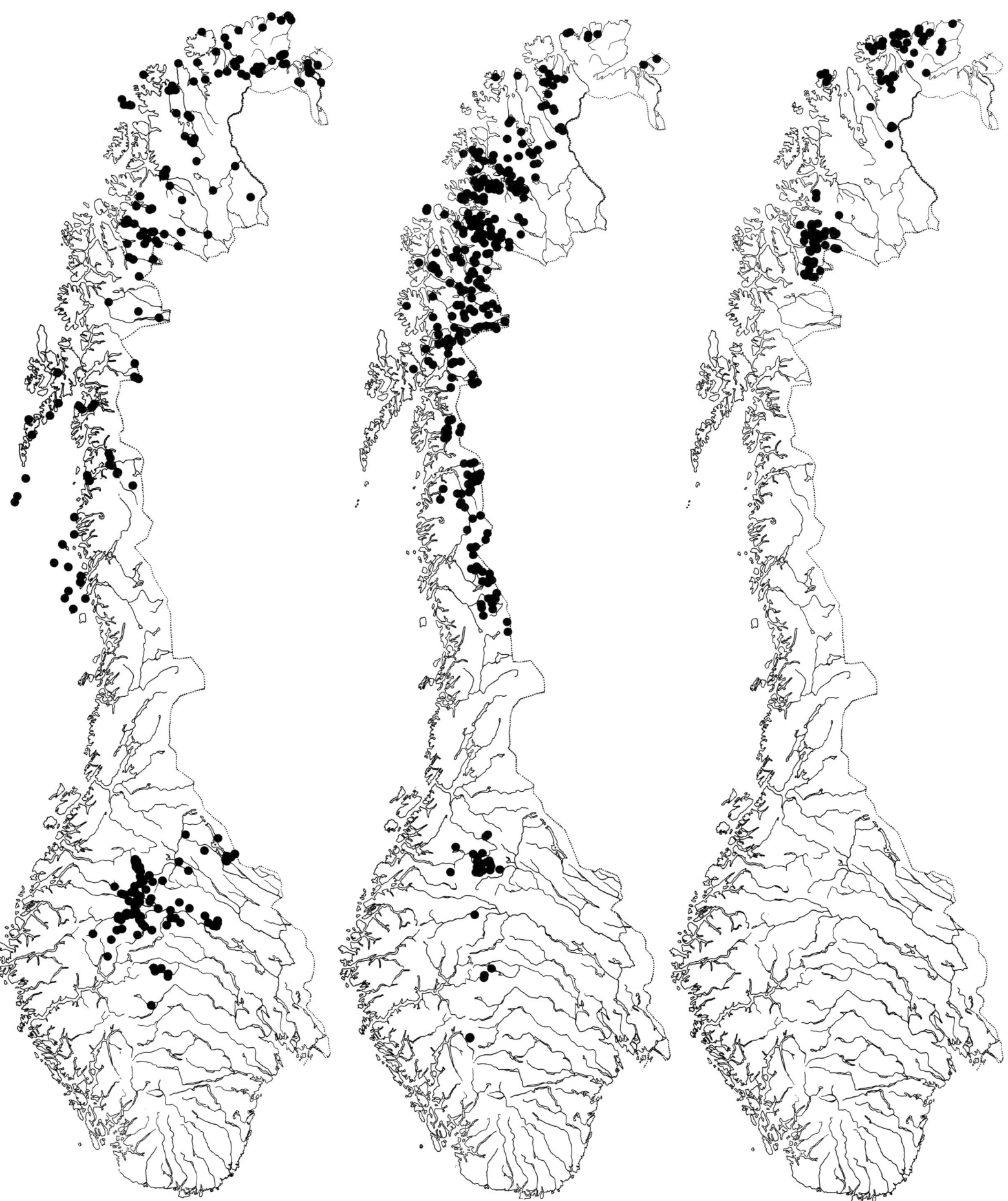

Rhododendron lapponicum
(L.) Wahlenb.

Roegneria borealis (Turcz.) Nevski

Sagina caespitosa
(J. Vahl) Lge.

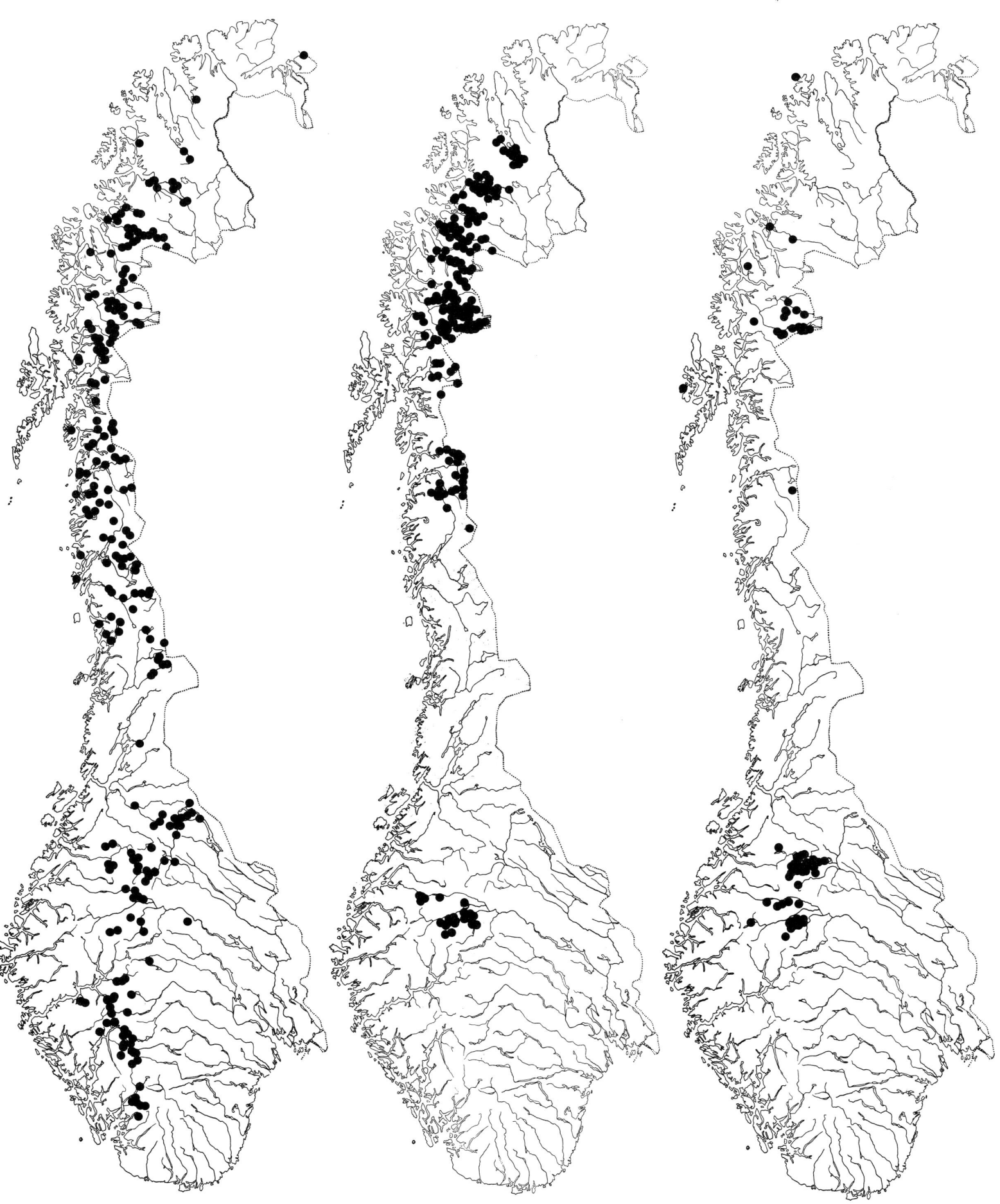

Sagina intermedia Fenzl   *Saxifraga adscendens L.*   *Saxifraga cernua L.*

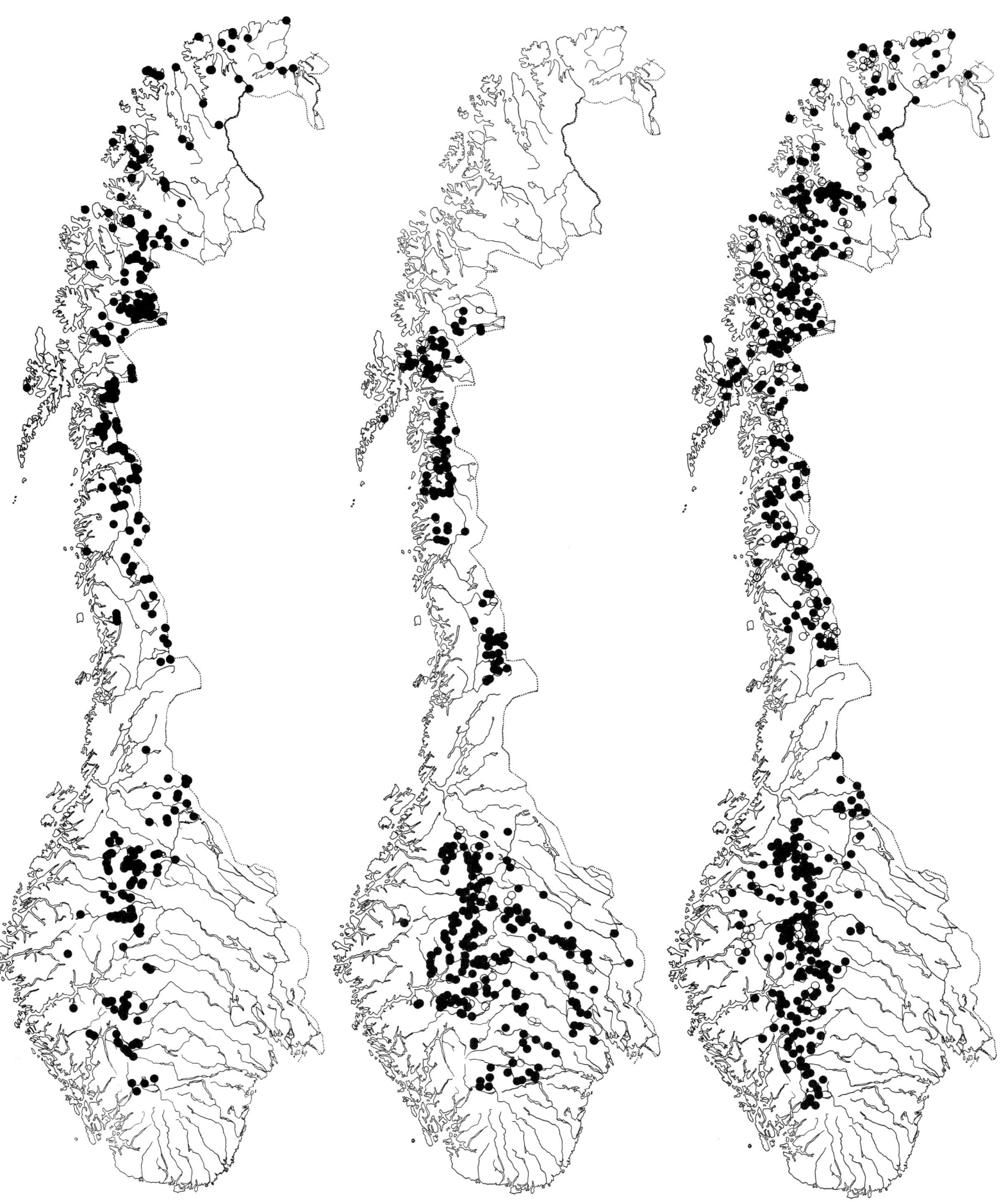

Saxifraga cotyledon L.   *Saxifraga foliolosa R.Br.*   *Saxifraga hieracifolia Waldst. & Kit.*

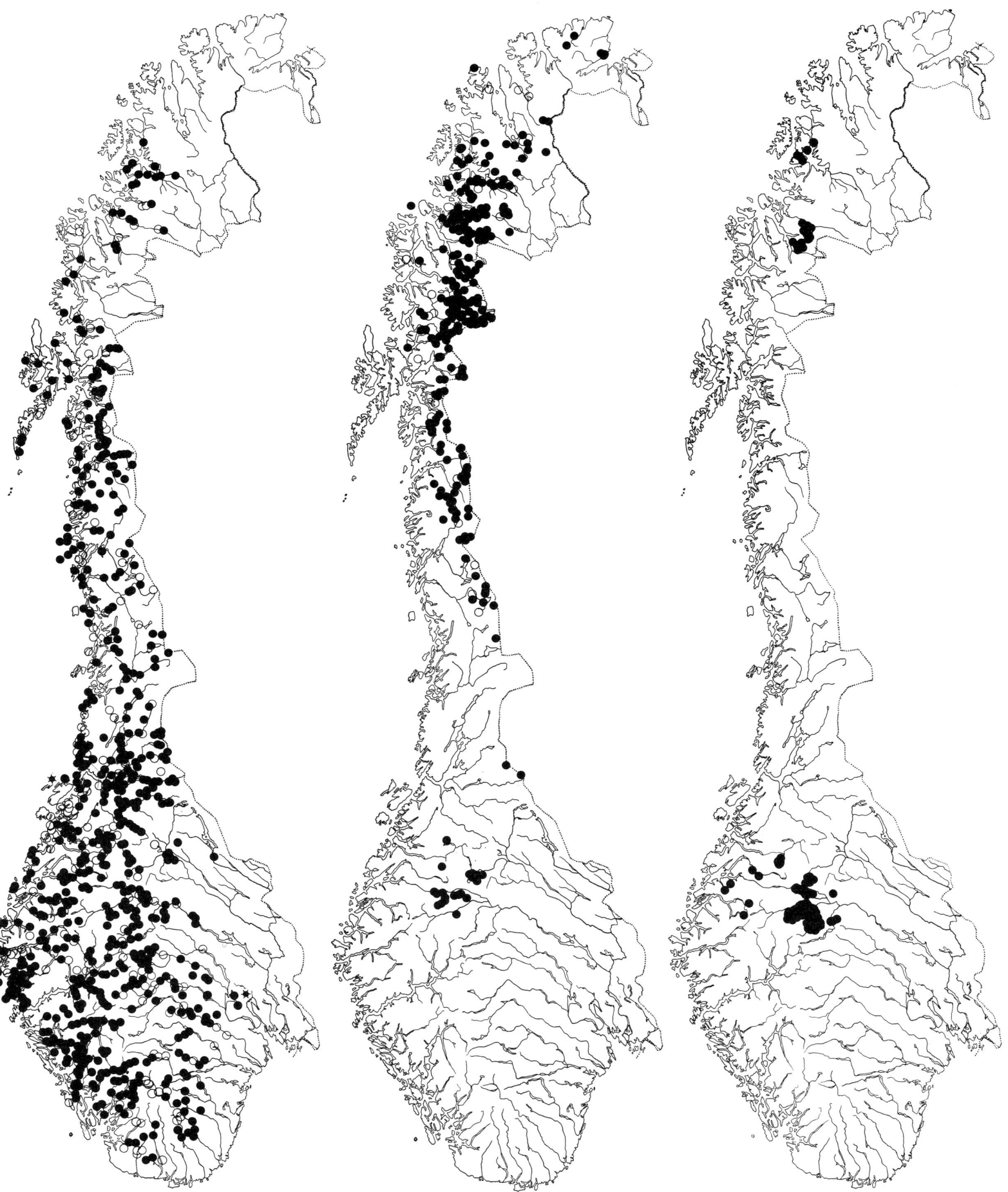

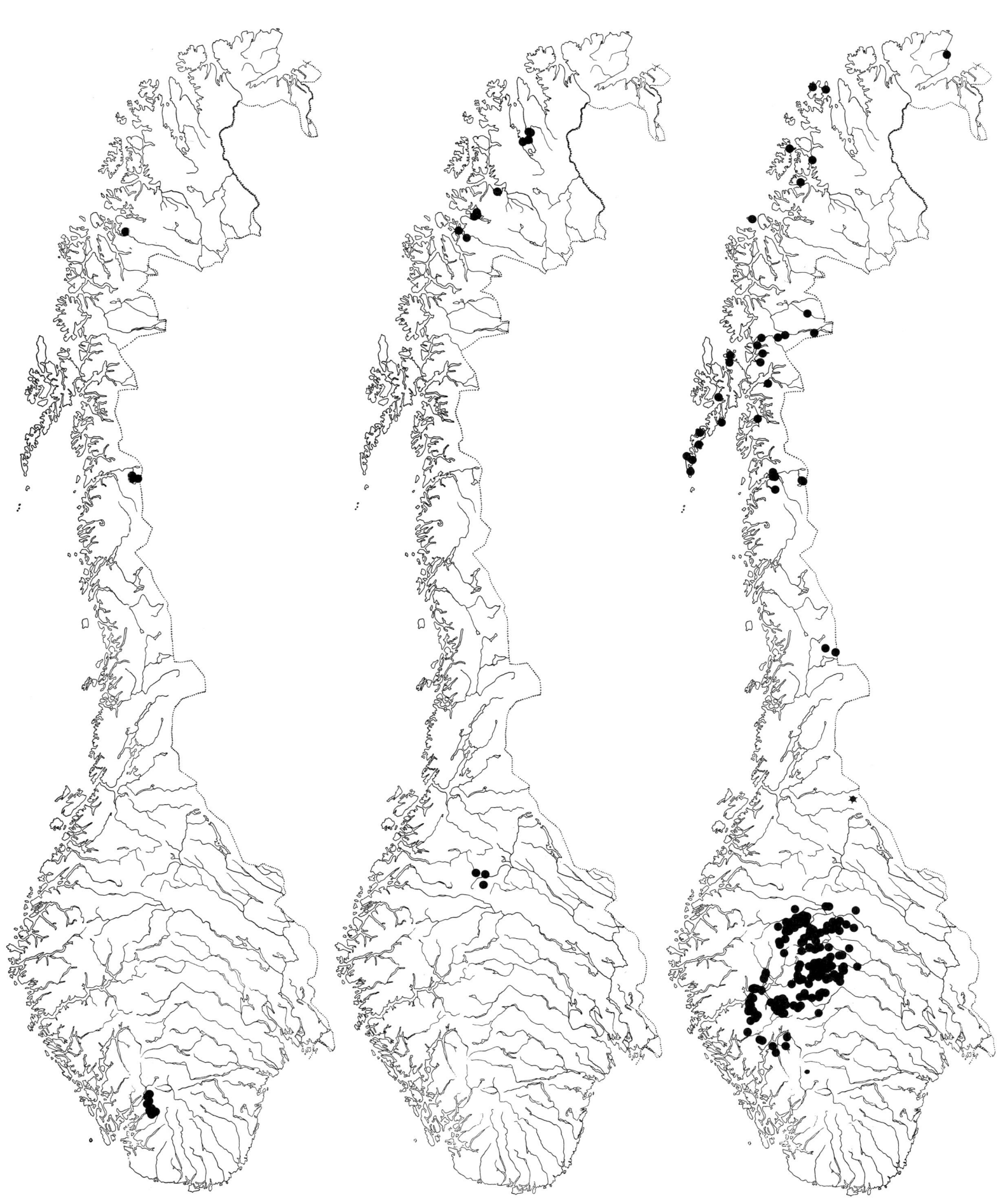

Saxifraga paniculata Miller   *Scirpus pumilus Vahl*   *Sedum villosum L.*

Silene acaulis (L.) Jacq. *Silene wahlbergella Chowd.* *Stellaria crassipes Hultén*

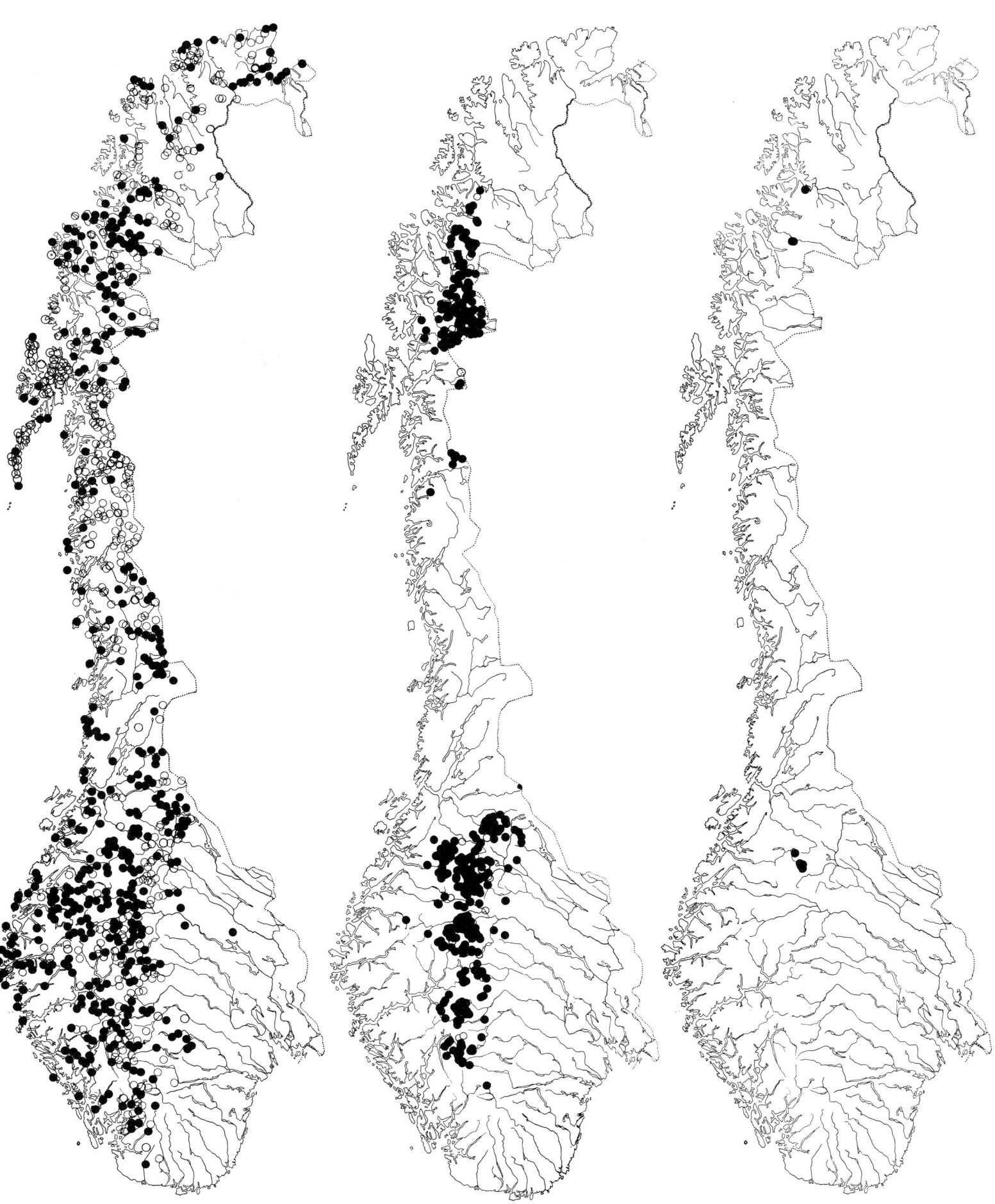

Taraxacum dovrense (Dt.) Dt. *Vahlodea atropurpurea* (Wahlenb.) Hartm. *Veronica fruticans* Jacq.

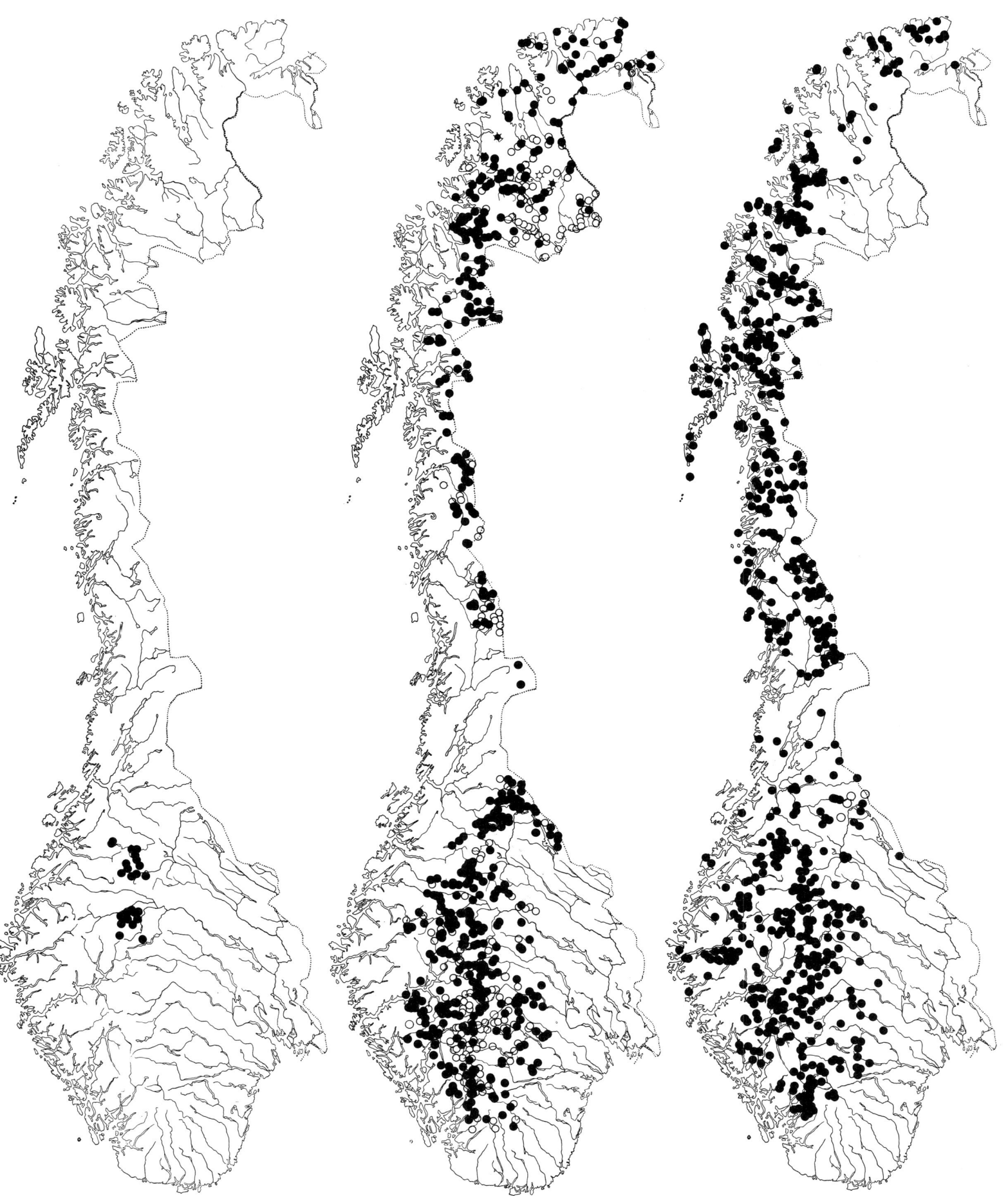

Woodsia glabella R.Br.